Ridge Functions and Applications in Neural Networks

Mathematical
Surveys
and
Monographs

Volume 263

Ridge Functions and Applications in Neural Networks

Vugar E. Ismailov

AMERICAN MATHEMATICAL SOCIETY
Providence, Rhode Island

2020 *Mathematics Subject Classification.* Primary 26B40, 41A30, 41A50, 41A63, 39B22, 46B28, 47A05, 65D15, 68T05, 92B20.

For additional information and updates on this book, visit
www.ams.org/bookpages/surv-263

Library of Congress Cataloging-in-Publication Data

Names: Ismailov, Vugar E., 1971- author.
Title: Ridge functions and applications in neural networks / Vugar E. Ismailov.
Description: Providence, Rhode Island : American Mathematical Society, [2021] | Series: Mathematical surveys and monographs, 0076-5376 ; volume 263 | Includes bibliographical references and index.
Identifiers: LCCN 2021035100 | ISBN 9781470467654 (paperback) | ISBN 9781470468002 (ebook)
Subjects: LCSH: Function spaces. | Multivariate analysis. | Approximation theory. | Linear operators. | Numbers, Real. | Neural networks (Computer science) | AMS: Real functions [See also 54C30] – Functions of several variables – Representation and superposition of functions. | Approximations and expansions {For all approximation theory in the complex domain, see 30E05 and 30E10; for all trigonometric approximation and interpolation, see 42A10 and 42A15; for numerical approx | Difference and functional equations – Functional equations and inequalities [See also 30D05] – Equations for real functions [See also 26A51, 26B25]. | Functional analysis {For manifolds modeled on topological linear spaces, see 57Nxx, 58Bxx} – Normed linear spaces and Banach spaces; Banach lattices {For function spaces, see 46Exx} | Operator theory – General theory of linear operators – General (adjoints, conjugates, products, inverses, domains, ranges, etc.). | Numerical analysis – Numerical approximation and computational geometry (primarily algorithms) {For theory, see 41-XX and 68Uxx} – Algorithms for functional approximation. | Computer science {For papers involving machine computations and programs in a specific mathematical area, see Section–04 in that area} – Artificial intelligence – Learning and adaptive systems | Biology and other natural sciences – Mathematical biology in general – Neural networks, artificial life and related topics [See also 68T05, 82C32, 94Cxx].
Classification: LCC QA323 .I76 2021 | DDC 515/.73–dc23
LC record available at https://lccn.loc.gov/2021035100

∞ The paper used in this book is acid-free and falls within the guidelines
established to ensure permanence and durability.
Visit the AMS home page at `https://www.ams.org/`

10 9 8 7 6 5 4 3 2 1   26 25 24 23 22 21

To the Memory of My Parents

Contents

Preface

This monograph is about *ridge functions.* Recent years have witnessed a flurry of interest in these functions. Ridge functions appear in various fields and under various guises. They appear in fields as diverse as partial differential equations (where they are called *plane waves*), computerized tomography and statistics. These functions are also the underpinnings of many central models in neural networks.

We are interested in ridge functions from the point of view of approximation theory. The basic goal in approximation theory is to approximate complicated objects by simpler objects. Among many classes of multivariate functions, linear combinations of ridge functions are a class of simpler functions. This book studies some problems of approximation of multivariate functions by linear combinations of ridge functions. We present here various properties of these functions. The questions we ask are as follows. When can a multivariate function be expressed as a linear combination of ridge functions from a certain class? When do such linear combinations represent each multivariate function? If a precise representation is not possible, can one approximate arbitrarily well? If well approximation fails, how can one compute/estimate the error of approximation, know that a best approximation exists? How can one characterize and construct best approximations? If a smooth function is a sum of arbitrarily behaved ridge functions, is it true that it can be expressed as a sum of smooth ridge functions? We also study properties of generalized ridge functions, which are very much related to linear superpositions and Kolmogorov's famous superposition theorem. This book ends with a few applications of ridge functions to the problem of approximation by single and two hidden layer neural networks with a restricted set of weights.

I hope that this book will be useful and interesting to both researchers and graduate students.

I would like to express my sincere gratitude to Eriko Hironaka at AMS for handling the review process and for providing professional support. I am also grateful to the Editorial Committee and the anonymous referees for their constructive suggestions which helped to improve the first version of this monograph.

Introduction

Recent years have seen a growing interest in the study of special multivariate functions called ridge functions. A *ridge function*, in its simplest format, is a multivariate function of the form $g(\mathbf{a}\cdot\mathbf{x})$, where $g:\mathbb{R}\to\mathbb{R}$, $\mathbf{a}=(a_1,\ldots,a_d)$ is a fixed vector (direction) in $\mathbb{R}^d\setminus\{\mathbf{0}\}$, $\mathbf{x}=(x_1,\ldots,x_d)$ is the variable and $\mathbf{a}\cdot\mathbf{x}$ is the standard inner product. In other words, a ridge function is a multivariate function constant on the parallel hyperplanes $\mathbf{a}\cdot\mathbf{x}=c$, $c\in\mathbb{R}$. These functions arise naturally in various fields. They arise in computerized tomography (see, e.g., [**109–111, 141, 149, 160**]), statistics (see, e.g., [**24, 25, 50, 56, 75**]), large-scale data analysis (see, e.g., [**35, 49, 53, 201, 205**]) and neural networks (see, e.g., [**42, 97, 99, 136, 145, 172, 178**]). These functions are also used in modern approximation theory as an effective and convenient tool for approximating complicated multivariate functions (see, e.g., [**66, 91, 95, 128, 146, 167, 171, 196**]).

It should be remarked that long before the appearance of the name "ridge", these functions were used in PDE theory under the name of *plane waves*. For example, see the book by F. John [**105**]. In general, sums of ridge functions with fixed directions occur in the study of hyperbolic constant coefficient partial differential equations. As an example, assume that (α_i,β_i), $i=1,\ldots,r$, are pairwise linearly independent vectors in $\mathbb{R}^2$. Then the general solution to the homogeneous partial differential equation

$$\prod_{i=1}^{r}\left(\alpha_i\frac{\partial}{\partial x}+\beta_i\frac{\partial}{\partial y}\right)u(x,y)=0$$

are all functions of the form

$$u(x,y)=\sum_{i=1}^{r}g_i\left(\beta_i x-\alpha_i y\right)$$

for arbitrary continuous univariate functions g_i, $i=1,\ldots,r$. Here the derivatives are understood in the sense of distributions.

The term "ridge function" was coined by Logan and Shepp in their seminal paper [**141**] devoted to the basic mathematical problem of computerized tomography. This problem consists of reconstructing a given multivariate function from values of its integrals along certain straight lines in the plane. The integrals along parallel lines can be considered as a ridge function. Thus, the problem is to reconstruct f from some set of ridge functions generated by the function f itself. In practice, one can consider only a finite number of directions along which the above integrals are taken. Obviously, reconstruction from such data needs some additional conditions to be unique, since there are many functions g having the same integrals. For uniqueness, Logan and Shepp [**141**] used the criterion of minimizing the L_2 norm of g. That is, they found a function $g(x,y)$ with the minimum L_2 norm among all functions, which has the same integrals as f. More precisely, let D be the unit disk in the plane and an unknown function $f(x,y)$ be square integrable and supported on D. We are given projections $P_f(t,\theta)$ (integrals of f along the lines $x\cos\theta + y\sin\theta = t$) and looking for a function $g = g(x,y)$ of minimum L_2 norm, which has the same projections as $f : P_g(t,\theta_j) = P_f(t,\theta_j)$, $j = 0,1,\dots,n-1$, where the angles θ_j generate equally spaced directions, i.e. $\theta_j = \frac{j\pi}{n}$, $j = 0,1,\dots,n-1$. The authors of [**141**] showed that this problem of tomography is equivalent to the problem of L_2-approximation of the function f by sums of ridge functions with the equally spaced directions $(\cos\theta_j, \sin\theta_j)$, $j = 0,1,\dots,n-1$. They gave a closed-form expression for the unique function $g(x,y)$ and showed that the unique polynomial $P(x,y)$ of degree $n-1$ which best approximates f in $L_2(D)$ is determined from the above n projections of f and can be represented as a sum of n ridge functions.

Kazantsev [**109**] solved the above problem of tomography without requiring that the considered directions are equally spaced. Marr [**149**] considered the problem of finding a polynomial of degree $n-2$, whose projections along lines joining each pair of n equally spaced points on the circumference of D best matches the given projections of f in the sense of minimizing the sum of squares of the differences. Thus we see that the problems of tomography give rise to an independent study of approximation theoretic properties of the following set of linear combinations of ridge functions:

$$\mathcal{R}\left(\mathbf{a}^1,\dots,\mathbf{a}^r\right) = \left\{\sum_{i=1}^{r} g_i\left(\mathbf{a}^i\cdot\mathbf{x}\right) : g_i : \mathbb{R}\to\mathbb{R}, i = 1,\dots,r\right\},$$

where directions $\mathbf{a}^1,\dots,\mathbf{a}^r$ are fixed and belong to the d-dimensional Euclidean space. Note that the set $\mathcal{R}\left(\mathbf{a}^1,\dots,\mathbf{a}^r\right)$ is a linear space.

Ridge function approximation also appears in statistics in *Projection Pursuit.* This term was introduced by Friedman and Tukey [**55**] to name a technique for the explanatory analysis of large and multivariate data sets. This technique seeks out "interesting" linear projections of the multivariate data onto a line or a plane. Projection Pursuit algorithms approximate a multivariate function f by sums of ridge functions with variable directions, that is, by functions from the set

$$\mathcal{R}_r = \left\{\sum_{i=1}^{r} g_i\left(\mathbf{a}^i\cdot\mathbf{x}\right) : \mathbf{a}^i\in\mathbb{R}^d\setminus\{\mathbf{0}\},\ g_i : \mathbb{R}\to\mathbb{R}, i = 1,\dots,r\right\}.$$

Here r is the only fixed parameter, directions $\mathbf{a}^1,\dots,\mathbf{a}^r$ and functions $g_1,\dots,g_r$ are free to choose. The first method of such approximation was developed by Friedman and Stuetzle [**56**]. Their approximation process called *Projection Pursuit Regression* (PPR) operates in a stepwise and greedy fashion. The process does not find a best approximation from $\mathcal{R}_r$, it algorithmically constructs functions $g_r \in \mathcal{R}_r$, such that $\|g_r - f\|_{L_2} \to 0$, as $r\to\infty$. At stage m, PPR looks for a univariate function g_m and direction $\mathbf{a}^m$ such that the ridge function $g_m\left(\mathbf{a}^m\cdot\mathbf{x}\right)$ best approximates the residual $f(x) - \sum_{j=1}^{m-1} g_j\left(\mathbf{a}^j\cdot\mathbf{x}\right)$. Projection pursuit regression has been proposed as an approach to bypass the curse of dimensionality and now is applied to prediction in applied sciences. In [**24, 25**], Candes developed a new approach based not on stepwise construction of approximation but on a new transform called the *ridgelet transform.* The ridgelet transform represents general functions as integrals of *ridgelets* – specifically chosen ridge functions.

Ridge functions are used in large-scale data analysis as a model, which efficiently solves the *recovery problem*. In the recovery problem, the aim is to compute an approximation of an unknown function $f : X \subseteq \mathbb{R}^d \to \mathbb{R}$ of some given class, where d is very large, from samples $f(x_1), \ldots, f(x_n)$ within an approximation error ε. A measure for the difficulty of the recovery problem is the so-called *information complexity* $n(\varepsilon, d)$, which is defined as the minimum number of information extraction operations (see [**199**]). The general question is if the above problem *tractable*, that is, if $n(\varepsilon, d)$ does not depend exponentially on ε^{-1} and d (see [**163**]). Note that if $n(\varepsilon, \mathcal{F}_d)$ depends exponentially on d, then an approximation problem suffers from the curse of dimensionality. Unfortunately, the number of samples to approximate even a C^∞-function grows exponentially with the dimension d (see [**163**]). That is, the approximation of such functions in high dimensions is intractable and suffers from the curse of dimensionality. This implies that if we want the recovery problem to be efficiently solvable, we must impose additional assumptions on f, different from just regularity. One popular approach to break the curse of dimensionality and make the recovery problem tractable is to assume a priory that f is a ridge function $g(\mathbf{a} \cdot \mathbf{x})$, or a generalized ridge function of the form $g(A\mathbf{x})$, where A is a $k \times d$ matrix, with k considerably smaller than d, and g is a function of k variables (see, e.g., [**35, 49, 53, 62, 112, 152, 201, 205**]). Typical results on the recovery of ridge functions give algorithms for computing the approximating function, whose complexity is polynomial both in the dimension d and in ε^{-1}.

The significance of approximation by ridge functions is well understood from its role in the theory of *neural networks*. Ridge functions appear in the definitions of many central neural network models. It is a broad knowledge that neural networks are being successfully applied across an extraordinary range of problem domains, in fields as diverse as finance, medicine, engineering, geology and physics. Generally speaking, neural networks are being introduced anywhere that there are problems of prediction, classification or control. Thus not surprisingly, there is a great interest to this powerful and very popular area of research (see, e.g., [**172**] and a great deal of references therein). An artificial neural network is a way to perform computations using networks of interconnected computational units vaguely analogous to neurons simulating how our brain solves them. An artificial neuron, which forms the basis for designing neural networks, is a device with d real inputs and an output. This output is generally a ridge function of the given inputs. In mathematical terms, a neuron may be described as

$$y = \sigma(\mathbf{w} \cdot \mathbf{x} - \theta),$$

where $\mathbf{x} = (\mathbf{x}_1, \ldots, x_d) \in \mathbb{R}^d$ are the input signals, $w = (w_1, \ldots, w_d) \in \mathbb{R}^d$ are the synaptic weights, $\theta \in \mathbb{R}$ is the bias, σ is the activation function and y is the output signal of the neuron. In a layered neural network the neurons are organized in the form of layers. We have at least two layers: an input and an output layer. The layers between the input and the output layers (if any) are called hidden layers, whose computation nodes are correspondingly called hidden neurons or hidden units. The output signals of the first layer are used as inputs to the second layer, the output signals of the second layer are used as inputs to the third layer, and so on for the rest of the network. Neural networks with this kind of architecture is called a *Multilayer Feedforward Perceptron* (MLP). This is the most popular model among other neural network models. In this model, a neural network with a single hidden layer and one output represents a function of the form

$$\sum_{i=1}^{r} c_i\sigma(\mathbf{w}^i{\cdot}\mathbf{x} - \theta_i).$$

Here the weights $\mathbf{w}^i$ are vectors in $\mathbb{R}^d$, the thresholds θ_i and the coefficients c_i are real numbers and the activation function σ is a univariate function. We fix only σ and r. Note that the functions $\sigma(\mathbf{w}^i{\cdot}\mathbf{x} - \theta_i)$ are ridge functions. Thus it is not surprising that some approximation theoretic problems related to neural networks have strong association with the corresponding problems of approximation by ridge functions.

It is clear that in the special case, linear combinations of ridge functions turn into sums of univariate functions. This is also the simplest case. The simplicity of the approximation apparatus itself guarantees its utility in applications where multivariate functions are constant obstacles. In mathematics, this type of approximation has arisen, for example, in connection with the classical functional equations [**19**], the numerical solution of certain PDE boundary value problems [**15**], dimension theory [**189, 190**], etc. In computer science, it arises in connection with the efficient storage of data in computer databases (see, e.g., [**200**]). There is an interesting interconnection between the theory of approximation by univariate functions and problems of equilibrium construction in economics (see [**195**]).

Linear combinations of ridge functions with fixed directions allow a natural generalization to functions of the form $g(\alpha_1(x_1) + \cdots + \alpha_d(x_d))$, where $\alpha_i(x_i)$, $i = \overline{1, d}$, are real univariate functions. Such a generalization has a strong association with linear superpositions. A linear superposition is a function expressed as the sum

$$\sum_{i=1}^{r} g_i(h_i(x)),\ x \in X,$$

where X is any set (in particular, a subset of $\mathbb{R}^d$), $h_i : X \to \mathbb{R}$, $i = 1, \ldots, r$, are arbitrarily fixed functions, and $g_i : \mathbb{R} \to \mathbb{R}$, $i = 1, \ldots, r$. Note that here we deal with more complicated composition than the composition of a univariate function with the inner product. A starting point in the study of linear superpositions was the well known superposition theorem of Kolmogorov [**122**] (see also the paper on Kolmogorov's works by Tikhomirov [**198**]). This theorem states that for the unit cube $\mathbb{I}^d$, $\mathbb{I} = [0, 1]$, $d \geq 2$, there exist $2d + 1$ functions $\{s_q\}_{q=1}^{2d+1} \subset C(\mathbb{I}^d)$ of the form

$$s_q(x_1, \ldots, x_d) = \sum_{p=1}^{d} \varphi_{pq}(x_p),\ \varphi_{pq} \in C(\mathbb{I}),\ p = 1, \ldots, d,\ q = 1, \ldots, 2d + 1$$

such that each function $f \in C(\mathbb{I}^d)$ admits the representation

$$f(x) = \sum_{q=1}^{2d+1} g_q(s_q(x)),\ x = (x_1, \ldots, x_d) \in \mathbb{I}^d,\ g_q \in C(\mathbb{R}).$$

Thus, any continuous function on the unit cube can be represented as a linear superposition with the fixed inner functions $s_1, \ldots, s_{2d+1}$. In literature, these functions are called universal functions or the Kolmogorov functions. Note that all the functions $g_q(s_q(x))$ in the Kolmogorov superposition formula are generalized ridge functions, since each s_q is a sum of univariate functions.

In this book, we consider some problems of approximation and/or representation of multivariate functions by linear combinations of ridge functions, generalized ridge functions and feedforward neural networks. The book consists of five chapters.

Chapter 1 is devoted to the approximation from some sets of ridge functions with arbitrarily fixed directions in C and L_2 metrics. First, we study problems of representation of multivariate functions by linear combinations of ridge functions. Then, in case of two fixed directions and under suitable conditions, we give complete solutions to three basic problems of uniform approximation, namely, problems on existence, characterization, and construction of a best approximation. We also study problems of well approximation (approximation with arbitrary accuracy) and representation of continuous multivariate functions by sums of two continuous ridge functions. The reader will see the main difficulties and remained open problems in the uniform approximation by sums of more than two ridge functions. For L_2 approximation, a number of summands does not play such an essential role as it plays in the uniform approximation. In this case, it is known that a best approximation always exists and unique. For some special domains in $\mathbb{R}^d$, we characterize and then construct the best approximation. We also give an explicit formula for the approximation error.

Chapter 2 explores the following open problem raised in Buhmann and Pinkus [**20**], and Pinkus [**175**, p. 14]. Assume we are given a function $f(\mathbf{x}) = f(x_1, \dots, x_n)$ of the form

$$f(\mathbf{x}) = \sum_{i=1}^{k} f_i(\mathbf{a}^i \cdot \mathbf{x}),$$

where the $\mathbf{a}^i$, $i = 1, \dots, k$, are pairwise linearly independent vectors (directions) in $\mathbb{R}^d$, f_i are arbitrarily behaved univariate functions and $\mathbf{a}^i \cdot \mathbf{x}$ are standard inner products. Assume, in addition, that f is of a certain smoothness class, that is, $f \in C^s(\mathbb{R}^d)$, where $s \geq 0$ (with the convention that $C^0(\mathbb{R}^d) = C(\mathbb{R}^d)$). Is it true that there will always exist $g_i \in C^s(\mathbb{R})$ such that

$$f(\mathbf{x}) = \sum_{i=1}^{k} g_i(\mathbf{a}^i \cdot \mathbf{x}) \ ?$$

In this chapter, we solve this problem up to some multivariate polynomial. We find various conditions on the directions $\mathbf{a}^i$ allowing to express this polynomial as a sum of smooth ridge functions with these directions. We also consider the question of constructing g_i using the information about the known functions f_i.

Chapter 3 is devoted to the simplest type of ridge functions – univariate functions. Note that a ridge function depends only on one variable if its direction coincides with the coordinate direction. Thus, in case of coincidence of all given directions with the coordinate directions, the problem of ridge function approximation turns into the problem of approximation of multivariate functions by sums of univariate functions. In this chapter, we first consider the approximation of a bivariate function $f(x, y)$ by sums $\varphi(x) + \psi(y)$ on a rectangular domain R. We construct special classes of continuous functions depending on a numerical parameter and characterize each class in terms of the approximation error calculation formulas. This parameter will show which points of R the calculation formula involves. We will also construct a best approximating sum $\varphi_0(x) + \psi_0(y)$ to a function from constructed classes. Then we develop a method for obtaining explicit formulas for the error of approximation of bivariate functions, defined on a union of rectangles, by sums of univariate functions. It should be remarked that formulas of such type were known only for functions defined on a rectangle with sides parallel to the coordinate axes. Our method, based on a maximization process over certain objects, called "closed bolts", allows the consideration of functions defined on hexagons, octagons and stairlike polygons with sides parallel to the coordinate axes. At the end of this chapter we discuss one important result from Golomb's paper [**64**]. This paper, published in 1959, made a start of a systematic study of approximation of multivariate functions by various compositions, including sums of univariate functions. In [**64**], along with many other results, Golomb obtained a duality formula for the error of approximation to a multivariate function from the set of sums of univariate functions. Unfortunately, his proof had a gap, which was 24 years later pointed out by Marshall and O'Farrell [**151**]. But the question if Golomb's formula was correct, remained unsolved. In Chapter 3, we show that Golomb's formula is correct, and moreover it holds in a stronger form.

Chapter 4 tells us about some problems concerning generalized ridge functions $g(\alpha_1(x_1) + \cdots + \alpha_d(x_d))$ and linear superpositions. We consider the problem of representation of general functions by linear superpositions. We show that if some representation by linear superpositions, in particular by linear combinations of generalized ridge functions, holds for continuous functions, then it holds for all functions. This leads us to extensions of many superpositions theorems (such as the well-known Kolmogorov superposition theorem, Ostrand's superposition theorem, etc.) from continuous to arbitrarily behaved multivariate functions. Concerning generalized ridge functions, we see that every multivariate function can be written as a generalized ridge function or as a sum of finitely many such functions. We also study the uniqueness of representation of functions by linear superpositions. At the end of this chapter we discuss a special algorithmic approximation method called the Diliberto-Straus algorithm. Under mild assumptions, we prove the convergence of this algorithm.

Chapter 5 is about neural network approximation. The analysis in this chapter is based on properties of ordinary and generalized ridge functions. We consider a single and two hidden layer feedforward neural network models with a restricted set of weights. Such network models are important from the point of view of practical applications. First, we discuss the universal approximation theorem. Then we study approximation properties of single hidden layer neural networks with weights varying on a finite set of directions and straight lines. We give several necessary and sufficient conditions for well approximation by such networks. For a set of weights consisting of two directions (and two straight lines), we show that there is a geometrically explicit solution to the problem. We also give an explicit formula for the approximation error. Regarding two hidden layer feedforward neural networks, we prove that two hidden layer neural networks with d inputs, d neurons in the first hidden layer, $2d + 2$ neurons in the second hidden layer and with a specifically constructed sigmoidal, infinitely differentiable and almost monotone activation function can approximate any continuous multivariate function with arbitrary precision. We show that for this approximation only a finite number of fixed weights (precisely, d fixed weights) suffice.

There are topics related to ridge functions that are not presented here. The glaring omission is that of interpolation at points and on straight lines by ridge functions. We also do not address, for example, questions of linear independence and spanning by linear combinations of ridge monomials in the spaces of homogeneous and algebraic polynomials of a fixed degree, integral representations of functions where the kernel is a ridge function, closure properties of finite sums of ridge functions. These and similar topics may be found in the monograph by Pinkus [**175**]. The reader may also consult the survey articles [**95, 127, 171**].

CHAPTER 1

Properties of linear combinations of ridge functions

In this chapter, we consider approximation-theoretic problems arising in ridge function approximation. First we briefly review some results on approximation by sums of ridge functions with both fixed and variable directions. Then we analyze the problem of representability of an arbitrary multivariate function by linear combinations of ridge functions with fixed directions. In the special case of two fixed directions, we characterize a best uniform approximation from the set of sums of ridge functions with these directions. For a class of bivariate functions we use this result to construct explicitly a best approximation. Questions on existence of a best approximation are also studied. We also study problems of well approximation (approximation with arbitrary accuracy) and representation of continuous multivariate functions by sums of two continuous ridge functions. The reader will see the main difficulties and remained open problems in the uniform approximation by sums of more than two ridge functions. For L_2 approximation, a number of summands does not play such an essential role as it plays in the uniform approximation. In this case, it is known that a best approximation always exists and unique. For some special domains in $\mathbb{R}^d$, we characterize and then construct the best approximation. We also give an explicit formula for the approximation error.

1.1. A brief excursion into the approximation theory of ridge functions

In this section we briefly review some results on approximation properties of the sets $\mathcal{R}\left(\mathbf{a}^1, \ldots, \mathbf{a}^r\right)$ and $\mathcal{R}_r$. These results are presented without proofs but with discussions and complete references. We hope this section will whet the reader's appetite for the rest of this book, where a more comprehensive study of concrete mathematical problems is provided.

1.1.1. $\mathcal{R}\left(\mathbf{a}^1, \ldots, \mathbf{a}^r\right)$ – ridge functions with fixed directions. It is clear that well approximation of a multivariate function $f : X \to \mathbb{R}$ from some normed space by using elements of the set $\mathcal{R}\left(\mathbf{a}^1, \ldots, \mathbf{a}^r\right)$ is not always possible. The value of the approximation error depends not only on the approximated function f but also on geometrical structure of the given set X. This poses challenging research problems on computing the error of approximation and constructing best approximations from $\mathcal{R}\left(\mathbf{a}^1, \ldots, \mathbf{a}^r\right)$. Serious difficulties arise when one attempts to solve these problems in continuous function spaces endowed with the uniform norm. For example, let us consider the algorithm for finding best approximations, called the *Diliberto-Straus algorithm* (see [**138**]). The essence of this algorithm is as follows. Let X be a compact subset of $\mathbb{R}^d$ and A_i be a best approximation operator from the space of continuous functions $C(X)$ to the subspace of ridge functions

$G_i = \{g_i(\mathbf{a}^i \cdot \mathbf{x}) : g_i \in C(\mathbb{R}),\ \mathbf{x} \in X\}$, $i = 1, \dots, r$. That is, for each function $f \in C(X)$, the function $A_i f$ is a best approximation to f from G_i. Set

$$Tf = (I - A_r)(I - A_{r-1}) \cdots (I - A_1)f,$$

where I is the identity operator. It is clear that

$$Tf = f - g_1 - g_2 - \cdots - g_r,$$

where g_k is a best approximation from G_k to the function $f - g_1 - g_2 - \cdots - g_{k-1}$, $k = 1, \dots, r$. Consider powers of the operator T: T^2, T^3 and so on. Is the sequence $\{T^n f\}_{n=1}^{\infty}$ convergent? In case of an affirmative answer, which function is the limit of $T^n f$, as $n \to \infty$? One may expect that the sequence $\{T^n f\}_{n=1}^{\infty}$ converges to $f - g^*$, where g^* is a best approximation from $\mathcal{R}(\mathbf{a}^1, \dots, \mathbf{a}^r)$ to f. This conjecture was first stated by Diliberto and Straus [**47**] in 1951 for the uniform approximation of a multivariate function, defined on the unit cube, by sums of univariate functions (that is, sums of ridge functions with the coordinate directions). But later it was shown by Aumann [**10**] that the sequence generated by this algorithm may not converge if $r > 2$. For $r = 2$ and certain convex compact sets X, the sequence $\{\|T^n f\|\}_{n=1}^{\infty}$ converges to the approximation error $\|f - g_0\|$, where g_0 is a best approximation from $\mathcal{R}(\mathbf{a}^1, \mathbf{a}^2)$ (see [**93, 175**]). However, it is not yet clear whether $\|T^n f - (f - g_0)\|$ converges to zero as $n \to \infty$. In the case $r > 2$ no efficient algorithm is known for finding a best uniform approximation from $\mathcal{R}(\mathbf{a}^1, \dots, \mathbf{a}^r)$. Note that in the L_2 metric, the Diliberto-Straus algorithm converges as desired for an arbitrary number of distinct directions. This also holds in the L_p space setting, provided that $p > 1$ and $\mathcal{R}(\mathbf{a}^1, \dots, \mathbf{a}^r)$ is closed (see [**171**]). But in the L_1 space setting, the alternating algorithm does not work even in the case of two directions (see [**175**]).

One of the basic problems concerning the approximation by sums of ridge functions with fixed directions is the problem of verifying if a given function f belongs to the space $\mathcal{R}(\mathbf{a}^1, \dots, \mathbf{a}^r)$. This problem has a simple solution if the space dimension $d = 2$ and a given function $f(x, y)$ has partial derivatives up to r-th order. For the representation of $f(x, y)$ in the form

$$f(x, y) = \sum_{i=1}^{r} g_i(a_i x + b_i y),$$

it is necessary and sufficient that

$$\prod_{i=1}^{r} \left(b_i \frac{\partial}{\partial x} - a_i \frac{\partial}{\partial y} \right) f = 0. \tag{1.1}$$

This recipe is also valid for continuous bivariate functions provided that the derivatives are understood in the sense of distributions. Unfortunately, such a simple characterization does not carry over to the case of more than two variables. Below we provide two results concerning the general case of arbitrarily many variables.

THEOREM 1.1 (Diaconis, Shahshahani [**46**]). *Let $\mathbf{a}^1, \dots, \mathbf{a}^r$ be pairwise linearly independent vectors in $\mathbb{R}^d$. Let for $i = 1, 2, \dots, r$, H^i denote the hyperplane $\{\mathbf{c} \in \mathbb{R}^d : \mathbf{c} \cdot \mathbf{a}^i = 0\}$. Then a function $f \in C^r(\mathbb{R}^d)$ can be represented in the form*

$$f(\mathbf{x}) = \sum_{i=1}^{r} g_i(\mathbf{a}^i \cdot \mathbf{x}) + P(\mathbf{x}),$$

where $P(\mathbf{x})$ is a polynomial of degree not more than r, if and only if

$$\prod_{i=1}^{r} \sum_{s=1}^{d} c_s^i \frac{\partial f}{\partial x_s} = 0,$$

for all vectors $\mathbf{c}^i = (c_1^i, c_2^i, \ldots, c_d^i) \in H^i$, $i = 1, 2, \ldots, r$.

There are examples showing that one cannot simply dispense with the polynomial $P(\mathbf{x})$ in the above theorem (see [**46**]). In fact, a polynomial term appears in the sufficiency part of the proof of this theorem.

Lin and Pinkus [**140**] obtained more general result on the representation by sums of ridge functions with fixed directions. We need some notation to present their result. Each multivariate polynomial $p(x_1, \ldots, x_d)$ generates the differential operator $p(\frac{\partial}{\partial x_1}, \ldots, \frac{\partial}{\partial x_d})$. Let $P(\mathbf{a}^1, \ldots, \mathbf{a}^r)$ denote the set of polynomials which vanish on all the lines $\{\lambda \mathbf{a}^i, \lambda \in \mathbb{R}\}$, $i = 1, \ldots, r$. Obviously, this is an ideal in the ring of all polynomials. Let Q be the set of polynomials $q = q(x_1, \ldots, x_d)$ such that $p(\frac{\partial}{\partial x_1}, \ldots, \frac{\partial}{\partial x_d})q = 0$, for all $p(x_1, \ldots, x_d) \in P(\mathbf{a}^1, \ldots, \mathbf{a}^r)$.

THEOREM 1.2 (Lin, Pinkus [**140**])). *Let $\mathbf{a}^1, \ldots, \mathbf{a}^r$ be pairwise linearly independent vectors in $\mathbb{R}^d$. A function $f \in C(\mathbb{R}^d)$ can be expressed in the form*

$$f(\mathbf{x}) = \sum_{i=1}^{r} g_i(\mathbf{a}^i \cdot \mathbf{x}),$$

if and only if f belongs to the closure of the linear span of Q.

In [**173**], A.Pinkus considered the problems of smoothness and uniqueness in ridge function representation. For a given function $f \in \mathcal{R}\left(\mathbf{a}^1, \ldots, \mathbf{a}^r\right)$, he posed and answered the following questions. If f belongs to some smoothness class, what can we say about the smoothness of the functions g_i? How many different ways can we write f as a linear combination of ridge functions? These and similar problems will be extensively discussed in Chapter 2.

The above problem of representation of fixed functions by sums of ridge functions gives rise to the problem of representation of some classes of functions by such sums. For example, one may consider the following problem. Let X be a subset of the d-dimensional Euclidean space. Let $C(X)$, $B(X)$, $T(X)$ denote the set of continuous, bounded and all real functions defined on X, respectively. In the first case, we additionally suppose that X is a compact set. Let $\mathcal{R}_c\left(\mathbf{a}^1, \ldots, \mathbf{a}^r\right)$ and $\mathcal{R}_b\left(\mathbf{a}^1, \ldots, \mathbf{a}^r\right)$ denote the subspaces of $\mathcal{R}\left(\mathbf{a}^1, \ldots, \mathbf{a}^r\right)$ comprising only sums with continuous and bounded terms $g_i\left(\mathbf{a}^i \cdot \mathbf{x}\right)$, $i = 1, \ldots, r$, respectively. The following questions naturally arise: For which sets X,

(1) $C(X) = \mathcal{R}_c\left(\mathbf{a}^1, \ldots, \mathbf{a}^r\right)$?
(2) $B(X) = \mathcal{R}_b\left(\mathbf{a}^1, \ldots, \mathbf{a}^r\right)$?
(3) $T(X) = \mathcal{R}\left(\mathbf{a}^1, \ldots, \mathbf{a}^r\right)$?

The first two questions in a more general setting were answered in Sternfeld [**190, 193**]. The third question will be answered in the next section. Let us briefly discuss some results of Sternfeld concerning ridge function representation. These results have been mostly overlooked in the corresponding ridge function literature, as they have to do with more general superpositions of functions and do not directly mention ridge functions. Assume we are given directions $\mathbf{a}^1, \ldots, \mathbf{a}^r \in \mathbb{R}^d \backslash \{\mathbf{0}\}$ and a set $X \subseteq \mathbb{R}^d$. Following Sternfeld, we say that a family $F = \{\mathbf{a}^1, \dots, \mathbf{a}^r\}$ *uniformly separates points* of X if there exists a number $0 < \lambda \leq 1$ such that for each pair $\{\mathbf{x}_j\}_{j=1}^m$, $\{\mathbf{z}_j\}_{j=1}^m$ of disjoint finite sequences in X, there exists some direction $\mathbf{a}^k \in F$ so that if from the two sequences $\{\mathbf{a}^k \cdot \mathbf{x}_j\}_{j=1}^m$ and $\{\mathbf{a}^k \cdot \mathbf{z}_j\}_{j=1}^m$ we remove a maximal number of pairs of points $\mathbf{a}^k \cdot \mathbf{x}_{j_1}$ and $\mathbf{a}^k \cdot \mathbf{z}_{j_2}$ with $\mathbf{a}^k \cdot \mathbf{x}_{j_1} = \mathbf{a}^k \cdot \mathbf{z}_{j_2}$, then there remains at least λm points in each sequence (or, equivalently, at most $(1-\lambda)m$ pairs can be removed). Sternfeld [**190**], in particular, proved that a family of directions $F = \{\mathbf{a}^1, \dots, \mathbf{a}^r\}$ uniformly separates points of X if and only if $\mathcal{R}_b\left(\mathbf{a}^1, \dots, \mathbf{a}^r\right) = B(X)$. In [**190**], he also obtained a practically convenient sufficient condition for the equality $\mathcal{R}_b\left(\mathbf{a}^1, \dots, \mathbf{a}^r\right) = B(X)$. To describe his condition, define the set functions

$$\tau_i(Z) = \{\mathbf{x} \in Z : \ |p_i^{-1}(p_i(\mathbf{x})) \bigcap Z| \geq 2\},$$

where $Z \subset X$, $p_i(\mathbf{x}) = \mathbf{a}^i \cdot \mathbf{x}$, $i = 1, \dots, r$, and $|Y|$ denotes the cardinality of a set Y. Define $\tau(Z)$ to be $\bigcap_{i=1}^k \tau_i(Z)$ and define $\tau^2(Z) = \tau(\tau(Z))$, $\tau^3(Z) = \tau(\tau^2(Z))$ and so on inductively.

THEOREM 1.3 (Sternfeld [**190**]). *If $\tau^n(X) = \emptyset$ holds for some natural n, then $\mathcal{R}_b\left(\mathbf{a}^1, \dots, \mathbf{a}^r\right) = B(X)$. If X is a compact subset of $\mathbb{R}^d$, and $\tau^n(X) = \emptyset$ for some n, then $\mathcal{R}_c\left(\mathbf{a}^1, \dots, \mathbf{a}^r\right) = C(X)$.*

If $r = 2$, the sufficient condition "$\tau^n(X) = \emptyset$ for some n" turns out to be also necessary. In this case, the equality $\mathcal{R}_b\left(\mathbf{a}^1, \mathbf{a}^2\right) = B(X)$ is equivalent to the equality $\mathcal{R}_c\left(\mathbf{a}^1, \mathbf{a}^2\right) = C(X)$. In another work [**193**], Sternfeld obtained a measure-theoretic necessary and sufficient condition for the equality $\mathcal{R}_c\left(\mathbf{a}^1, \dots, \mathbf{a}^r\right) = C(X)$. Let $p_i(\mathbf{x}) = \mathbf{a}^i \cdot \mathbf{x}$, $i = 1, \dots, r$, X be a compact set in $\mathbb{R}^d$ and $M(X)$ be a class of measures defined on some field of subsets of X. Following Sternfeld, we say that a family $F = \{\mathbf{a}^1, \dots, \mathbf{a}^r\}$ *uniformly separates measures* of the class $M(X)$ if there exists a number $0 < \lambda \leq 1$ such that for each measure μ in $M(X)$ the equality $\left\|\mu \circ p_k^{-1}\right\| \geq \lambda \left\|\mu\right\|$ holds for some direction $\mathbf{a}^k \in F$. Sternfeld [**191, 193**], in particular, proved that the equality $\mathcal{R}_c\left(\mathbf{a}^1, \dots, \mathbf{a}^r\right) = C(X)$ holds if and only if the family of directions $\{\mathbf{a}^1, \dots, \mathbf{a}^r\}$ uniformly separates measures of the class $C(X)^*$ (that is, the class of regular Borel measures). In addition, he proved that $\mathcal{R}_b\left(\mathbf{a}^1, \dots, \mathbf{a}^r\right) = B(X)$ if and only if the family of directions $\{\mathbf{a}^1, \dots, \mathbf{a}^r\}$ uniformly separates measures of the class $l_1(X)$ (that is, the class of finite measures defined on countable subsets of X). Since $l_1(X) \subset C(X)^*$, the first equality $\mathcal{R}_c\left(\mathbf{a}^1, \dots, \mathbf{a}^r\right) = C(X)$ implies the second equality $\mathcal{R}_b\left(\mathbf{a}^1, \dots, \mathbf{a}^r\right) = B(X)$. The inverse is not true (see [**193**]). We emphasize again that the above results of Sternfeld were obtained for more general functions, than linear combinations of ridge functions, namely for functions of the form $\sum_{i=1}^r g_i(h_i(x))$, where h_i arbitrarily fixed functions (bounded or continuous) defined on X. Such functions will be discussed in Chapter 4.

1.1.2. $\mathcal{R}_r$ – ridge functions with variable directions. Obviously, the set $\mathcal{R}_c\left(\mathbf{a}^1, \dots, \mathbf{a}^r\right)$ is not dense in $C(\mathbb{R}^d)$ in the topology of uniform convergence on compact subsets of $\mathbb{R}^d$. Density here does not hold because the number of considered directions is finite. If consider all the possible directions, then the set $\mathcal{R} = span\{g(\mathbf{a} \cdot \mathbf{x}) : \ g \in C(\mathbb{R}), \ \mathbf{a} \in \mathbb{R}^d \backslash \{\mathbf{0}\}\}$ will certainly be dense in the space $C(\mathbb{R}^d)$ in the above mentioned topology. In order to be sure, it is enough to consider only the functions $e^{\mathbf{a} \cdot \mathbf{x}} \in \mathcal{R}$, the linear span of which is dense in $C(\mathbb{R}^d)$ by the Stone-Weierstrass theorem (for this theorem see [**177**, p.122]). In fact, for density it is not necessary to comprise all directions. The following theorem shows how many directions in totality satisfy the density requirements.

THEOREM 1.4 (Vostrecov and Kreines [**203**], Lin and Pinkus [**140**]). *For density of the set*

$$\mathcal{R}(\mathcal{A}) = span\{g(\mathbf{a}\cdot\mathbf{x}) : \ g \in C(\mathbb{R}), \ \mathbf{a} \in \mathcal{A} \subset \mathbb{R}^d\}$$

in $C(\mathbb{R}^d)$ (in the topology of uniform convergence on compact sets) it is necessary and sufficient that the only homogeneous polynomial which vanishes identically on $\mathcal{A}$ is the zero polynomial.

Since in the definition of $\mathcal{R}(\mathcal{A})$ we vary over all univariate functions g, allowing one direction $\mathbf{a}$ is equivalent to allowing all directions $k\mathbf{a}$ for every real k. Thus it is sufficient to consider only the set $\mathcal{A}$ of directions normalized to the unit sphere S^{n-1}. For example, if $\mathcal{A}$ is a subset of the sphere S^{n-1}, which contains an interior point (interior point with respect to the induced topology on S^{n-1}), then $\mathcal{R}(\mathcal{A})$ is dense in the space $C(\mathbb{R}^d)$. The proof of Theorem 1.4 highlights an important fact that the set $\mathcal{R}(\mathcal{A})$ is dense in $C(\mathbb{R}^d)$ in the topology of uniform convergence on compact subsets if and only if $\mathcal{R}(\mathcal{A})$ contains all the polynomials (see [**140**]).

Representability of polynomials by sums of ridge functions is a building block for many results. In many works (see, e.g., [**172**]), the following fact is fundamental: Every multivariate polynomial $h(\mathbf{x}) = h(x_1, \dots, x_d)$ of degree k can be represented in the form

$$h(\mathbf{x}) = \sum_{i=1}^{l} p_i(\mathbf{a}^i \cdot \mathbf{x}),$$

where p_i is a univariate polynomial, $\mathbf{a}^i \in \mathbb{R}^d$, and $l = \binom{d-1+k}{k}$.

For example, for the representation of a bivariate polynomial of degree k, it is needed $k+1$ univariate polynomials and $k+1$ directions (see [**141**]). The proof of this fact is organized so that the directions $\mathbf{a}^i$, $i = 1, \dots, k+1$, are chosen once for all multivariate polynomials of k-th degree. At one of the seminars in the Technion – Israel Institute of Technology in 2007, A. Pinkus posed two problems:

1) Can every multivariate polynomial of degree k be represented by less than l ridge functions?

2) How large is the set of polynomials represented by $l-1$, $l-2, \dots$ ridge functions?

Note that for bivariate polynomials the 1-st problem is solved positively, that is, the number $l = k+1$ can be reduced. Indeed, for a bivariate polynomial $P(x, y)$ of k-th degree, there exist many combinations of real numbers $c_0, \dots, c_k$ such that

$$\sum_{i=0}^{k} c_i \frac{\partial^k}{\partial x^i \partial y^{k-i}} P(x, y) = 0.$$

Further the numbers c_i, $i = 0, \dots, k$, can be selected to enjoy the property that the polynomial $\sum_{i=0}^{k} c_i t^i$ has distinct real zeros. Then it is not difficult to verify that the differential operator $\sum_{i=0}^{k} c_i \frac{\partial^k}{\partial x^i \partial y^{k-i}}$ can be written in the form

$$\prod_{i=1}^{k} \left(b_i \frac{\partial}{\partial x} - a_i \frac{\partial}{\partial y} \right),$$

for some pairwise linearly independent vectors (a_i, b_i), $i = 1, \dots, k$. Now from the above criterion (1.1) we obtain that the polynomial $P(x, y)$ can be represented as a sum of k ridge functions. Note that the problem of representation of a multivariate algebraic polynomial $P(\mathbf{x})$ in the form $\sum_{i=0}^{r} g_i(\mathbf{a}^i \cdot \mathbf{x})$ with minimal r was extensively studied in the monograph by Pinkus [**175**].

In connection with the 2-nd problem of Pinkus, V. Maiorov [**144**] studied certain geometrical properties of the manifold $\mathcal{R}_r$. Namely, he estimated the ε-entropy numbers in terms of smaller ε-covering numbers of the compact class formed by the intersection of the class $\mathcal{R}_r$ with the unit ball in the space of polynomials of degree at most s on $\mathbb{R}^d$. Let E be a Banach space and let for $x \in E$ and $\delta > 0$, $S(x, \delta)$ denote the ball of radius δ centered at the point x. For any positive number ε, the ε-covering number of a set F in the space E represents the quantity

$$L_\varepsilon(F, E) = \min\left\{ N : \exists x_1, \dots, x_N \in F \text{ such that } F \subset \bigcup_{i=1}^{N} S(x_i, \varepsilon) \right\}.$$

The ε-entropy of F is defined as the number $H_\varepsilon(F, E) \overset{def}{=} \log_2 L_\varepsilon(F, E)$. The notion of ε-entropy has been devised by A.N.Kolmogorov (see [**121, 123**]) to classify compact metric sets according to their massivity.

In order to formulate Maiorov's result, let $\mathcal{P}_s^d$ be the space of all polynomials of degree at most s on $\mathbb{R}^d$, $L_q = L_q(I)$, $1 \le q \le \infty$, be the space of q-integrable functions on the unit cube $I = [0, 1]^d$ with the norm $\|f\|_q = \left(\int_I |f(x)|^q \, dx\right)^{1/q}$, BL_q be the unit ball in the space L_q, and $B_q\mathcal{P}_s^d = BL_q \cap \mathcal{P}_s^d$ be the unit ball in the space $\mathcal{P}_s^d$ equipped with the L_q metric.

THEOREM 1.5 (Maiorov [**144**]). *Let $r, s \in \mathbb{N}$, $1 \le q \le \infty$, $0 < \varepsilon < 1$. The ε-entropy of the class $B_q\mathcal{P}_s^d \cap \mathcal{R}_r$ in the space L_q satisfies the inequalities*

1)

$$c_1 r s \le \frac{H_\varepsilon(B_q\mathcal{P}_s^d \cap \mathcal{R}_r, L_q)}{\log_2 \frac{1}{\varepsilon}} \le c_2 r s \log_2 \frac{2es^{d-1}}{r},$$

for $r \le s^{d-1}$.

2)

$$c_1' s^d \le \frac{H_\varepsilon(B_q\mathcal{P}_s^d \cap \mathcal{R}_r, L_q)}{\log_2 \frac{1}{\varepsilon}} \le c_2' s^d,$$

for $r > s^{d-1}$. In these inequalities c_1, c_2, c_1', c_2' are constants depending only on d.

Let us consider $\mathcal{R}_r$ as a subspace of some normed linear space X endowed with the norm $\|\cdot\|_X$. The error of approximation of a given function $f \in X$ by functions $g \in \mathcal{R}_r$ is defined as follows

$$E(f, \mathcal{R}_r, X) \overset{def}{=} \inf_{g \in \mathcal{R}_r} \|f - g\|_X.$$

Let B^d denote the unit ball in the space $\mathbb{R}^d$. Besides, let $\mathbb{Z}_+^d$ denote the lattice of nonnegative multi-integers in $\mathbb{R}^d$. For $k = (k_1, \dots, k_d) \in \mathbb{Z}_+^d$, set $|k| = k_1 + \dots + k_d$, $\mathbf{x}^\mathbf{k} = x_1^{k_1} \cdots x_d^{k_d}$ and

$$D^\mathbf{k} = \frac{\partial^{|k|}}{\partial^{k_1} x_1 \cdots \partial^{k_d} x_d}$$

The Sobolev space $W_p^m(B^d)$ is the space of functions defined on B^d with the norm

$$\|f\|_{m,p} = \begin{cases} \left(\sum_{0\le|\mathbf{k}|\le m} \left\|D^{\mathbf{k}}f\right\|_p^p\right)^{1/p}, & \text{if } 1 \le p < \infty \\ \max_{0\le|\mathbf{k}|\le m} \left\|D^{\mathbf{k}}f\right\|_\infty, & \text{if } p = \infty. \end{cases}$$

Here

$$\|h(\mathbf{x})\|_p = \begin{cases} \left(\int_{B^n} |h(\mathbf{x})|^p \, d\mathbf{x}\right)^{1/p}, & \text{if } 1 \le p < \infty \\ ess\sup_{\mathbf{x}\in B^d} |h(\mathbf{x})|, & \text{if } p = \infty. \end{cases}$$

Let $S_p^m(B^d)$ be the unit ball in $W_p^m(B^d)$:

$$S_p^m(B^d) = \{f \in W_p^m(B^d) : \|f\|_{m,p} \le 1\}.$$

In 1999, Maiorov [**143**] proved the following result

THEOREM 1.6 (Maiorov [**143**]). *Assume $m \ge 1$ and $d \ge 2$. Then for each $r \in \mathbb{N}$ there exists a function $f \in S_2^m(B^d)$ such that*

$$E(f, \mathcal{R}_r, L_2) \ge Cr^{-m/(d-1)}, \tag{1.2}$$

where C is a constant independent of f and r.

For $d = 2$, this inequality was proved by Oskolkov [**167**]. In [**143**], Maiorov also proved that for each function $f \in S_2^m(B^d)$

$$E(f, \mathcal{R}_r, L_2) \le Cr^{-m/(d-1)}. \tag{1.3}$$

Thus he established the following order for the error of approximation to functions in $S_2^m(B^d)$ from the class $\mathcal{R}_r$:

$$E(S_2^m(B^d), \mathcal{R}_r, L_2) \overset{def}{=} \sup_{f\in S_2^m(B^d)} E(f, \mathcal{R}_r, L_2) \asymp r^{-m/(d-1)}.$$

Pinkus [**172**] revealed that the upper bound (1.3) is also valid in the L_p metric ($1 \le p \le \infty$). In other words, for every function $f \in S_p^m(B^d)$

$$E(f, \mathcal{R}_r, L_p) \le Cr^{-m/(d-1)}.$$

These inequalities were successfully applied to some problems of approximation of multivariate functions by neural networks with a single hidden layer. Recall that such networks are given by the formula $\sum_{i=1}^r c_i\sigma(\mathbf{w}^i{\cdot}\mathbf{x}-\theta_i)$. By $\mathcal{M}_r(\sigma)$ let us denote the set of all single hidden layer networks with the activation function σ. That is,

$$\mathcal{M}_r(\sigma) = \left\{\sum_{i=1}^r c_i\sigma(\mathbf{w}^i{\cdot}\mathbf{x} - \theta_i) : \ c_i, \theta_i \in \mathbb{R}, \ \mathbf{w}^i \in \mathbb{R}^d\right\}.$$

The above results on ridge approximation from $\mathcal{R}_r$ enable us to estimate the rate with which the approximation error $E(f, \mathcal{M}_r(\sigma), L_2)$ tends to zero. First note that $\mathcal{M}_r(\sigma) \subset \mathcal{R}_r$, since each function of the form $\sigma(\mathbf{w}\cdot\mathbf{x} - \theta)$ is a ridge function with the direction $\mathbf{w}$. Thus the lower bound (1.2) holds also for the set $\mathcal{M}_r(\sigma)$: there exists a function $f \in S_2^m(B^d)$ for which

$$E(f, \mathcal{M}_r(\sigma), L_2) \ge Cr^{-m/(d-1)}.$$

It remains to see whether the upper bound (1.3) is valid for $\mathcal{M}_r(\sigma)$. Clearly, it cannot be valid if σ is an arbitrary continuous function. Here we are dealing with the question if there exists a function $\sigma^* \in C(\mathbb{R})$, for which

$$E(f, \mathcal{M}_r(\sigma^*), L_2) \le Cr^{-m/(d-1)}.$$

This question is answered affirmatively by the following result.

THEOREM 1.7 (Maiorov, Pinkus [**147**]). *There exists a function $\sigma^* \in C(\mathbb{R})$ with the following properties*

1) σ^* is infinitely differentiable and strictly increasing;

2) $\lim_{t\to\infty} \sigma^*(t) = 1$ and $\lim_{t\to-\infty} \sigma^*(t) = 0$;

3) for every $g \in \mathcal{R}_r$ and $\varepsilon > 0$ there exist $c_i, \theta_i \in \mathbb{R}$ and $\mathbf{w}^i \in \mathbb{R}^d$ satisfying

$$\sup_{\mathbf{x}\in B^d} \left| g(\mathbf{x}) - \sum_{i=1}^{r+d+1} c_i \sigma^*(\mathbf{w}^i \cdot \mathbf{x} - \theta_i) \right| < \varepsilon.$$

Temlyakov [**197**] considered the approximation from some certain subclass of $\mathcal{R}_r$ in L_2 metric. More precisely, he considered the approximation of a function $f \in L_2(D)$, where D is the unit disk in $\mathbb{R}^2$, by functions $\sum_{i=1}^{r} g_i(\mathbf{a}^i \cdot \mathbf{x}) \in \mathcal{R}_r \cap L_2(D)$, which satisfy the additional condition $\left\| g_i(\mathbf{a}^i \cdot \mathbf{x}) \right\|_2 \leq B \left\| f \right\|_2$, $i = 1, \dots, r$ (B is a given positive number). Let $\sigma_r^B(f)$ be the error of this approximation. For this approximation error, the author of [**197**] obtained upper and lower bounds. Let, for $\alpha > 0$, $H^\alpha(D)$ denote the set of all functions $f \in L_2(D)$, which can be represented in the form

$$f = \sum_{n=1}^{\infty} P_n,$$

where P_n are bivariate algebraic polynomials of total degree $2^n - 1$ satisfying the inequalities

$$\left\| P_n \right\|_2 \leq 2^{-\alpha n}, \; n = 1, 2, \dots$$

THEOREM 1.8 (Temlyakov [**197**]). *1) For every $f \in H^\alpha(D)$, we have*

$$\sigma_r^1(f) \leq C(\alpha) r^{-\alpha}.$$

2) For any given $\alpha > 0$, $B > 0$, $r > 1$, there exists a function $f \in H^\alpha(D)$ such that

$$\sigma_r^B(f) \geq C(\alpha, B)(r \ln r)^{-\alpha}.$$

Petrushev [**170**] proved the following interesting result: Let X_k be the k dimensional linear space of univariate functions in $L_2[-1,1]$, $k = 1, 2, \dots$. Besides, let B^d and S^{d-1} denote correspondingly the unit ball and unit sphere in the space $\mathbb{R}^d$. If X_k provides order of approximation $O(k^{-m})$ for univariate functions with m derivatives in $L_2[-1,1]$ and Ω_k are appropriately chosen finite sets of directions distributed on S^{d-1}, then the space $Y_k = span\{p_k(\mathbf{a} \cdot \mathbf{x}) : \; p_k \in X_k, \; \mathbf{a} \in \Omega_k\}$ will provide approximation of order $O(k^{-m-d/2+1/2})$ for every function $f \in L_2(B^d)$ with smoothness of order $m + d/2 - 1/2$. Thus, Petrushev showed that the above form of ridge approximation has the same efficiency of approximation as the traditional multivariate polynomial approximation.

Many other results concerning the approximation of multivariate functions by functions from the set $\mathcal{R}_r$ and their applications in neural network theory may be found in [**89, 136, 147, 172, 178**].

At the end of this section we want to draw the reader's attention to the problem of identifying ridge functions with variable directions. Assume that $f \in \mathcal{R}_r$, i.e.,

$$f(\mathbf{x}) = \sum_{i=1}^{r} g_i\left(\mathbf{a}^i \cdot \mathbf{x}\right) \tag{1.4}$$

for some set of unknown functions g_i and unknown directions $\mathbf{a}^i \in \mathbb{R}^d \backslash \{\mathbf{0}\}$, $i = 1, \ldots, r$. Can we determine the g_i and $\mathbf{a}^i$? In other words, we want to find some g_i and $\mathbf{a}^i$ satisfying (1.4). Note that g_i and $\mathbf{a}^i$ are not uniquely determined. Specifically, we can always replace $\mathbf{a}^i$ by $c\mathbf{a}^i$ for any constant $c \neq 0$, and appropriately alter g_i. This problem was discussed in Buhmann and Pinkus [**20**] (see also [**175**, p.30]). They gave a general recipe for finding the unknowns under certain assumptions. The suggested method first finds the directions $\mathbf{a}^i$, and then the functions g_i. Note that we have a simple method of determining the $\mathbf{a}^i$ only in the case $r = 1$. Let us focus on this case. Assume that

$$f(\mathbf{x}) = g\left(\mathbf{a} \cdot \mathbf{x}\right)$$

for some unknown g and $\mathbf{a} = (a_1, \ldots, a_d)$. Assume also that f, and hence g, is continuously differentiable. Then

$$\frac{\partial f}{\partial x_j}(\mathbf{x}) = a_j g'\left(\mathbf{a} \cdot \mathbf{x}\right),$$

for $j = 1, \ldots, d$. Taking ratios we have

$$\frac{\frac{\partial f}{\partial x_j}(\mathbf{x})}{\frac{\partial f}{\partial x_k}(\mathbf{x})} = \frac{a_j}{a_k},$$

for $j, k = 1, \ldots, d$, provided that $a_k \neq 0$ and $g'\left(\mathbf{a} \cdot \mathbf{x}\right) \neq 0$. Thus, knowing all the ratios a_j/a_k we can find the vector $\mathbf{a}$ up to multiplication by a constant.

Unfortunately, for $r > 1$ no simple and practically useful recipe for determining $\mathbf{a}^i$ is known. However, there is an approach, which tells us that it is generally theoretically possible to determine the $\mathbf{a}^i$ from a knowledge of f, under various restrictive assumptions (see [**175**, p.32]). It should be remarked that once we find the appropriate $\mathbf{a}^i$, we can find the associated g_i by applying the known methods (see [**175**, p.28] and Section 2.3).

The above problem is related to the recovery of ridge functions in the analysis of large multivariate data sets. In the recovery problem, the aim is to construct an approximation f_{app} of an unknown function $f : X \subseteq \mathbb{R}^d \to \mathbb{R}$, where d is very large, from samples $f(x_1), \ldots, f(x_n)$ within a small approximation error $\|f - f_{app}\|_{\infty}$. About f it is only known that f belongs to some class of functions $\mathcal{F}_d$, which describes certain model assumptions on f. A measure for the difficulty of the recovery problem is the so-called *information complexity* $n(\varepsilon, \mathcal{F}_d)$ (see [**163, 199**]). To be more precise, let $M_n : \mathcal{F}_d \to \mathbb{R}^n$ be a suitable measurement operator and $H_n : \mathbb{R}^n \to L_\infty(X)$ be a recovery map. For example, M_n can give n samples $f(x_1), \ldots, f(x_n)$ of f and H_n can be an interpolation operator. For each n we define the approximation numbers

$$e(n, \mathcal{F}_d) = \inf_{H_n, M_n} \sup_{f \in \mathcal{F}_d} \|f - H_n[M_n(f)]\|_\infty$$

and the information complexity

$$n(\varepsilon, \mathcal{F}_d) = \inf\{n : e(n, \mathcal{F}_d) \leq \varepsilon\},$$

where $\varepsilon > 0$. Roughly speaking, the information complexity is the smallest number n such that there is an algorithm which evaluates n samples with worst case error at most ε, irrespective of which function $f \in \mathcal{F}_d$ is chosen as an input for the algorithm. The general question is which classes $\mathcal{F}_d$ make the recovery problem efficiently solvable or, in *information-based complexity* terminology, *tractable*. Following Novak and Wozniakowski [**163**], we say that a problem is tractable if $n(\varepsilon, \mathcal{F}_d)$ does not depend exponentially on ε^{-1} and d. If $n(\varepsilon, \mathcal{F}_d)$ depends exponentially on d, then we say that a problem suffers from the *curse of dimensionality*. Note that there are also other notions of tractability (see [**163**]). The main result of Novak and Woźniakowski [**163**] says that if $\mathcal{F}_d$ is a linear space of functions $f : [0,1]^d \to \mathbb{R}$ that are infinitely differentiable with respect to all variables and the partial derivatives of f are uniformly bounded, then $n(\varepsilon, \mathcal{F}_d) \geq 2^{\lfloor d/2 \rfloor}$ for all $\varepsilon \in (0,1)$ and $d \in \mathbb{N}$. Thus the number of samples to approximate even a C^∞-function grows exponentially with the dimension d. This means that for such functions there is no efficient solution to the recovery problem in high dimensions, more precisely, the problem is intractable and suffers from the curse of dimensionality. It should be remarked that the tractability of approximation problems is an actively studied topic in the developing area of *information-based complexity* (see, e.g., [**164, 165, 199**]).

The above arguments show that the approximation of functions depending on a large number of variables is particularly difficult even with smoothness assumptions on f. If we want the recovery problem to be efficiently solvable in high dimensions, we need additional assumptions, different from just regularity. One popular approach to break the curse of dimensionality and make the recovery problem tractable is to assume a priory that f is a ridge function $g(\mathbf{a}\cdot\mathbf{x})$, or a generalized ridge function of the form $g(A\mathbf{x})$, where A is a $k \times d$ matrix, with k considerably smaller than d, and g is a function of k variables. Such assumption is reasonable, since ridge function structure is fundamental to any empirical model as a consequence of the dimensional analysis (see, e.g., [**36, 37**]).

Recent papers on this topic propose and analyze algorithms for ridge recovery. For example, Cohen et al. [**35**] investigate the recovery problem for ridge functions defined on the unit cube, with condition $g \in C^s[0,1]$ for $s \geq 1$, $\|\mathbf{a}\|_1 = 1$ and $a_i \geq 0$, $i = 1, \ldots d$. Doerr and Mayer [**49**] continues the research from [**35**] and impose the following condition on the vector $\mathbf{a}$: $\|\mathbf{a}\|_p \leq M$ for some $p \in (0,1)$ and $M \geq 1$. Fornasier et al. [**53**] considers the recovery of generalized ridge functions $g(A\mathbf{x})$ under certain conditions on A and assuming that $g \in C^s[0,1]^k$ for $s \geq 2$. Zaitseva et al. [**205**] propose a polynomial time algorithm for recovering $g(\mathbf{a}\cdot\mathbf{x})$ provided that g is analytic. Some other related results may be found, e.g., in [**62, 112, 152, 201**].

1.2. Representation of multivariate functions by linear combinations of ridge functions

In this section we develop a technique for verifying if a multivariate function can be expressed as a sum of ridge functions with given directions. We also obtain a necessary and sufficient condition for the representation of all multivariate functions on a subset X of $\mathbb{R}^d$ by sums of ridge functions with fixed directions.

1.2.1. Two representation problems. Let X be a subset of $\mathbb{R}^d$ and $\{\mathbf{a}^i\}_{i=1}^r$ be arbitrarily fixed nonzero directions (vectors) in $\mathbb{R}^d$. Consider the following set of linear combinations of ridge functions.

$$\mathcal{R}(\mathbf{a}^1,\dots,\mathbf{a}^r;X)=\left\{\sum_{i=1}^{r} g_i(\mathbf{a}^i\cdot\mathbf{x}),\ \mathbf{x}\in X,\ g_i:\mathbb{R}\to\mathbb{R},\ i=1,\dots,r\right\}$$

In this section, we are going to deal with the following two problems:

PROBLEM 1. *What conditions imposed on $f: X\to\mathbb{R}$ are necessary and sufficient for the inclusion $f\in\mathcal{R}(\mathbf{a}^1,\dots,\mathbf{a}^r;X)$?*

PROBLEM 2. *What conditions imposed on X are necessary and sufficient that every function defined on X belongs to the space $\mathcal{R}(\mathbf{a}^1,\dots,\mathbf{a}^r;X)$?*

As noticed in Section 1.1, Problem 1 was considered for continuous functions in [**140**] and a theoretical result was obtained. It was also noticed there that the similar problem of representation of f in the form $\sum_{i=1}^{r} g_i(\mathbf{a}^i\cdot\mathbf{x})+P(\mathbf{x})$ with polynomial $P(\mathbf{x})$ was solved for continuously differentiable functions in [**46**]. Problem 2 was solved in [**18**] for finite subsets X of $\mathbb{R}^d$ and in [**119**] for the case when $r=d$ and $\mathbf{a}^i$ are the coordinate directions.

Here we consider both Problem 1 and Problem 2 without imposing on X, f and r any conditions. In fact, we solve these problems for more general, than $\mathcal{R}(\mathbf{a}^1,\dots,\mathbf{a}^r;X)$, set of functions. Namely, we solve them for the set

$$\mathcal{B}(X)=\mathcal{B}(h_1,\dots,h_r;X)=\left\{\sum_{i=1}^{r} g_i(h_i(x)),\ x\in X,\ g_i:\mathbb{R}\to\mathbb{R},\ i=1,\dots,r\right\},$$

where $h_i: X\to\mathbb{R}$, $i=1,\dots,r$, are arbitrarily fixed functions. In particular, the functions h_i, $i=1,\dots,r$, may be equal to scalar products of the variable $\mathbf{x}$ with some vectors $\mathbf{a}^i$, $i=1,\dots,r$. Only in this special case, we have $\mathcal{B}(h_1,\dots,h_r;X)=\mathcal{R}(\mathbf{a}^1,\dots,\mathbf{a}^r;X)$.

1.2.2. Cycles. The main idea leading to solutions of the above problems is in using new objects called *cycles* with respect to r functions $h_i: X\to\mathbb{R}$, $i=1,\dots,r$ (and in particular, with respect to r directions $\mathbf{a}^1,\dots,\mathbf{a}^r$). In the sequel, by δ_A we will denote the characteristic function of a set $A\subset\mathbb{R}$. That is,

$$\delta_A(y)=\begin{cases}1, & if\ y\in A\\ 0, & if\ y\notin A.\end{cases}$$

DEFINITION 1.9. Given a subset $X\subset\mathbb{R}^d$ and functions $h_i: X\to\mathbb{R}$, $i=1,\dots,r$. A set of points $\{x_1,\dots,x_n\}\subset X$ is called a cycle with respect to the functions $h_1,\dots,h_r$ (or, concisely, a cycle if there is no confusion), if there exists a vector $\lambda=(\lambda_1,\dots,\lambda_n)$ with the nonzero real coordinates λ_i, $i=1,\dots,n$, such that

$$\sum_{j=1}^{n}\lambda_j\delta_{h_i(x_j)}=0,\ i=1,\dots,r. \tag{1.5}$$

If $h_i=\mathbf{a}^i\cdot\mathbf{x}$, $i=1,\dots,r$, where $\mathbf{a}^1,\dots,\mathbf{a}^r$ are some directions in $\mathbb{R}^d$, a cycle, with respect to the functions $h_1,\dots,h_r$, is called a cycle with respect to the directions $\mathbf{a}^1,\dots,\mathbf{a}^r$.

Let for $i=1,\dots,r$, the set $\{h_i(x_j),\ j=1,\dots,n\}$ have k_i different values. Then it is not difficult to see that Eq. (1.5) stands for a system of $\sum_{i=1}^{r}k_i$ homogeneous linear equations in unknowns $\lambda_1,\dots,\lambda_n$. If this system has any solution with the nonzero components, then the given set $\{x_1, \dots, x_n\}$ is a cycle. In the last case, the system has also a solution $m = (m_1, \dots, m_n)$ with the nonzero integer components m_i, $i = 1, \dots, n$. Thus, in Definition 1.9, the vector $\lambda = (\lambda_1, \dots, \lambda_n)$ can be replaced with a vector $m = (m_1, \dots, m_n)$ with $m_i \in \mathbb{Z}\backslash\{0\}$.

For example, the set $l = \{(0,0,0),\ (0,0,1),\ (0,1,0),\ (1,0,0),\ (1,1,1)\}$ is a cycle in $\mathbb{R}^3$ with respect to the functions $h_i(z_1, z_2, z_3) = z_i,\ i = 1,2,3$. The vector λ in Definition 1.9 can be taken as $(2,1,1,1,-1)$.

In case $r = 2$, the picture of cycles becomes more clear. Let, for example, h_1 and h_2 be the coordinate functions on $\mathbb{R}^2$. In this case, a cycle is the union of some sets A_k with the property: each A_k consists of vertices of a closed broken line with the sides parallel to the coordinate axis. These objects (sets A_k) have been exploited in practically all works devoted to the approximation of bivariate functions by univariate functions, although under various different names (see "bolt of lightning" in Section 1.3. If the functions h_1 and h_2 are arbitrary, the sets A_k can be described as a trace of some point traveling alternatively in the level sets of h_1 and h_2, and then returning to its primary position. It should be remarked that in the case $r > 2$, cycles do not admit such a simple geometric description. We refer the reader to Braess and Pinkus [**18**] for the description of cycles when $r = 3$ and $h_i(\mathbf{x}) = \mathbf{a}^i \cdot \mathbf{x}$, $\mathbf{x} \in \mathbb{R}^2$, $\mathbf{a}^i \in \mathbb{R}^2\backslash\{\mathbf{0}\}$, $i = 1,2,3$.

Let $T(X)$ denote the set of all functions on X. With each pair $\langle p, \lambda \rangle$, where $p = \{x_1, \dots, x_n\}$ is a cycle in X and $\lambda = (\lambda_1, \dots, \lambda_n)$ is a vector known from Definition 1.9, we associate the functional

$$G_{p,\lambda} : T(X) \to \mathbb{R},\ G_{p,\lambda}(f) = \sum_{j=1}^{n} \lambda_j f(x_j).$$

In the following, such pairs $\langle p, \lambda \rangle$ will be called *cycle-vector pairs* of X. It is clear that the functional $G_{p,\lambda}$ is linear and $G_{p,\lambda}(g) = 0$ for all functions $g \in \mathcal{B}(h_1, \dots, h_r; X)$.

LEMMA 1.10. *Let X have cycles and $h_i(X) \cap h_j(X) = \varnothing$, for all $i, j \in \{1, \dots, r\}$, $i \neq j$. Then a function $f : X \to \mathbb{R}$ belongs to the set $\mathcal{B}(h_1, \dots, h_r; X)$ if and only if $G_{p,\lambda}(f) = 0$ for any cycle-vector pair $\langle p, \lambda \rangle$ of X.*

PROOF. The necessity is obvious, since the functional $G_{p,\lambda}$ annihilates all members of $\mathcal{B}(h_1, \dots, h_r; X)$. Let us prove the sufficiency. Introduce the notation

$$\begin{aligned} Y_i &= h_i(X),\ i = 1, \dots, r; \\ \Omega &= Y_1 \cup \dots \cup Y_r. \end{aligned}$$

Consider the following set

(1.6)
$$\mathcal{L} = \{Y = \{y_1, \dots, y_r\} : \text{if there is } x \in X \text{ such that } h_i(x) = y_i,\ i = 1, \dots, r\}.$$

Note that $\mathcal{L}$ is not a subset of Ω. It is a set of some certain subsets of Ω. Each element of $\mathcal{L}$ is a set $Y = \{y_1, \dots, y_r\} \subset \Omega$ with the property that there exists $x \in X$ such that $h_i(x) = y_i,\ i = 1, \dots, r$.

In what follows, all the points x associated with Y by (1.6) will be called $(*)$-points of Y. It is clear that the number of such points depends on Y as well as on the functions $h_1, \dots, h_r$, and may be greater than 1. But note that if any two points x_1 and x_2 are $(*)$-points of Y, then the set $\{x_1,\ x_2\}$ necessarily forms a cycle with the associated vector $\lambda_0 = (1; -1)$. Indeed, if x_1 and x_2 are $(*)$-points of Y, then $h_i(x_1) = h_i(x_2)$, $i = 1, \ldots, r$, whence

$$1 \cdot \delta_{h_i(x_1)} + (-1) \cdot \delta_{h_i(x_2)} \equiv 0, \ i = 1, \ldots, r.$$

The last identity means that the set $p_0 = \{x_1, \ x_2\}$ forms a cycle and $\lambda_0 = (1; -1)$ is an associated vector. Then by the the sufficiency condition, $G_{p_0, \lambda_0}(f) = 0$, whcih yields that $f(x_1) = f(x_2)$.

Let now Y^* be the set of all $(*)$-points of Y. Since we have already known that $f(Y^*)$ is a single number, we can define the function

$$t : \mathcal{L} \to \mathbb{R}, \ t(Y) = f(Y^*).$$

Or, equivalently, $t(Y) = f(x)$, where x is an arbitrary $(*)$-point of Y.

Consider now a class $\mathcal{S}$ of functions of the form $\sum_{j=1}^{k} r_j \delta_{D_j}$, where k is a positive integer, r_j are real numbers and D_j are elements of $\mathcal{L}$, $j = 1, \ldots, k$. We fix neither the numbers k, r_j, nor the sets D_j. Clearly, $\mathcal{S}$ is a linear space. Over $\mathcal{S}$, we define the functional

$$F : \mathcal{S} \to \mathbb{R}, \ F\left(\sum_{j=1}^{k} r_j \delta_{D_j}\right) = \sum_{j=1}^{k} r_j t(D_j).$$

First of all, we must show that this functional is well defined. That is, the equality

$$\sum_{j=1}^{k_1} r_j' \delta_{D_j'} = \sum_{j=1}^{k_2} r_j'' \delta_{D_j''}$$

always implies the equality

$$\sum_{j=1}^{k_1} r_j' t(D_j') = \sum_{j=1}^{k_2} r_j'' t(D_j'').$$

In fact, this is equivalent to the implication

$$\sum_{j=1}^{k} r_j \delta_{D_j} = 0 \Longrightarrow \sum_{j=1}^{k} r_j t(D_j) = 0, \ \text{for all } k \in \mathbb{N}, \ r_j \in \mathbb{R}, \ D_j \subset \mathcal{L}. \tag{1.7}$$

Suppose that the left-hand side of the implication (1.7) be satisfied. Each set D_j consists of r real numbers $y_1^j, \ldots, y_r^j$, $j = 1, \ldots, k$. By the hypothesis of the lemma, all these numbers are different. Therefore,

$$\delta_{D_j} = \sum_{i=1}^{r} \delta_{y_i^j}, \ j = 1, \ldots, k. \tag{1.8}$$

Eq. (1.8) together with the left-hand side of (1.7) gives

$$\sum_{i=1}^{r} \sum_{j=1}^{k} r_j \delta_{y_i^j} = 0. \tag{1.9}$$

Since the sets $\{y_i^1, y_i^2, \ldots, y_i^k\}$, $i = 1, \ldots, r$, are pairwise disjoint, we obtain from (1.9) that

$$\sum_{j=1}^{k} r_j \delta_{y_i^j} = 0, \ i = 1, \ldots, r. \tag{1.10}$$

Let now $x_1, \dots, x_k$ be some $(*)$-points of the sets $D_1, \dots, D_k$ respectively. Since by (1.6), $y_i^j = h_i(x_j)$, for $i = 1, \dots, r$ and $j = 1, \dots, k$, it follows from (1.10) that the set $\{x_1, \dots, x_k\}$ is a cycle. Then by the condition of the sufficiency, $\sum_{j=1}^k r_j f(x_j) = 0$. Hence $\sum_{j=1}^k r_j t(D_j) = 0$. We have proved the implication (1.7) and hence the functional F is well defined. Note that the functional F is linear (this can be easily seen from its definition).

Consider now the following space:

$$\mathcal{S}' = \left\{ \sum_{j=1}^{k} r_j \delta_{\omega_j} \right\},$$

where $k \in \mathbb{N}$, $r_j \in \mathbb{R}$, $\omega_j \subset \Omega$. As above, we do not fix the parameters k, r_j and ω_j. Clearly, the space $\mathcal{S}'$ is larger than $\mathcal{S}$. Let us prove that the functional F can be linearly extended to the space $\mathcal{S}'$. So, we must prove that there exists a linear functional $F' : \mathcal{S}' \to \mathbb{R}$ such that $F'(x) = F(x)$, for all $x \in \mathcal{S}$. Let H denote the set of all linear extensions of F to subspaces of $\mathcal{S}'$ containing $\mathcal{S}$. The set H is not empty, since it contains a functional F. For each functional $v \in H$, let $dom(v)$ denote the domain of v. Consider the following partial order in H: $v_1 \leq v_2$, if v_2 is a linear extension of v_1 from the space $dom(v_1)$ to the space $dom(v_2)$. Let now P be any chain (linearly ordered subset) in H. Consider the following functional u defined on the union of domains of all functionals $p \in P$:

$$u : \bigcup_{p \in P} dom(p) \to \mathbb{R}, \ u(x) = p(x), \text{ if } x \in dom(p)$$

Obviously, this functional is well defined and linear. Besides, the functional u provides an upper bound for P. We see that the arbitrarily chosen chain P has an upper bound. Then by Zorn's lemma, there is a maximal element $F' \in H$. We claim that the functional F' must be defined on the whole space $\mathcal{S}'$. Indeed, if F' is defined on a proper subspace $\mathcal{D} \subset \mathcal{S}'$, then it can be linearly extended to a space larger than $\mathcal{D}$ by the following way: take any point $x \in \mathcal{S}' \backslash \mathcal{D}$ and consider the linear space $\mathcal{D}' = \{\mathcal{D} + \alpha x\}$, where α runs through all real numbers. For an arbitrary point $y + \alpha x \in \mathcal{D}'$, set $F''(y + \alpha x) = F'(y) + \alpha b$, where b is any real number considered as the value of F'' at x. Thus, we constructed a linear functional $F'' \in H$ satisfying $F' \leq F''$. The last contradicts the maximality of F'. This means that the functional F' is defined on the whole $\mathcal{S}'$ and $F \leq F'$ (F' is a linear extension of F).

Define the following functions by means of the functional F':

$$g_i : Y_i \to \mathbb{R}, \ g_i(y_i) \stackrel{def}{=} F'(\delta_{y_i}), \ i = 1, \dots, r.$$

Let x be an arbitrary point in X. Obviously, x is a $(*)$-point of some set $Y = \{y_1, \dots, y_r\} \subset \mathcal{L}$. Thus,

$$\begin{aligned} f(x) &= t(Y) = F(\delta_Y) = F\left(\sum_{i=1}^{r} \delta_{y_i}\right) = F'\left(\sum_{i=1}^{r} \delta_{y_i}\right) = \\ \sum_{i=1}^{r} F'(\delta_{y_i}) &= \sum_{i=1}^{r} g_i(y_i) = \sum_{i=1}^{r} g_i(h_i(x)). \end{aligned}$$

□

1.2.3. Minimal cycles and the main results.

DEFINITION 1.11. A cycle $p = \{x_1, \dots, x_n\}$ is said to be minimal if p does not contain any cycle as its proper subset.

For example, the set $l = \{(0,0,0),\ (0,0,1),\ (0,1,0),\ (1,0,0),\ (1,1,1)\}$ considered above is a minimal cycle with respect to the functions $h_i(z_1, z_2, z_3) = z_i$, $i = 1, 2, 3$. Adding the point $(0,1,1)$ to l, we will have a cycle, but not minimal. The vector λ associated with $l \cup \{(0,1,1)\}$ can be taken as $(3, -1, -1, -2, 2, -1)$.

A minimal cycle $p = \{x_1, \dots, x_n\}$ has the following obvious properties:

(a) *The vector λ associated with p through Eq.* (1.5) *is unique up to multiplication by a constant;*
(b) *If in* (1.5), $\sum_{j=1}^{n} |\lambda_j| = 1$, *then all the numbers* λ_j, $j = 1, \dots, n$, *are rational.*

Thus, a minimal cycle p uniquely (up to a sign) defines the functional

$$G_p(f) = \sum_{j=1}^{n} \lambda_j f(x_j), \quad \sum_{j=1}^{n} |\lambda_j| = 1.$$

LEMMA 1.12. *The functional $G_{p,\lambda}$ is a linear combination of the functionals $G_{p_1}, \dots, G_{p_k}$, where $p_1, \dots, p_k$ are minimal cycles in p.*

PROOF. Let $\langle p, \lambda \rangle$ be a cycle-vector pair of X, where $p = \{x_1, \dots, x_n\}$ and $\lambda = (\lambda_1, \dots, \lambda_n)$. Let $p_1 = \{y_1^1, \dots, y_{s_1}^1\}$, $s_1 < n$, be a minimal cycle in p and

$$G_{p_1}(f) = \sum_{j=1}^{s_1} \nu_j^1 f(y_j^1), \quad \sum_{j=1}^{s_1} |\nu_j^1| = 1.$$

Without loss of generality, we may assume that $y_1^1 = x_1$. Put

$$t_1 = \frac{\lambda_1}{\nu_1^1}.$$

Then the functional $G_{p,\lambda} - t_1 G_{p_1}$ has the form

$$G_{p,\lambda} - t_1 G_{p_1} = \sum_{j=1}^{n_1} \lambda_j^1 f(x_j^1),$$

where $x_j^1 \in p$, $\lambda_j^1 \neq 0$, $j = 1, \dots, n_1$. Clearly, the set $l_1 = \{x_1^1, \dots, x_{n_1}^1\}$ is a cycle in p with the associated vector $\lambda^1 = (\lambda_1^1, \dots, \lambda_{n_1}^1)$. Besides, $x_1 \notin l_1$. Thus, $n_1 < n$ and $G_{l_1,\lambda^1} = G_{p,\lambda} - t_1 G_{p_1}$. If l_1 is minimal, then the proof is completed. Assume l_1 is not minimal. Let $p_1 = \{y_1^2, \dots, y_{s_2}^2\}$, $s_2 < n_1$, be a minimal cycle in l_1 and

$$G_{p_2}(f) = \sum_{j=1}^{s_2} \nu_j^2 f(y_j^2), \quad \sum_{j=1}^{s_2} |\nu_j^2| = 1.$$

Without loss of generality, we may assume that $y_1^2 = x_1^1$. Put

$$t_2 = \frac{\lambda_1^1}{\nu_1^2}.$$

Then the functional $G_{l_1,\lambda^1} - t_2 G_{p_2}$ has the form

$$G_{l_1,\lambda^1} - t_2 G_{p_2} = \sum_{j=1}^{n_2} \lambda_j^2 f(x_j^2),$$

where $x_j^2 \in l_1$, $\lambda_j^2 \neq 0$, $j = 1, \ldots, n_2$. Clearly, the set $l_2 = \{x_1^2, \ldots, x_{n_2}^2\}$ is a cycle in l_1 with the associated vector $\lambda^2 = (\lambda_1^2, \ldots, \lambda_{n_2}^2)$. Besides, $x_1^1 \notin l_2$. Thus, $n_2 < n_1$ and $G_{l_2,\lambda^2} = G_{l_1,\lambda^1} - t_2 G_{p_2}$. If l_2 is minimal, then the proof is completed. Let l_2 be not minimal. Repeating the above process for l_2, then for l_3, etc., after some $k-1$ steps we will come to a minimal cycle l_{k-1} and the functional

$$G_{l_{k-1},\lambda^{k-1}} = G_{l_{k-2},\lambda^{k-2}} - t_{k-1} G_{p_{k-1}} = \sum_{j=1}^{n_{k-1}} \lambda_j^{k-1} f(x_j^{k-1}).$$

Since the cycle l_{k-1} is minimal,

$$G_{l_{k-1},\lambda^{k-1}} = t_k G_{l_{k-1}}, \quad \text{where } t_k = \sum_{j=1}^{n_{k-1}} \left|\lambda_j^{k-1}\right|.$$

Now putting $p_k = l_{k-1}$ and considering the above chain relations between the functionals G_{l_i,λ^i}, $i = 1, \ldots, k-1$, we obtain that

$$G_{p,\lambda} = \sum_{i=1}^{k} t_i G_{p_i}.$$

□

THEOREM 1.13. *Assume $X \subset \mathbb{R}^d$ and $h_1, \ldots, h_r$ are arbitrarily fixed real functions on X. The following assertions are valid.*

1) Let X have cycles with respect to the functions $h_1, \ldots, h_r$. A function $f : X \to \mathbb{R}$ belongs to the space $\mathcal{B}(h_1, \ldots, h_r; X)$ if and only if $G_p(f) = 0$ for any minimal cycle $p \subset X$.

2) Let X have no cycles. Then $\mathcal{B}(h_1, \ldots, h_r; X) = T(X)$.

PROOF. 1) The necessity is clear. Let us prove the sufficiency. On the strength of Lemma 1.12, it is enough to prove that if $G_{p,\lambda}(f) = 0$ for any cycle-vector pair $\langle p, \lambda \rangle$ of X, then $f \in \mathcal{B}(X)$.

Consider a system of intervals $\{(a_i, b_i) \subset \mathbb{R}\}_{i=1}^r$ such that $(a_i, b_i) \cap (a_j, b_j) = \varnothing$ for all the indices $i, j \in \{1, \ldots, r\}$, $i \neq j$. For $i = 1, \ldots, r$, let τ_i be one-to-one mappings of $\mathbb{R}$ onto (a_i, b_i). Introduce the following functions on X:

$$h_i'(x) = \tau_i(h_i(x)), \ i = 1, \ldots, r.$$

It is clear that any cycle with respect to the functions $h_1, \ldots, h_r$ is also a cycle with respect to the functions $h_1', \ldots, h_r'$, and vice versa. Besides, $h_i'(X) \cap h_j'(X) = \varnothing$, for all $i, j \in \{1, \ldots, r\}$, $i \neq j$. Then by Lemma 1.10,

$$f(x) = g_1'(h_1'(x)) + \cdots + g_r'(h_r'(x)),$$

where $g_1', \ldots, g_r'$ are univariate functions depending on f. From the last equality we obtain that

$$f(x) = g_1'(\tau_1(h_1(x))) + \cdots + g_r'(\tau_r(h_r(x))) = g_1(h_1(x)) + \cdots + g_r(h_r(x)).$$

That is, $f \in \mathcal{B}(X)$.

2) Let $f : X \to \mathbb{R}$ be an arbitrary function. First suppose that $h_i(X) \cap h_j(X) = \varnothing$, for all $i, j \in \{1, \dots, r\}$, $i \neq j$. In this case, the proof is similar to and even simpler than that of Lemma 1.10. Indeed, the set of all $(*)$-points of Y consists of a single point, since otherwise we would have a cycle with two points, which contradicts the hypothesis of the 2-nd part of the theorem. Further, well definition of the functional F becomes obvious, since the left-hand side of (1.7) also contradicts the nonexistence of cycles. Thus, as in the proof of Lemma 1.10, we can extend F to the space $\mathcal{S}'$ and then obtain the desired representation for the function f. Since f is arbitrary, $T(X) = \mathcal{B}(X)$.

Using the techniques from the proof of the 1-st part of the theorem, one can easily generalize the above argument to the case when the functions $h_1, \dots, h_r$ have arbitrary ranges. □

THEOREM 1.14. *$\mathcal{B}(h_1, \dots, h_r; X) = T(X)$ if and only if X has no cycles with respect to the functions $h_1, \dots, h_r$.*

PROOF. The sufficiency immediately follows from Theorem 1.13. To prove the necessity, assume that X has a cycle $p = \{x_1, \dots, x_n\}$. Let $\lambda = (\lambda_1, \dots, \lambda_n)$ be a vector associated with p by Eq. (1.5). Consider a function f_0 on X with the property: $f_0(x_i) = 1$, for indices i such that $\lambda_i > 0$ and $f_0(x_i) = -1$, for indices i such that $\lambda_i < 0$. For this function, $G_{p,\lambda}(f_0) \neq 0$. Then by Theorem 1.13, $f_0 \notin \mathcal{B}(X)$. Hence $\mathcal{B}(X) \neq T(X)$. The contradiction shows that X does not admit cycles. □

1.2.4. Corollaries. From Theorems 1.13 and 1.14 we obtain the following corollaries for the ridge function representation.

COROLLARY 1.15. *Assume $X \subset \mathbb{R}^d$ and $\mathbf{a}^1, \dots, \mathbf{a}^r \in \mathbb{R}^d \setminus \{\mathbf{0}\}$. The following assertions are valid.*

1) Let X have cycles with respect to the directions $\mathbf{a}^1, \dots, \mathbf{a}^r$. A function $f : X \to \mathbb{R}$ belongs to the space $\mathcal{R}(\mathbf{a}^1, \dots, \mathbf{a}^r; X)$ if and only if $G_p(f) = 0$ for any minimal cycle $p \subset X$.

2) Let X have no cycles. Then every function $f : X \to \mathbb{R}$ belongs to the space $\mathcal{R}(\mathbf{a}^1, \dots, \mathbf{a}^r; X)$.

COROLLARY 1.16. *$\mathcal{R}(\mathbf{a}^1, \dots, \mathbf{a}^r; X) = T(X)$ if and only if X has no cycles with respect to the directions $\mathbf{a}^1, \dots, \mathbf{a}^r$.*

Note that solutions to Problems 1 and 2 are given by Corollaries 1.15 and 1.16, correspondingly. Although it is not always easy to find all cycles of a given set X and even to know if X possesses a single cycle, Corollaries 1.15 and 1.16 are of more practical than theoretical character. Particular cases of Problems 1 and 2 evidence in favor of our opinion. For example, for the problem of representation by sums of two ridge functions, the picture of cycles is completely describable (see the beginning of this section). The interpretation of cycles with respect to three directions in the plane can be found in Braess and Pinkus [**18**]. A geometric description of cycles with respect to 4 and more directions is quite complicated and requires deep techniques from geometry and graph theory. This is not within the aim of our study.

From the last corollary, it follows that if representation by sums of ridge functions with fixed directions $\mathbf{a}^1, \ldots, \mathbf{a}^r$ is valid in the class of continuous functions (or in the class of bounded functions), then such representation is valid in the class of all functions. For a rigid mathematical formulation of this result, let us introduce the notation:

$$\mathcal{R}_c(\mathbf{a}^1, \ldots, \mathbf{a}^r; X) = \left\{ \sum_{i=1}^{r} g_i(\mathbf{a}^i \cdot \mathbf{x}), \ \mathbf{x} \in X, \ g_i(\mathbf{a}^i \cdot \mathbf{x}) \in C(X), \ i = 1, \ldots, r \right\}$$

and

$$\mathcal{R}_b(\mathbf{a}^1, \ldots, \mathbf{a}^r; X) = \left\{ \sum_{i=1}^{r} g_i(\mathbf{a}^i \cdot \mathbf{x}), \ \mathbf{x} \in X, \ g_i(\mathbf{a}^i \cdot \mathbf{x}) \in B(X), \ i = 1, \ldots, r \right\}$$

Here $C(X)$ and $B(X)$ denote the spaces of continuous and bounded functions defined on $X \subset \mathbb{R}^d$ correspondingly (for the first space, the set X is supposed to be compact). As we know (see Section 1.1 from the results of Sternfeld it follows that the equality $\mathcal{R}_c(\mathbf{a}^1, \ldots, \mathbf{a}^r; X) = C(X)$ implies the equality $\mathcal{R}_b(\mathbf{a}^1, \ldots, \mathbf{a}^r; X) = B(X)$. In other words, if every continuous function is represented by sums of ridge functions (with fixed directions!), then every bounded function also obeys such representation (with bounded summands). Corollaries 1.15 and 1.16 allow us to obtain the following result.

COROLLARY 1.17. *Let X be a compact subset of $\mathbb{R}^d$ and $\mathbf{a}^1, \ldots, \mathbf{a}^r$ be given directions in $\mathbb{R}^d \backslash \{\mathbf{0}\}$. If $\mathcal{R}_c(\mathbf{a}^1, \ldots, \mathbf{a}^r; X) = C(X)$, then $\mathcal{R}(\mathbf{a}^1, \ldots, \mathbf{a}^r; X) = T(X)$.*

PROOF. If every continuous function defined on $X \subset \mathbb{R}^d$ is represented by sums of ridge functions with the directions $\mathbf{a}^1, \ldots, \mathbf{a}^r$, then it can be shown by applying the same idea (as in the proof of Theorem 1.14 that the set X has no cycles with respect to the given directions. Only, because of continuity, Urysohn's lemma should be taken into account. That is, it should be taken into account that, by assuming the existence of a cycle $p_0 = \{x_1, \ldots, x_n\}$ with an associated vector $\lambda_0 = (\lambda_1, \ldots, \lambda_n)$, we can deduce from Urysohn's lemma the existence of a continuous function $u : X \to \mathbb{R}$ satisfying

1) $u(x_i) = 1$, for indices i such that $\lambda_i > 0$

2) $u(x_j) = -1$, for indices j such that $\lambda_j < 0$,

3) $-1 < u(x) < 1$, for all $x \in X \backslash p_0$.

These properties mean that $G_{p_0, \lambda_0}(u) \neq 0 \Longrightarrow u \notin \mathcal{R}_c(\mathbf{a}^1, \ldots, \mathbf{a}^r; X) \Longrightarrow \mathcal{R}_c(\mathbf{a}^1, \ldots, \mathbf{a}^r; X) \neq C(X)$.

But if X has no cycles with respect to the directions $\mathbf{a}^1, \ldots, \mathbf{a}^r$, then by Corollary 1.16, $\mathcal{R}(\mathbf{a}^1, \ldots, \mathbf{a}^r; X) = T(X)$. □

Let us now give some examples of sets over which the representation by linear combinations of ridge functions is possible.

(1) Let $r = 2$ and X be the union of two parallel lines not perpendicular to the directions $\mathbf{a}^1$ and $\mathbf{a}^2$. Then X has no cycles with respect to $\{\mathbf{a}^1, \mathbf{a}^2\}$. Therefore, by Corollary 1.16, $\mathcal{R}\left(\mathbf{a}^1, \mathbf{a}^2; X\right) = T(X)$.

(2) Let $r = 2$, $\mathbf{a}^1 = (1, 1)$, $\mathbf{a}^2 = (1, -1)$ and X be the graph of the function $y = \arcsin(\sin x)$. Then X has no cycles and hence $\mathcal{R}\left(\mathbf{a}^1, \mathbf{a}^2; X\right) = T(X)$.

(3) Assume we are given r directions $\{\mathbf{a}^j\}_{j=1}^r$ and $r+1$ points $\{\mathbf{x}^i\}_{i=1}^{r+1} \subset \mathbb{R}^d$ such that

$$\begin{aligned} \mathbf{a}^1 \cdot \mathbf{x}^i &= \mathbf{a}^1 \cdot \mathbf{x}^j \neq \mathbf{a}^1 \cdot \mathbf{x}^2, \text{ for } 1 \le i,j \le r+1,\ i,j \neq 2 \\ \mathbf{a}^2 \cdot \mathbf{x}^i &= \mathbf{a}^2 \cdot \mathbf{x}^j \neq \mathbf{a}^2 \cdot \mathbf{x}^3, \text{ for } 1 \le i,j \le r+1,\ i,j \neq 3 \\ &\ldots\ldots\ldots\ldots\ldots\ldots\ldots\ldots\ldots\ldots \\ \mathbf{a}^r \cdot \mathbf{x}^i &= \mathbf{a}^r \cdot \mathbf{x}^j \neq \mathbf{a}^r \cdot \mathbf{x}^{r+1}, \text{ for } 1 \le i,j \le r. \end{aligned}$$

The simplest data realizing these equations are the basis directions in $\mathbb{R}^d$ and the points $(0,0,\ldots,0)$, $(1,0,\ldots,0)$, $(0,1,\ldots,0)$,..., $(0,0,\ldots,1)$. From the first equation we obtain that $\mathbf{x}^2$ cannot be a point of any cycle in $X = \{\mathbf{x}^1,\ldots,\mathbf{x}^{r+1}\}$. Sequentially, from the second, third, ..., r-th equations it follows that the points $\mathbf{x}^3, \mathbf{x}^4, \ldots, \mathbf{x}^{r+1}$ also cannot be points of cycles in X, respectively. Thus the set X does not contain cycles at all. By Corollary 1.16, $\mathcal{R}\left(\mathbf{a}^1,\ldots,\mathbf{a}^r;X\right) = T(X)$.

(4) Assume we are given directions $\{\mathbf{a}^j\}_{j=1}^r$ and a curve γ in $\mathbb{R}^d$ such that for any $c \in \mathbb{R}$, γ has at most one common point with at least one of the hyperplanes $\mathbf{a}^j \cdot \mathbf{x} = c$, $j = 1,\ldots,r$. Clearly, the curve γ has no cycles and hence $\mathcal{R}\left(\mathbf{a}^1,\ldots,\mathbf{a}^r;\gamma\right) = T(\gamma)$.

Braess and Pinkus [**18**] considered the partial case of Problem 2: characterize a set of points $\left(\mathbf{x}^1,\ldots,\mathbf{x}^k\right) \subset \mathbb{R}^d$ such that for any data $\{\alpha_1,\ldots,\alpha_k\} \subset \mathbb{R}$ there exists a function $g \in \mathcal{R}\left(\mathbf{a}^1,\ldots,\mathbf{a}^r;\mathbb{R}^d\right)$ satisfying $g(\mathbf{x}^i) = \alpha_i$, $i = 1,\ldots,k$. In connection with this problem, they introduced the notion of the *NI*-property (non interpolation property) and *MNI*-property (minimal non interpolation property) of a finite set of points as follows:

Given directions $\{\mathbf{a}^j\}_{j=1}^r \subset \mathbb{R}^d \backslash \{\mathbf{0}\}$, we say that a set of points $\{\mathbf{x}^i\}_{i=1}^k \subset \mathbb{R}^d$ has the *NI*-property with respect to $\{\mathbf{a}^j\}_{j=1}^r$, if there exists $\{\alpha_i\}_{i=1}^k \subset \mathbb{R}$ such that we cannot find a function $g \in \mathcal{R}\left(\mathbf{a}^1,\ldots,\mathbf{a}^r;\mathbb{R}^d\right)$ satisfying $g(\mathbf{x}^i) = \alpha_i$, $i = 1,\ldots,k$. We say that the set $\{\mathbf{x}^i\}_{i=1}^k \subset \mathbb{R}^d$ has the *MNI*-property with respect to $\{\mathbf{a}^j\}_{j=1}^r$, if $\{\mathbf{x}^i\}_{i=1}^k$ but no proper subset thereof has the *NI*-property.

It follows from Corollary 1.16 that a set $\{\mathbf{x}^i\}_{i=1}^k$ has the *NI*-property if and only if $\{\mathbf{x}^i\}_{i=1}^k$ contains a cycle with respect to the functions $h_i = \mathbf{a}^i \cdot \mathbf{x}$, $i = 1,\ldots,r$ (or, simply, to the directions $\mathbf{a}^i$, $i = 1,\ldots,r$) and the *MNI*-property if and only if the set $\{\mathbf{x}^i\}_{i=1}^k$ itself is a minimal cycle with respect to the given directions. Taking into account this argument and Definitions 1.9 and 1.11, we obtain that the set $\{\mathbf{x}^i\}_{i=1}^k$ has the *NI*-property if and only if there is a vector $\mathbf{m} = (m_1,\ldots,m_k) \in \mathbb{Z}^k \backslash \{\mathbf{0}\}$ such that

$$\sum_{j=1}^k m_j g(\mathbf{a}^i \cdot \mathbf{x}^j) = 0,$$

for $i = 1,\ldots,r$ and all functions $g : \mathbb{R} \to \mathbb{R}$. This set has the *MNI*-property if and only if the vector $\mathbf{m}$ has the additional properties: it is unique up to multiplication by a constant and all its components are different from zero. This special consequence of Corollary 1.16 was proved in [**18**].

1.3. Characterization of an extremal sum of ridge functions

The approximation problem considered in this section is to approximate a continuous multivariate function $f(\mathbf{x}) = f(x_1,\ldots,x_d)$ by sums of two ridge functions in the uniform norm. We give a necessary and sufficient condition for a sum of two ridge functions to be a best approximation to $f(\mathbf{x})$. This main result is next used in a special case to obtain an explicit formula for the approximation error and to construct a best approximation. The problem of well approximation by such sums is also considered.

1.3.1. Exposition of the problem. Consider the following set of sums of ridge functions

$$\mathcal{R} = \mathcal{R}\left(\mathbf{a}^1, \mathbf{a}^2\right) = \left\{g_1\left(\mathbf{a}^1 \cdot \mathbf{x}\right) + g_2\left(\mathbf{a}^2 \cdot \mathbf{x}\right) : g_i \in C(\mathbb{R}), i = 1, 2\right\}.$$

That is, we fix directions $\mathbf{a}^1$ and $\mathbf{a}^2$ and consider linear combinations of ridge functions with these directions.

Assume $f(\mathbf{x})$ is a continuous function on a compact subset Q of $\mathbb{R}^d$. We want to find conditions that are necessary and sufficient for a function $g_0 \in \mathcal{R}\left(\mathbf{a}^1, \mathbf{a}^2\right)$ to be an extremal element (or a best approximation) to f. In other words, we want to characterize such sums $g_0(\mathbf{x}) = g_1\left(\mathbf{a}^1 \cdot \mathbf{x}\right) + g_2\left(\mathbf{a}^2 \cdot \mathbf{x}\right)$ of ridge functions that

$$\|f - g_0\| = \max_{\mathbf{x} \in Q} |f(\mathbf{x}) - g_0(\mathbf{x})| = E(f),$$

where

$$E(f) = E(f, \mathcal{R}) \stackrel{def}{=} \inf_{g \in \mathcal{R}(\mathbf{a}^1, \mathbf{a}^2)} \|f - g\|$$

is the error in approximating from $\mathcal{R}\left(\mathbf{a}^1, \mathbf{a}^2\right)$. The other related problem is how to construct these sums of ridge functions. We also want to know if we can approximate well, i.e. for which compact sets Q, $\mathcal{R}\left(\mathbf{a}^1, \mathbf{a}^2\right)$ is dense in $C(Q)$ in the topology of uniform convergence. It should be remarked that solutions to these problems may be useful in connection with the study of partial differential equations. For example, assume that (a_1, b_1) and (a_2, b_2) are linearly independent vectors in $\mathbb{R}^2$. Then the general solution to the homogeneous partial differential equation

$$\left(a_1 \frac{\partial}{\partial x} + b_1 \frac{\partial}{\partial y}\right)\left(a_2 \frac{\partial}{\partial x} + b_2 \frac{\partial}{\partial y}\right) u(x, y) = 0 \tag{1.11}$$

are all functions of the form

$$u(x, y) = g_1(b_1 x - a_1 y) + g_2(b_2 x - a_2 y) \tag{1.12}$$

for arbitrary g_1 and g_2. In [**63**], Golitschek and Light described an algorithm that computes the error of approximation of a continuous function $f(x, y)$ by solutions of equation (1.11), provided that $a_1 = b_2 = 1$, $a_2 = b_1 = 0$. Using our result (see Theorem 1.22), one can characterize those solutions 1.12 that are extremal to a given function $f(x, y)$. For a certain class of functions $f(x, y)$, one can also easily calculate the approximation error and construct an extremal solution (see Theorems 1.24 and 1.27 below).

The problem of approximating by functions from the set $\mathcal{R}\left(\mathbf{a}^1, \mathbf{a}^2\right)$ arises in other contexts too. Buck [**19**] studied the classical functional equation: given $\beta(t) \in C[0,1]$, $0 \le \beta(t) \le 1$, for which $u \in C[0,1]$ does there exist $\varphi \in C[0,1]$ such that

$$\varphi(t) = \varphi(\beta(t)) + u(t)?$$

He proved that the set of all u satisfying this condition is dense in the set

$$\{v \in C[0,1] : \ v(t) = 0 \text{ whenever } \beta(t) = t\}$$

if and only if $\mathcal{R}\left(\mathbf{a}^1, \mathbf{a}^2\right)$ with the unit directions $\mathbf{a}^1 = (1;0)$ and $\mathbf{a}^2 = (0,1)$ is dense in $C(K)$, where $K = \{(x,y) : y = x \text{ or } y = \beta(x),\ 0 \leq x \leq 1\}$.

Although there are enough reasons to consider approximation problems associated with the set $\mathcal{R}\left(\mathbf{a}^1, \mathbf{a}^2\right)$ in an independent way, one may ask why sums of only two ridge functions are considered instead of sums with an arbitrary number of terms. We will try to answer this fair question at the end of this section.

1.3.2. The characterization theorem. Let Q be a compact subset of $\mathbb{R}^d$ and $\mathbf{a}^1, \mathbf{a}^2 \in \mathbb{R}^d \backslash \{\mathbf{0}\}$.

DEFINITION 1.18. A finite or infinite ordered set $p = (\mathbf{p}_1, \mathbf{p}_2, \ldots) \subset Q$ with $\mathbf{p}_i \neq \mathbf{p}_{i+1}$, and either $\mathbf{a}^1 \cdot \mathbf{p}_1 = \mathbf{a}^1 \cdot \mathbf{p}_2, \mathbf{a}^2 \cdot \mathbf{p}_2 = \mathbf{a}^2 \cdot \mathbf{p}_3, \mathbf{a}^1 \cdot \mathbf{p}_3 = \mathbf{a}^1 \cdot \mathbf{p}_4, \ldots$ or $\mathbf{a}^2 \cdot \mathbf{p}_1 = \mathbf{a}^2 \cdot \mathbf{p}_2,\ \mathbf{a}^1 \cdot \mathbf{p}_2 = \mathbf{a}^1 \cdot \mathbf{p}_3, \mathbf{a}^2 \cdot \mathbf{p}_3 = \mathbf{a}^2 \cdot \mathbf{p}_4, \ldots$ is called a path with respect to the directions $\mathbf{a}^1$ and $\mathbf{a}^2$.

This notion (in the two-dimensional case) was introduced by Braess and Pinkus [**18**]. They showed that paths give geometric means of deciding if a set of points $\{\mathbf{x}^i\}_{i=1}^m \subset \mathbb{R}^2$ has the *NI* property (see Section 1.2.4). Ismailov and Pinkus [**98**] used these objects to study the problem of interpolation on straight lines by linear combinations of a finite number of ridge functions with fixed directions. In [**80**, **82**, **90**] paths were generalized to those with respect to two functions. The last objects turned out to be useful in problems of approximation and representation by sums of compositions of fixed multivariate functions with univariate functions.

If $\mathbf{a}^1$ and $\mathbf{a}^2$ are the coordinate vectors in $\mathbb{R}^2$, then Definition 1.18 defines a *bolt of lightning*. The idea of bolts was first introduced in Diliberto and Straus [**47**], where these objects are called *permissible lines*. They appeared further in a number of papers, although under several different names (see, e.g., [**51**, **60**, **63**, **78**, **79**, **114**, **115**, **117**, **118**, **137**, **150**, **151**, **166**]). Note that the term "bolt of lightning" is due to Arnold [**8**].

For the sake of brevity, we use the term "path" instead of the long expression "path with respect to the directions $\mathbf{a}^1$ and $\mathbf{a}^2$".

The length of a path is the number of its points. A single point is a path of the unit length. A finite path $(\mathbf{p}_1, \mathbf{p}_2, \ldots, \mathbf{p}_{2n})$ is said to be closed if $(\mathbf{p}_1, \mathbf{p}_2, \ldots, \mathbf{p}_{2n}, \mathbf{p}_1)$ is a path.

We associate each closed path $p = (\mathbf{p}_1, \mathbf{p}_2, \ldots, \mathbf{p}_{2n})$ with the functional

$$G_p(f) = \frac{1}{2n} \sum_{k=1}^{2n} (-1)^{k+1} f(\mathbf{p}_k).$$

This functional has the following obvious properties:

(a) If $g \in \mathcal{R}\left(\mathbf{a}^1, \mathbf{a}^2\right)$, then $G_p(g) = 0$.
(b) $\|G_p\| \leq 1$ and if $\mathbf{p}_i \neq \mathbf{p}_j$ for all $i \neq j$, $1 \leq i, j \leq 2n$, then $\|G_p\| = 1$.

LEMMA 1.19. *Let a compact set Q have closed paths. Then*

$$\sup_{p \subset Q} |G_p(f)| \leq E(f), \tag{1.13}$$

where the sup is taken over all closed paths. Moreover, inequality (1.13) *is sharp, i.e. there exist functions for which* (1.13) *turns into equality.*

PROOF. Let p be a closed path in Q and g be any function from $\mathcal{R}\left(\mathbf{a}^1, \mathbf{a}^2\right)$. By the linearity of G_p and properties (a) and (b),

$$|G_p(f)| = |G_p(f-g)| \leq \|f-g\|. \tag{1.14}$$

Since the left-hand and the right-hand sides of (1.14) do not depend on g and p respectively, it follows from (1.14) that

$$\sup_{p\subset Q} |G_p(f)| \leq \inf_{g\in\mathcal{R}(\mathbf{a}^1,\mathbf{a}^2)} \|f-g\|. \tag{1.15}$$

Now we prove the sharpness of (1.13). By assumption Q has closed paths. Then Q has a closed path $p' = (\mathbf{p}_1', \dots, \mathbf{p}_{2m}')$ with distinct points $\mathbf{p}_1', \dots, \mathbf{p}_{2m}'$. In fact, such a special path can be obtained from any closed path $p = (\mathbf{p}_1, \dots, \mathbf{p}_{2n})$ by the following simple algorithm: if the points of the path p are not all distinct, let i and $k > 0$ be the minimal indices such that $\mathbf{p}_i = \mathbf{p}_{i+2k}$; delete from p the subsequence $\mathbf{p}_{i+1}, \dots, \mathbf{p}_{i+2k}$ and call p the obtained path; repeat the above step until all points of p are all distinct; set $p' := p$. On the other hand there exist continuous functions $h = h(\mathbf{x})$ on Q such that $h(\mathbf{p}_i') = 1$, $i = 1, 3, \dots, 2m-1$, $h(\mathbf{p}_i') = -1$, $i = 2, 4, \dots, 2m$ and $-1 < h(\mathbf{x}) < 1$ elsewhere. For such functions we have

$$G_{p'}(h) = \|h\| = 1 \tag{1.16}$$

and

$$E(h) \leq \|h\|, \tag{1.17}$$

where the last inequality follows from the fact that $0 \in \mathcal{R}\left(\mathbf{a}^1, \mathbf{a}^2\right)$. From (1.15)-(1.17) it follows that

$$\sup_{p\subset Q} |G_p(h)| = E(h).$$

□

LEMMA 1.20. *Let Q be a convex compact subset of $\mathbb{R}^d$ and $f \in C(Q)$. For a vector $\mathbf{e} \in \mathbb{R}^d \backslash \{\mathbf{0}\}$ and a real number t set*

$$Q_t = \{\mathbf{x} \in Q : \mathbf{e}\cdot\mathbf{x} = t\}, \quad T_h = \{t \in \mathbb{R} : Q_t \neq \emptyset\}.$$

The functions

$$g_1(t) = \max_{\mathbf{x}\in Q_t} f(\mathbf{x}), \quad t \in T_h \quad \textit{and} \quad g_2(t) = \min_{\mathbf{x}\in Q_t} f(\mathbf{x}), \quad t \in T_h$$

are defined and continuous on T_h.

The proof of this lemma is not difficult and can be obtained by the well-known elementary methods of mathematical analysis.

DEFINITION 1.21. A finite or infinite path $(\mathbf{p}_1, \mathbf{p}_2, \dots)$ is said to be extremal for a function $u \in C(Q)$ if $u(\mathbf{p}_i) = (-1)^i \|u\|, i = 1, 2, \dots$ or $u(\mathbf{p}_i) = (-1)^{i+1} \|u\|$, $i = 1, 2, \dots$

THEOREM 1.22. *Let $Q \subset \mathbb{R}^d$ be a convex compact set satisfying the following condition*

Condition (A): For any path $q = (\mathbf{q}_1, \mathbf{q}_2, \dots, \mathbf{q}_n) \subset Q$ there exist some points $\mathbf{q}_{n+1}, \mathbf{q}_{n+2}, \dots, \mathbf{q}_{n+s} \in Q$ such that $(\mathbf{q}_1, \mathbf{q}_2, \dots, \mathbf{q}_{n+s})$ is a closed path and s is not more than some positive integer N_0 independent of q.

Then a necessary and sufficient condition for a function $g_0 \in \mathcal{R}\left(\mathbf{a}^1, \mathbf{a}^2\right)$ to be an extremal element to the given function $f \in C(Q)$ is the existence of a closed or infinite path $l = (\mathbf{p}_1, \mathbf{p}_2, \dots)$ extremal for the function $f_1 = f - g_0$.

It should be remarked that the above condition (A) strongly depends on the fixed directions $\mathbf{a}^1$ and $\mathbf{a}^2$. For example, in the familiar case of a square $S \subset \mathbb{R}^2$ there are many directions which are not allowed. If it is possible to reach a corner of S with not more than one of the two directions orthogonal to $\mathbf{a}^1$ and $\mathbf{a}^2$, respectively (we don't differentiate between directions $\mathbf{c}$ and $-\mathbf{c}$), the triple $(S, \mathbf{a}^1, \mathbf{a}^2)$ does not satisfy condition (A) of the theorem. Here are simple examples: Let $S = [0;1]^2$, $\mathbf{a}^1 = (1;0)$, $\mathbf{a}^2 = (1;1)$. Then the ordered set $\{(0;1), (1;0), (1;1)\}$ is a path in S which can not be made closed. In this case, $(1;1)$ is not reached with the direction orthogonal to $\mathbf{b}$. Let now $\mathbf{a}^1 = (1; \frac{1}{2})$, $\mathbf{a}^2 = (1;1)$. Then the corner $(1;1)$ is reached with none of the directions orthogonal to $\mathbf{a}^1$ and $\mathbf{a}^2$ respectively. In this case, for any positive integer N_0 and any point $\mathbf{q}_0$ in S one can chose a point $\mathbf{q}_1 \in S$ from a sufficiently small neighborhood of the corner $(1;1)$ so that any path containing $\mathbf{q}_0$ and $\mathbf{q}_1$ has the length more than N_0. These examples and a little geometry show that if a convex compact set $Q \subset \mathbb{R}^2$ satisfies condition (A) of Theorem 1.22, then any point in the boundary of Q must be reached with each of the two directions orthogonal to $\mathbf{a}^1$ and $\mathbf{a}^2$ respectively. If $Q \subset \mathbb{R}^d, \mathbf{a}^1, \mathbf{a}^2 \in \mathbb{R}^d \backslash \{\mathbf{0}\}$, $d > 2$, there are many directions orthogonal to $\mathbf{a}^1$ and $\mathbf{a}^2$. In this case, condition (A) requires that any point in the boundary of Q should be reached with at least two directions orthogonal to $\mathbf{a}^1$ and $\mathbf{a}^2$, respectively.

PROOF. *Necessity.* Let $g_0(\mathbf{x}) = g_{1,0}\left(\mathbf{a}^1 \cdot \mathbf{x}\right) + g_{2,0}\left(\mathbf{a}^2 \cdot \mathbf{x}\right)$ be an extremal element from $\mathcal{R}\left(\mathbf{a}^1, \mathbf{a}^2\right)$ to f. We must show that if there is not a closed path extremal for f_1, then there exists a path extremal for f_1 with the infinite length (number of points). Suppose the contrary. Suppose that there exists a positive integer N such that the length of each path extremal for f_1 is not more than N. Set the following functions:

$$f_n = f_{n-1} - g_{1,n-1} - g_{2,n-1}, \quad n = 2, 3, \dots,$$

where

$$g_{1,n-1} = g_{1,n-1}\left(\mathbf{a}^1 \cdot \mathbf{x}\right) = \frac{1}{2}\left(\max_{\substack{\mathbf{y} \in Q \\ \mathbf{a}^1 \cdot \mathbf{y} = \mathbf{a}^1 \cdot \mathbf{x}}} f_{n-1}(\mathbf{y}) + \min_{\substack{\mathbf{y} \in Q \\ \mathbf{a}^1 \cdot \mathbf{y} = \mathbf{a}^1 \cdot \mathbf{x}}} f_{n-1}(\mathbf{y})\right)$$

$$g_{2,n-1} = g_{2,n-1}(\mathbf{a}^2 \cdot \mathbf{x}) = \frac{1}{2}\left(\max_{\substack{\mathbf{y} \in Q \\ \mathbf{a}^2 \cdot \mathbf{y} = \mathbf{a}^2 \cdot \mathbf{x}}} (f_{n-1}(\mathbf{y}) - g_{1,n-1}(\mathbf{a}^1 \cdot \mathbf{y})) \right.$$

$$\left. + \min_{\substack{\mathbf{y} \in Q \\ \mathbf{a}^2 \cdot \mathbf{y} = \mathbf{a}^2 \cdot \mathbf{x}}} (f_{n-1}(\mathbf{y}) - g_{1,n-1}(\mathbf{a}^1 \cdot \mathbf{y}))\right).$$

By Lemma 1.20, all the functions $f_n(\mathbf{x})$, $n = 2, 3, \dots$, are continuous on Q. By assumption g_0 is a best approximation to f. Hence $\|f_1\| = E(f)$. Now let us show that $\|f_2\| = E(f)$. Indeed, for any $\mathbf{x} \in Q$

$$f_1(\mathbf{x}) - g_{1,1}(\mathbf{a}^1 \cdot \mathbf{x}) \leq \frac{1}{2}\left(\max_{\substack{\mathbf{y} \in Q \\ \mathbf{a}^1 \cdot \mathbf{y} = \mathbf{a}^1 \cdot \mathbf{x}}} f_1(\mathbf{y}) - \min_{\substack{\mathbf{y} \in Q \\ \mathbf{a}^1 \cdot \mathbf{y} = \mathbf{a}^1 \cdot \mathbf{x}}} f_1(\mathbf{y})\right) \leq E(f) \tag{1.18}$$

and

$$f_1(\mathbf{x}) - g_{1,1}(\mathbf{a}^1{\cdot}\mathbf{x}) \geq \frac{1}{2}\left(\min_{\substack{\mathbf{y}\in Q\\ \mathbf{a}^1\cdot\mathbf{y}=\mathbf{a}^1\cdot\mathbf{x}}} f_1(\mathbf{y}) - \max_{\substack{\mathbf{y}\in Q\\ \mathbf{a}^1\cdot\mathbf{y}=\mathbf{a}^1\cdot\mathbf{x}}} f_1(\mathbf{y})\right) \geq -E(f). \tag{1.19}$$

Using the definition of $g_{2,1}(\mathbf{a}^2 \cdot \mathbf{x})$, for any $\mathbf{x} \in Q$ we have

$$f_1(\mathbf{x}) - g_{1,1}(\mathbf{a}^1 \cdot \mathbf{x}) - g_{2,1}(\mathbf{a}^2 \cdot \mathbf{x})$$

$$\leq \frac{1}{2}\left(\max_{\substack{\mathbf{y}\in Q\\ \mathbf{a}^2\cdot\mathbf{y}=\mathbf{a}^2\cdot\mathbf{x}}} (f_1(\mathbf{y}) - g_{1,1}(\mathbf{a}^1 \cdot \mathbf{y})) - \min_{\substack{\mathbf{y}\in Q\\ \mathbf{a}^2\cdot\mathbf{y}=\mathbf{a}^2\cdot\mathbf{x}}} (f_1(\mathbf{y}) - g_{1,1}(\mathbf{a}^1 \cdot \mathbf{y}))\right)$$

and

$$f_1(\mathbf{x}) - g_{1,1}(\mathbf{a}^1 \cdot \mathbf{x}) - g_{2,1}(\mathbf{a}^2 \cdot \mathbf{x})$$

$$\leq \frac{1}{2}\left(\min_{\substack{\mathbf{y}\in Q\\ \mathbf{a}^2\cdot\mathbf{y}=\mathbf{a}^2\cdot\mathbf{x}}} (f_1(\mathbf{y}) - g_{1,1}(\mathbf{a}^1 \cdot \mathbf{y})) - \max_{\substack{\mathbf{y}\in Q\\ \mathbf{a}^2\cdot\mathbf{y}=\mathbf{a}^2\cdot\mathbf{x}}} (f_1(\mathbf{y}) - g_{1,1}(\mathbf{a}^1 \cdot \mathbf{y}))\right).$$

Using (1.18) and (1.19) in the last two inequalities, we obtain that for any $\mathbf{x} \in Q$

$$-E(f) \leq f_2(\mathbf{x}) = f_1(\mathbf{x}) - g_{1,1}(\mathbf{a}^1{\cdot}\mathbf{x}) - g_{2,1}(\mathbf{a}^2{\cdot}\mathbf{x}) \leq E(f).$$

Therefore,

$$\|f_2\| \leq E(f). \tag{1.20}$$

Since $f_2(\mathbf{x}) - f(\mathbf{x})$ belongs to $\mathcal{R}\left(\mathbf{a}^1, \mathbf{a}^2\right)$, we deduce from (1.20) that

$$\|f_2\| = E(f).$$

By the same way, one can show that $\|f_3\| = E(f)$, $\|f_4\| = E(f)$, and so on. Thus we can write

$$\|f_n\| = E(f), \text{ for any } n.$$

Let us now prove the implications

$$f_1(\mathbf{p}_0) < E(f) \Rightarrow f_2(\mathbf{p}_0) < E(f) \tag{1.21}$$

and

$$f_1(\mathbf{p}_0) > -E(f) \Rightarrow f_2(\mathbf{p}_0) > -E(f), \tag{1.22}$$

where $\mathbf{p}_0 \in Q$. First, we are going to prove the implication

$$f_1(\mathbf{p}_0) < E(f) \Rightarrow f_1(\mathbf{p}_0) - g_{1,1}(\mathbf{a} \cdot \mathbf{p}_0) < E(f). \tag{1.23}$$

There are two possible cases.

1) $\max\limits_{\substack{\mathbf{y}\in Q\\ \mathbf{a}^1\cdot\mathbf{y}=\mathbf{a}^1\cdot\mathbf{p}_0}} f_1(\mathbf{y}) = E(f)$ and $\min\limits_{\substack{\mathbf{y}\in Q\\ \mathbf{a}^1\cdot\mathbf{y}=\mathbf{a}^1\cdot\mathbf{p}_0}} f_1(\mathbf{y}) = -E(f)$.

In this case, $g_{1,1}(\mathbf{a}^1 \cdot \mathbf{p}_0) = 0$. Hence

$$f_1(\mathbf{p}_0) - g_{1,1}(\mathbf{a}^1 \cdot \mathbf{p}_0) < E(f).$$

2) $\max\limits_{\substack{\mathbf{y}\in Q\\ \mathbf{a}^1\cdot\mathbf{y}=\mathbf{a}^1\cdot\mathbf{p}_0}} f_1(\mathbf{y}) = E(f)-\varepsilon_1$ and $\min\limits_{\substack{\mathbf{y}\in Q\\ \mathbf{a}^1\cdot\mathbf{y}=\mathbf{a}^1\cdot\mathbf{p}_0}} f_1(\mathbf{y}) = -E(f)+\varepsilon_2$,

where ε_1, ε_2 are nonnegative real numbers with the sum $\varepsilon_1+\varepsilon_2\neq 0$. In this case,

$$\begin{aligned} f_1(\mathbf{p}_0)-g_{1,1}(\mathbf{a}^1\cdot\mathbf{p}_0) &\leq \max_{\substack{\mathbf{y}\in Q\\ \mathbf{a}^1\cdot\mathbf{y}=\mathbf{a}^1\cdot\mathbf{p}_0}} f_1(\mathbf{y})-g_{1,1}(\mathbf{a}^1\cdot\mathbf{p}_0)= \\ &= \frac{1}{2}\left(\max_{\substack{\mathbf{y}\in Q\\ \mathbf{a}^1\cdot\mathbf{y}=\mathbf{a}^1\cdot\mathbf{p}_0}} f_1(\mathbf{y})-\min_{\substack{\mathbf{y}\in Q\\ \mathbf{a}^1\cdot\mathbf{y}=\mathbf{a}^1\cdot\mathbf{p}_0}} f_1(\mathbf{y})\right)= \\ &= E(f)-\frac{\varepsilon_1+\varepsilon_2}{2}<E(f). \end{aligned}$$

Thus we have proved (1.23). Using this method, we can also prove that

$$(1.24)\quad f_1(\mathbf{p}_0)-g_{1,1}(\mathbf{a}^1\cdot\mathbf{p}_0)<E(f)\Rightarrow f_1(\mathbf{p}_0)-g_{1,1}(\mathbf{a}^1\cdot\mathbf{p}_0)-g_{2,1}(\mathbf{a}^2\cdot\mathbf{p}_0)<E(f).$$

Now (1.21) follows from (1.23) and (1.24). By the same way we can prove (1.22). It follows from implications (1.21) and (1.22) that if $f_2(\mathbf{p}_0)=E(f)$, then $f_1(\mathbf{p}_0)=E(f)$ and if $f_2(\mathbf{p}_0)=-E(f)$, then $f_1(\mathbf{p}_0)=-E(f)$. This simply means that each path extremal for f_2 will be extremal for f_1.

Now we show that if any path extremal for f_1 has the length not more than N, then any path extremal for f_2 has the length not more than $N-1$. Suppose the contrary. Suppose that there is a path extremal for f_2 with the length equal to N. Denote it by $q=(\mathbf{q}_1,\mathbf{q}_2,\dots,\mathbf{q}_N)$. Without loss of generality we may assume that $\mathbf{a}^2\cdot\mathbf{q}_{N-1}=\mathbf{a}^2\cdot\mathbf{q}_N$. As it has been shown above, the path q is also extremal for f_1. Assume that $f_1(\mathbf{q}_N)=E(f)$. Then there is not any $\mathbf{q}_0\in Q$ such that $\mathbf{q}_0\neq\mathbf{q}_N$, $\mathbf{a}^1\cdot\mathbf{q}_0=\mathbf{a}^1\cdot\mathbf{q}_N$ and $f_1(\mathbf{q}_0)=-E(f)$. Indeed, if there was such $\mathbf{q}_0$ and $\mathbf{q}_0\notin q$, the path $(\mathbf{q}_1,\mathbf{q}_2,\dots,\mathbf{q}_N,\mathbf{q}_0)$ would be extremal for f_1. But this would contradict our assumption that any path extremal for f_1 has the length not more than N. Besides, if there was such $\mathbf{q}_0$ and $\mathbf{q}_0\in q$, we could form some closed path extremal for f_1. This also would contradict our assumption that there does not exist a closed path extremal for f_1.

Hence

$$\max_{\substack{\mathbf{y}\in Q\\ \mathbf{a}^1\cdot\mathbf{y}=\mathbf{a}^1\cdot\mathbf{q}_N}} f_1(\mathbf{y})=E(f),\qquad \min_{\substack{\mathbf{y}\in Q\\ \mathbf{a}^1\cdot\mathbf{y}=\mathbf{a}^1\cdot\mathbf{q}_N}} f_1(\mathbf{y})>-E(f).$$

Therefore,

$$\left|f_1(\mathbf{q}_N)-g_{1,1}(\mathbf{a}^1\cdot\mathbf{q}_N)\right|<E(f).$$

From the last inequality it is easy to obtain that (see the proof of implications (1.21) and (1.22)

$$|f_2(\mathbf{q}_N)|<E(f).$$

This means, on the contrary to our assumption, that the path $(\mathbf{q}_1,\mathbf{q}_2,\dots,\mathbf{q}_N)$ can not be extremal for f_2. Hence any path extremal for f_2 has the length not more than $N-1$.

By the same way, it can be shown that any path extremal for f_3 has the length not more than $N-2$, any path extremal for f_4 has the length not more than $N-3$ and so on. Finally, we will obtain that there is not a path extremal for f_{N+1}. Hence there is not a point $\mathbf{p}_0\in Q$ such that $|f_{N+1}(\mathbf{p}_0)|=\|f_{N+1}\|$. But by Lemma 1.20, all the functions f_2, $f_3,\dots,f_{N+1}$ are continuous on the compact set Q; hence the norm $\|f_{N+1}\|$ must be attained. This contradiction means that there exists a path extremal for f_1 with the infinite length.

Sufficiency. Let a path $p = (\mathbf{p}_1, \mathbf{p}_2, \dots, \mathbf{p}_{2n})$ be closed and extremal for f_1. Then

$$|G_p(f)| = \|f - g_0\|. \tag{1.25}$$

By Lemma 1.19,

$$|G_p(f)| \leq E(f). \tag{1.26}$$

It follows from (1.25) and (1.26) that g_0 is a best approximation.

Let now a path $p = (\mathbf{p}_1, \mathbf{p}_2, \dots, \mathbf{p}_n, \dots)$ be infinite and extremal for f_1. Consider the sequence $p_n = (\mathbf{p}_1, \mathbf{p}_2, \dots, \mathbf{p}_n)$, $n = 1, 2, \dots$, of finite paths. By condition (A) of the theorem, for each path p_n there exists a closed path $p_n^{m_n} = (\mathbf{p}_1, \mathbf{p}_2, \dots, \mathbf{p}_n, \mathbf{q}_{n+1}, \dots, \mathbf{q}_{n+m_n})$, where $m_n \leq N_0$. Then for any positive integer n,

$$\left|G_{p_n^{m_n}}(f)\right| = \left|G_{p_n^{m_n}}(f - g_0)\right| \leq \frac{n\|f - g_0\| + m_n\|f - g_0\|}{n + m_n} = \|f - g_0\|$$

and

$$\left|G_{p_n^{m_n}}(f)\right| \geq \frac{n\|f - g_0\| - m_n\|f - g_0\|}{n + m_n} = \frac{n - m_n}{n + m_n}\|f - g_0\|.$$

It follows from the above two inequalities for $\left|G_{p_n^{m_n}}(f)\right|$ that

$$\sup_{p_n^{m_n}} \left|G_{p_n^{m_n}}(f)\right| = \|f - g_0\|.$$

This together with Lemma 1.19 give that

$$\|f - g_0\| \leq E(f).$$

Hence g_0 is a best approximation. □

Theorem 1.22 has been proved by using only methods of classical analysis. By implementing more deep techniques from functional analysis we will see below that condition (A) and the convexity assumption on a compact set Q can be dropped.

THEOREM 1.23. *Assume Q is a compact subset of $\mathbb{R}^d$. A function $g_0 \in \mathcal{R}$ is a best approximation to a function $f \in C(Q)$ if and only if there exists a closed or infinite path $p = (\mathbf{p}_1, \mathbf{p}_2, \dots)$ extremal for the function $f - g_0$.*

PROOF. *Sufficiency.* There are two possible cases. The first case happens when there exists a closed path $(\mathbf{p}_1, \dots, \mathbf{p}_{2n})$ extremal for the function $f - g_0$. Let us check that in this case, $f - g_0$ is a best approximation. Indeed, on the one hand, the following equalities are valid

$$\left|\sum_{i=1}^{2n}(-1)^i f(\mathbf{p}_i)\right| = \left|\sum_{i=1}^{2n}(-1)^i [f - g_0](\mathbf{p}_i)\right| = 2n\|f - g_0\|.$$

On the other hand, for any function $g \in \mathcal{R}$, we have

$$\left|\sum_{i=1}^{2n}(-1)^i f(\mathbf{p}_i)\right| = \left|\sum_{i=1}^{2n}(-1)^i [f - g](\mathbf{p}_i)\right| \leq 2n\|f - g\|.$$

Therefore, $\|f - g_0\| \leq \|f - g\|$ for any $g \in \mathcal{R}$. That is, g_0 is a best approximation.

The second case happens when we do not have closed paths extremal for $f-g_0$, but there exists an infinite path $(\mathbf{p}_1, \mathbf{p}_2, \dots)$ extremal for $f-g_0$. To analyze this case, consider the following linear functional

$$L_q : C(Q) \to \mathbb{R}, \; L_q(F) = \frac{1}{n}\sum_{i=1}^{n}(-1)^i F(\mathbf{q}_i),$$

where $q = \{\mathbf{q}_1, \dots, \mathbf{q}_n\}$ is a finite path in Q. It is easy to see that the norm $\|L_q\| \le 1$ and $\|L_q\| = 1$ if and only if the set of points of q with odd indices $O = \{\mathbf{q}_i \in q : i \textit{ is an odd number}\}$ do not intersect with the set of points of q with even indices $E = \{\mathbf{q}_i \in q : i \textit{ is an even number}\}$. Indeed, from the definition of L_q it follows that $|L_q(F)| \le \|F\|$ for all functions $F \in C(Q)$, whence $\|L_q\| \le 1$. If $O \cap E = \varnothing$, then for a function F_0 with the property $F_0(\mathbf{q}_i) = -1$ if i is odd, $F_0(\mathbf{q}_i) = 1$ if i is even and $-1 < F_0(x) < 1$ elsewhere on Q, we have $|L_q(F_0)| = \|F_0\|$. Hence, $\|L_q\| = 1$. Recall that such a function F_0 exists on the basis of Urysohn's lemma.

Note that if q is a closed path, then L_q annihilates all members of the class $\mathcal{R}$. But in general, when q is not closed, we do not have the equality $L_q(g) = 0$, for all members $g \in \mathcal{R}$. Nonetheless, this functional has the important property that

$$|L_q(g_1 + g_2)| \le \frac{2}{n}(\|g_1\| + \|g_2\|), \tag{1.27}$$

where g_1 and g_2 are ridge functions with the directions $\mathbf{a}_1$ and $\mathbf{a}_2$, respectively, that is, $g_1 = g_1(\mathbf{a}_1 \cdot \mathbf{x})$ and $g_2 = g_2(\mathbf{a}_2 \cdot \mathbf{x})$. This property is important in the sense that if n is sufficiently large, then the functional L_q is close to an annihilating functional. To prove (1.27), note that $|L_q(g_1)| \le \frac{2}{n}\|g_1\|$ and $|L_q(g_2)| \le \frac{2}{n}\|g_2\|$. These estimates become obvious if consider the chain of equalities $g_1(\mathbf{a}_1 \cdot \mathbf{q}_1) = g_1(\mathbf{a}_1 \cdot \mathbf{q}_2)$, $g_1(\mathbf{a}_1 \cdot \mathbf{q}_3) = g_1(\mathbf{a}_1 \cdot \mathbf{q}_4), \dots$ (or $g_1(\mathbf{a}_1 \cdot \mathbf{q}_2) = g_1(\mathbf{a}_1 \cdot \mathbf{q}_3)$, $g_1(\mathbf{a}_1 \cdot \mathbf{q}_4) = g_1(\mathbf{a}_1 \cdot \mathbf{q}_5), \dots$) for $g_1(\mathbf{a}_1 \cdot \mathbf{x})$ and the corresponding chain of equalities for $g_2(\mathbf{a}_2 \cdot \mathbf{x})$.

Now consider the infinite path $p = (\mathbf{p}_1, \mathbf{p}_2, \dots)$ and form the finite paths $p_k = (\mathbf{p}_1, \dots, \mathbf{p}_k)$, $k = 1, 2, \dots$. For ease of notation, let us set $L_k = L_{p_k}$. The sequence $\{L_k\}_{k=1}^{\infty}$ is a subset of the unit ball of the conjugate space $C^*(Q)$. By the Banach-Alaoglu theorem, the unit ball is weak* compact in the weak* topology of $C^*(Q)$ (see [**177**, p.68]). It follows from this theorem that the sequence $\{L_k\}_{k=1}^{\infty}$ must have weak* cluster points. Suppose L^* denotes one of them. Without loss of generality we may assume that $L_k \overset{weak^*}{\longrightarrow} L^*$, as $k \to \infty$. From (1.27) it follows that $L^*(g_1 + g_2) = 0$. That is, $L^* \in \mathcal{R}^{\perp}$, where the symbol $\mathcal{R}^{\perp}$ stands for the annihilator of $\mathcal{R}$. Since in addition $\|L^*\| \le 1$, we can write that

$$|L^*(f)| = |L^*(f-g)| \le \|f-g\|, \tag{1.28}$$

for all functions $g \in \mathcal{R}$. On the other hand, since the infinite path p is extremal for $f - g_0$

$$|L_k(f-g_0)| = \|f-g_0\|, \; k = 1, 2, \dots$$

Therefore,

$$|L^*(f)| = |L^*(f-g_0)| = \|f-g_0\|. \tag{1.29}$$

From (1.28) and (1.29) we conclude that

$$\|f-g_0\| \le \|f-g\|,$$

for all $g \in \mathcal{R}$. In other words, g_0 is a best approximation to f. We proved the sufficiency of the theorem.

Necessity. The proof of this part is mainly based on the following result of Singer [**182**]: Let X be a compact space, U be a linear subspace of $C(X)$, $f \in C(X)\backslash U$ and $u_0 \in U$. Then u_0 is a best approximation to f if and only if there exists a regular Borel measure μ on X such that

(1) The total variation $\|\mu\| = 1$;

(2) μ is orthogonal to the subspace U, that is, $\int_X u d\mu = 0$ for all $u \in U$;

(3) For the Jordan decomposition $\mu = \mu^+ - \mu^-$,

$$f(x) - u_0(x) = \begin{cases} \|f - u_0\| & \text{for } x \in S^+, \\ -\|f - u_0\| & \text{for } x \in S^-, \end{cases}$$

where S^+ and S^- are closed supports of the positive measures μ^+ and μ^-, respectively.

Let us show how we use this theorem in the proof of necessity part of our theorem. Assume $g_0 \in \mathcal{R}$ is a best approximation. For the subspace $\mathcal{R}$, the existence of a measure μ satisfying the conditions (1)-(3) is a direct consequence of Singer's result. Let $\mathbf{x}_0$ be any point in S^+. Consider the point $y_0 = \mathbf{a}_1 \cdot \mathbf{x}_0$ and a δ-neighborhood of y_0. That is, choose an arbitrary $\delta > 0$ and consider the set $I_\delta = (y_0 - \delta, y_0 + \delta) \cap \mathbf{a}_1 \cdot Q$. Here, $\mathbf{a}_1 \cdot Q = \{\mathbf{a}_1 \cdot \mathbf{x} : \mathbf{x} \in Q\}$. For any subset $E \subset \mathbb{R}$, put

$$E^i = \{\mathbf{x} \in Q : \mathbf{a}_i \cdot \mathbf{x} \in E\},\ i = 1, 2.$$

Clearly, for some sets E, one or both the sets E^i may be empty. Since $I_\delta^1 \cap S^+$ is not empty (note that $\mathbf{x}_0 \in I_\delta^1$), it follows that $\mu^+(I_\delta^1) > 0$. At the same time $\mu(I_\delta^1) = 0$, since μ is orthogonal to all functions $g_1(\mathbf{a}_1 \cdot \mathbf{x})$. Therefore, $\mu^-(I_\delta^1) > 0$. We conclude that $I_\delta^1 \cap S^-$ is not empty. Denote this intersection by A_δ. Tending δ to 0, we obtain a set A which is a subset of S^- and has the property that for each $\mathbf{x} \in A$, we have $\mathbf{a}_1 \cdot \mathbf{x} = \mathbf{a}_1 \cdot \mathbf{x}_0$. Fix any point $\mathbf{x}_1 \in A$. Changing $\mathbf{a}_1$, μ^+, S^+ to $\mathbf{a}_2$, μ^- and S^- correspondingly, repeat the above process with the point $y_1 = \mathbf{a}_2 \cdot \mathbf{x}_1$ and a δ-neighborhood of y_1. Then we obtain a point $\mathbf{x}_2 \in S^+$ such that $\mathbf{a}_2 \cdot \mathbf{x}_2 = \mathbf{a}_2 \cdot \mathbf{x}_1$. Continuing this process, one can construct points $\mathbf{x}_3$, $\mathbf{x}_4$, and so on. Note that the set of all constructed points $\mathbf{x}_i$, $i = 0, 1, \ldots$, forms a path. By Singer's above result, this path is extremal for the function $f - g_0$. We have proved the necessity and hence Theorem 1.23. □

Theorem 1.23, in a more general setting, was proven in Pinkus [**175**, p.99] under additional assumption that Q is convex. Convexity assumption was made to guarantee continuity of the following functions

$$g_{1,i}(t) = \max_{\substack{\mathbf{x} \in Q \\ \mathbf{a}_i \cdot \mathbf{x} = t}} F(\mathbf{x}) \text{ and } g_{2,i}(t) = \min_{\substack{\mathbf{x} \in Q \\ \mathbf{a}_i \cdot \mathbf{x} = t}} F(\mathbf{x}),\ i = 1, 2,$$

where F is an arbitrary continuous function on Q. Note that in the proof of Theorem 1.23 we did not need continuity of these functions.

It is well known that characterization theorems of this type are very essential in approximation theory. Chebyshev was the first to prove a similar result for polynomial approximation. Khavinson [**114**] characterized extremal elements in the special case of the problem considered here. His case allows the approximation of a continuous bivariate function $f(x, y)$ by functions of the form $\varphi(x) + \psi(y)$.

1.3.3. Construction of an extremal element. In 1951 Diliberto and Straus [**47**] established a formula for the error of approximation of a bivariate function by sums of univariate functions. Their formula contains the supremum over all closed bolts (see Section 3.3). Although the mentioned formula is valid for all continuous functions, it is not easily calculable. Therefore, it cannot give a desired effect if one is interested in the precise value of the approximation error. After this general result some authors started to seek easily calculable formulas for the approximation error by considering not the whole space of continuous functions, but some subsets thereof (see, for example, [**10, 13, 78, 79, 114, 176**]). These subsets were chosen so that they could provide precise and easy computation of the approximation error. Since the set of ridge functions contains univariate functions, one may ask for explicit formulas for the error of approximation of a multivariate function by sums of ridge functions.

In this subsection, we see how with the use of Theorem 1.22 (or 1.23) it is possible to find the approximation error and construct an extremal element in the problem of approximation by sums of ridge functions. We restrict ourselves to $\mathbb{R}^2$. To make the problem more precise, let Ω be a compact set in $\mathbb{R}^2$, $f \in C(\Omega)$, $\mathbf{a} = (a_1, a_2)$ and $\mathbf{b} = (b_1, b_2)$ be linearly independent vectors. Consider the approximation of f by functions from $\mathcal{R} = \mathcal{R}(\mathbf{a}, \mathbf{b})$. We want, under some suitable conditions on f and Ω, to establish a formula for an easy and direct computation of the approximation error $E(f, \mathcal{R})$.

THEOREM 1.24. *Let*

$$\Omega = \left\{\mathbf{x} \in \mathbb{R}^2 : c_1 \leq \mathbf{a}\cdot\mathbf{x} \leq d_1, \quad c_2 \leq \mathbf{b}\cdot\mathbf{x} \leq d_2\right\},$$

where $c_1 < d_1$ and $c_2 < d_2$. Let a function $f(\mathbf{x}) \in C(\Omega)$ have the continuous partial derivatives $\frac{\partial^2 f}{\partial x_1^2}, \frac{\partial^2 f}{\partial x_1 \partial x_2}, \frac{\partial^2 f}{\partial x_2^2}$ and for any $\mathbf{x} \in \Omega$

$$\frac{\partial^2 f}{\partial x_1 \partial x_2}(a_1 b_2 + a_2 b_1) - \frac{\partial^2 f}{\partial x_1^2} a_2 b_2 - \frac{\partial^2 f}{\partial x_2^2} a_1 b_1 \geq 0.$$

Then

$$E(f, \mathcal{R}) = \frac{1}{4}\left(f_1(c_1, c_2) + f_1(d_1, d_2) - f_1(c_1, d_2) - f_1(d_1, c_2)\right),$$

where

$$f_1(y_1, y_2) = f\left(\frac{y_1 b_2 - y_2 a_2}{a_1 b_2 - a_2 b_1}, \frac{y_2 a_1 - y_1 b_1}{a_1 b_2 - a_2 b_1}\right). \tag{1.30}$$

PROOF. Introduce the new variables

$$y_1 = a_1 x_1 + a_2 x_2, \quad y_2 = b_1 x_1 + b_2 x_2. \tag{1.31}$$

Since the vectors (a_1, a_2) and (b_1, b_2) are linearly independent, for any $(y_1, y_2) \in Y$, where $Y = [c_1, d_1] \times [c_2, d_2]$, there exists only one solution $(x_1, x_2) \in \Omega$ of the system (1.31). The coordinates of this solution are

$$x_1 = \frac{y_1 b_2 - y_2 a_2}{a_1 b_2 - a_2 b_1}, \qquad x_2 = \frac{y_2 a_1 - y_1 b_1}{a_1 b_2 - a_2 b_1}. \tag{1.32}$$

The linear transformation (1.32) transforms the function $f(x_1, x_2)$ to the function $f_1(y_1, y_2)$. Consider the approximation of $f_1(y_1, y_2)$ from the set

$$\mathcal{Z} = \left\{z_1(y_1) + z_2(y_2) : z_i \in C(\mathbb{R}),\ i = 1, 2\right\}.$$

It is easy to see that

$$E(f,\mathcal{R}) = E(f_1,\mathcal{Z}). \tag{1.33}$$

With each rectangle $S = [u_1,v_1] \times [u_2,v_2] \subset Y$ we associate the functional

$$L(h,S) = \frac{1}{4}\left(h(u_1,u_2) + h(v_1,v_2) - h(u_1,v_2) - h(v_1,u_2)\right), \quad h \in C(Y).$$

This functional has the following obvious properties:
(i) $L(z,S) = 0$ for any $z \in \mathcal{Z}$ and $S \subset Y$.

(ii) For any point $(y_1,y_2) \in Y$, $L(f_1,Y) = \sum_{i=1}^{4} L(f_1,S_i)$, where $S_1 = [c_1,y_1] \times [c_2,y_2]$, $S_2 = [y_1,d_1] \times [y_2,d_2]$, $S_3 = [c_1,y_1] \times [y_2,d_2]$, $S_4 = [y_1,d_1] \times [c_2,y_2]$.

By the conditions of the theorem, it is not difficult to verify that

$$\frac{\partial^2 f_1}{\partial y_1 \partial y_2} \geq 0 \text{ for any } (y_1,y_2) \in Y.$$

Integrating both sides of the last inequality over arbitrary rectangle $S=[u_1,v_1]\times[u_2,v_2] \subset Y$, we obtain that

$$L(f_1,S) \geq 0. \tag{1.34}$$

Set the function

$$f_2(y_1,y_2) = L(f_1,S_1) + L(f_1,S_2) - L(f_1,S_3) - L(f_1,S_4). \tag{1.35}$$

It is not difficult to verify that the function $f_1 - f_2$ belongs to $\mathcal{Z}$. Hence

$$E(f_1,\mathcal{Z}) = E(f_2,\mathcal{Z}). \tag{1.36}$$

Calculate the norm $\|f_2\|$. From the property (ii), it follows that

$$f_2(y_1,y_2) = L(f_1,Y) - 2(L(f_1,S_3) + L(f_1,S_4))$$

and

$$f_2(y_1,y_2) = 2\left(L(f_1,S_1) + L(f_1,S_2)\right) - L(f_1,Y).$$

From the last equalities and (1.34), we obtain that

$$|f_2(y_1,y_2)| \leq L(f_1,Y), \text{ for any } (y_1,y_2) \in Y.$$

On the other hand, one can check that

$$f_2(c_1,c_2) = f_2(d_1,d_2) = L(f_1,Y) \tag{1.37}$$

and

$$f_2(c_1,d_2) = f_2(d_1,c_2) = -L(f_1,Y). \tag{1.38}$$

Therefore,

$$\|f_2\| = L(f_1,Y). \tag{1.39}$$

Note that the points $(c_1,c_2), (c_1,d_2), (d_1,d_2), (d_1,c_2)$ in the given order form a closed path with respect to the directions $(0;1)$ and $(1;0)$. We conclude from (1.37)-(1.39) that this path is extremal for f_2. By Theorem 1.22, $z_0 = 0$ is a best approximation to f_2. Hence

$$E(f_2,\mathcal{Z}) = L(f_1,Y). \tag{1.40}$$

Now from (1.33),(1.36) and (1.40) we finally conclude that

$$E(f,\mathcal{R}) = L(f_1,Y) = \frac{1}{4}\left(f_1(c_1,c_2) + f_1(d_1,d_2) - f_1(c_1,d_2) - f_1(d_1,c_2)\right),$$

which is the desired result. □

COROLLARY 1.25. *Let all the conditions of Theorem* 1.24 *hold and* $f_1(y_1,y_2)$ *is the function defined in* (1.30). *Then the function* $g_0(y_1,y_2) = g_{1,0}(y_1) + g_{2,0}(y_2)$, *where*

$$g_{1,0}(y_1) = \frac{1}{2}f_1(y_1,c_2) + \frac{1}{2}f_1(y_1,d_2) - \frac{1}{4}f_1(c_1,c_2) - \frac{1}{4}f_1(d_1,d_2),$$

$$g_{2,0}(y_2) = \frac{1}{2}f_1(c_1,y_2) + \frac{1}{2}f_1(d_1,y_2) - \frac{1}{4}f_1(c_1,d_2) - \frac{1}{4}f_1(d_1,c_2)$$

and $y_1 = a_1x_1 + a_2x_2$, $y_2 = b_1x_1 + b_2x_2$, *is a best approximation from the set* $\mathcal{R}(a,b)$ *to the function* f.

PROOF. It is not difficult to verify that the function $f_2(y_1,y_2)$ defined in (1.35) has the form

$$f_2(y_1,y_2) = f_1(y_1,y_2) - g_{1,0}(y_1) - g_{2,0}(y_2).$$

On the other hand, we know from the proof of Theorem 1.24 that

$$E(f_1,\mathcal{Z}) = \|f_2\|.$$

Therefore, the function $g_{1,0}(y_1) + g_{2,0}(y_2)$ is a best approximation to f_1. Then the function $g_{1,0}(\mathbf{a}\cdot\mathbf{x}) + g_{2,0}(\mathbf{b}\cdot\mathbf{x})$ is an extremal element from $\mathcal{R}(\mathbf{a},\mathbf{b})$ to $f(\mathbf{x})$. □

REMARK 1.26. Rivlin and Sibner [**176**], and Babaev [**13**] proved Theorem 1.24 for the case in which $\mathbf{a}$ and $\mathbf{b}$ are the coordinate vectors. Our proof of Theorem 1.24 is different, short and elementary. Moreover, it has turned out to be useful in constructing an extremal element (see the proof of Corollary 1.25).

1.3.4. Density of ridge functions and some problems. Let $\mathbf{a}^1$ and $\mathbf{a}^2$ be nonzero directions in $\mathbb{R}^d$. One may ask the following question: are there cases in which the set $\mathcal{R}\left(\mathbf{a}^1,\mathbf{a}^2\right)$ is dense in the space of all continuous functions? Undoubtedly, a positive answer depends on the geometrical structure of compact sets over which all the considered functions are defined. This problem may be interesting in the theory of partial differential equations. Take, for example, equation (1.11). A positive answer to the problem means that for any continuous function f there exist solutions of the given equation uniformly converging to f.

It should be remarked that our problem is a special case of the problem considered by Marshall and O'Farrell. In [**151**], they obtained a necessary and sufficient condition for a sum $A_1 + A_2$ of two subalgebras to be dense in $C(U)$, where $C(U)$ denotes the space of real-valued continuous functions on a compact Hausdorff space U. Below we describe Marshall and O' Farrell's result for sums of ridge functions.

Let X be a compact subset of $\mathbb{R}^d$. The relation on X, defined by setting $\mathbf{x} \approx \mathbf{y}$ if $\mathbf{x}$ and $\mathbf{y}$ belong to some path in X, is an equivalence relation. The equivalence classes we call orbits.

THEOREM 1.27. *Let* X *be a compact subset of* $\mathbb{R}^d$ *with all its orbits closed. The set* $\mathcal{R}\left(\mathbf{a}^1,\mathbf{a}^2\right)$ *is dense in* $C(X)$ *if and only if* X *contains no closed path with respect to the directions* $\mathbf{a}^1$ *and* $\mathbf{a}^2$.

The proof immediately follows from proposition 2 in [**150**] established for the sum of two algebras. Since that proposition was given without proof, for completeness of the exposition we give the proof of Theorem 1.27.

PROOF. *Necessity.* If X has closed paths, then X has a closed path $p' = (\mathbf{p}'_1, \ldots, \mathbf{p}'_{2m})$ such that all points $\mathbf{p}'_1, \ldots, \mathbf{p}'_{2m}$ are distinct. In fact, such a special path can be obtained from any closed path $p = (\mathbf{p}_1, \ldots, \mathbf{p}_{2n})$ by the following simple algorithm: if the points of the path p are not all distinct, let i and $k > 0$ be the minimal indices such that $\mathbf{p}_i = \mathbf{p}_{i+2k}$; delete from p the subsequence $\mathbf{p}_{i+1}, \ldots, \mathbf{p}_{i+2k}$ and call p the obtained path; repeat the above step until all points of p are all distinct; set $p' := p$. By Urysohn's lemma, there exist continuous functions $h = h(\mathbf{x})$ on X such that $h(\mathbf{p}'_i) = 1$, $i = 1, 3, \ldots, 2m-1$, $h(\mathbf{p}'_i) = -1$, $i = 2, 4, \ldots, 2m$ and $-1 < h(\mathbf{x}) < 1$ elsewhere. Consider the measure

$$\mu_{p'} = \frac{1}{2m} \sum_{i=1}^{2m} (-1)^{i-1} \delta_{\mathbf{p}'_i} ,$$

where $\delta_{\mathbf{p}'_i}$ is a point mass at $\mathbf{p}'_i$. For this measure, $\int_X h d\mu_{p'} = 1$ and $\int_X g d\mu_{p'} = 0$ for all functions $g \in \mathcal{R}\left(\mathbf{a}^1, \mathbf{a}^2\right)$. Thus the set $\mathcal{R}\left(\mathbf{a}^1, \mathbf{a}^2\right)$ cannot be dense in $C(X)$.

Sufficiency. We are going to prove that the only annihilating regular Borel measure for $\mathcal{R}\left(\mathbf{a}^1, \mathbf{a}^2\right)$ is the zero measure. Suppose, contrary to this assumption, there exists a nonzero annihilating measure on X for $\mathcal{R}\left(\mathbf{a}^1, \mathbf{a}^2\right)$. The class of such measures with total variation not more than 1 we denote by S. Clearly, S is weak-* compact and convex. By the Krein-Milman theorem (see [**177**, p.75]), there exists an extreme measure μ in S. Since the orbits are closed, μ must be supported on a single orbit. Denote this orbit by T.

For $i = 1, 2$, let X_i be the quotient space of X obtained by identifying the points $\mathbf{y}$ and $\mathbf{z}$ whenever $\mathbf{a}^i \cdot \mathbf{y} = \mathbf{a}^i \cdot \mathbf{z}$. Let π_i be the natural projection of X onto X_i. For a fixed point $t \in X$ set $T_1 = \{t\}$, $T_2 = \pi_1^{-1}(\pi_1 T_1)$, $T_3 = \pi_2^{-1}(\pi_2 T_2)$, $T_4 = \pi_1^{-1}(\pi_1 T_3)$, ... Obviously, $T_1 \subset T_2 \subset T_3 \subset \cdots$. Therefore, for some $k \in \mathbb{N}$, $|\mu|(T_{2k}) > 0$, where $|\mu|$ is a total variation measure of μ. Since μ is orthogonal to every continuous function of the form $g\left(\mathbf{a}^1 \cdot \mathbf{x}\right)$, $\mu(T_{2k}) = 0$. From the Jordan decomposition $\mu(T_{2k}) = \mu^+(T_{2k}) - \mu^-(T_{2k})$ it follows that $\mu^+(T_{2k}) = \mu^-(T_{2k}) > 0$. Fix a Borel subset $S_0 \subset T_{2k}$ such that $\mu^+(S_0) > 0$ and $\mu^-(S_0) = 0$. Since μ is orthogonal to every continuous function of the form $g\left(\mathbf{a}^2 \cdot \mathbf{x}\right)$, $\mu(\pi_2^{-1}(\pi_2 S_0)) = 0$. Therefore, one can chose a Borel set S_1 such that $S_1 \subset \pi_2^{-1}(\pi_2 S_0) \subset T_{2k+1}$, $S_1 \cap S_0 = \varnothing$, $\mu^+(S_1) = 0$, $\mu^-(S_1) \geqslant \mu^+(S_0)$. By the same way one can chose a Borel set S_2 such that $S_2 \subset \pi_1^{-1}(\pi_1 S_1) \subset T_{2k+2}$, $S_2 \cap S_1 = \varnothing$, $\mu^-(S_2) = 0$, $\mu^+(S_2) \geqslant \mu^-(S_1)$, and so on.

The sets $S_0, S_1, S_2, \ldots$ are pairwise disjoint. For otherwise, there would exist positive integers n and m, with $n < m$ and a path $(y_n, y_{n+1}, \ldots, y_m)$ such that $y_i \in S_i$ for $i = n, \ldots, m$ and $y_m \in S_m \cap S_n$. But then there would exist paths $(z_1, z_2, \ldots, z_{n-1}, y_n)$ and $(z_1, z'_2, \ldots, z'_{n-1}, y_m)$ with z_i and z'_i in T_i for $i = 2, \ldots, n-1$. Hence, the set

$$\{z_1, z_2, \ldots, z_{n-1}, y_n, y_{n+1}, \ldots, y_m, z'_{n-1}, \ldots, z'_2, z_1\}$$

would contain a closed path. This would contradict our assumption on X.

Now, since the sets $S_0, S_1, S_2, \ldots,$ are pairwise disjoint and $|\mu|(S_i) \geqslant \mu^+(S_0) > 0$ for each $i = 1, 2, \ldots,$ it follows that the total variation of μ is infinite. This contradiction completes the proof. $\square$

The following corollary concerns the problem considered by Colitschek and Light [**63**].

COROLLARY 1.28. *Let D be a compact subset of $\mathbb{R}^2$ with all its orbits closed. Let W denote the set of all solutions of the wave equation*

$$\frac{\partial^2 w}{\partial s \partial t}(s,t) = 0, \qquad (s,t) \in D.$$

Then

$$\inf_{w \in W} \|f - w\| = 0$$

for any continuous function $f(s,t)$ on D if and only if D contains no closed bolt of lightning.

PROOF. Let π_1 and π_2 denote the usual coordinate projections, viz: $\pi_1(s,t) = s$ and $\pi_2(s,t) = t$, $(s,t) \in \mathbb{R}^2$. Set $S = \pi_1(D)$ and $T = \pi_2(D)$. It is easy to see that

$$W = \left\{w \in C(D) : w(s,t) = x(s) + y(t), \quad x \in C^2(S), \ \ y \in C^2(T)\right\}.$$

Set

$$\widetilde{W} = \{w \in C(D) : w(s,t) = x(s) + y(t), \quad x \in C(S), \ \ y \in C(T)\}.$$

Since the set W is dense in $\widetilde{W}$,

$$\inf_{w \in W} \|f - w\| = \inf_{w \in \widetilde{W}} \|f - w\|.$$

But by Theorem 1.27, the equality

$$\inf_{w \in \widetilde{W}} \|f - w\| = 0$$

holds for any $f \in C(D)$ if and only if D contains no closed bolt of lightning. $\square$

Let us discuss some difficulties that arise when studying sums of more than two ridge functions. Consider the set

$$\mathcal{R}\left(\mathbf{a}^1, \ldots, \mathbf{a}^r\right) = \left\{\sum_{i=1}^{r} g_i\left(\mathbf{a}^i \cdot \mathbf{x}\right), g_i \in C\left(\mathbb{R}\right), i = 1, \ldots, r\right\},$$

where $\mathbf{a}^1, \ldots, \mathbf{a}^r$ are pairwise linearly independent vectors in $\mathbb{R}^d \backslash \{\mathbf{0}\}$. Let $r \geq 3$. How can one define a path in this general case? Recall that in the case when $r = 2$, a path is an ordered set of points $(\mathbf{p}_1, \mathbf{p}_2, \ldots, \mathbf{p}_n)$ in $\mathbb{R}^d$ with edges $\mathbf{p}_i\mathbf{p}_{i+1}$ in alternating hyperplanes. The first, the third, the fifth,... hyperplanes (also the second, the fourth, the sixth,... hyperplanes) are parallel. If not differentiate between parallel hyperplanes, the path $(\mathbf{p}_1, \mathbf{p}_2, \ldots, \mathbf{p}_n)$ can be considered as a trace of some point traveling in two alternating hyperplanes. In this case, if the point starts and stops at the same location (i.e., if $\mathbf{p}_n = \mathbf{p}_1$) and n is an odd number, then the path functional

$$G(f) = \frac{1}{n-1} \sum_{i=1}^{n-1} (-1)^{i+1} f(\mathbf{p}_i),$$

annihilates sums of ridge functions with the corresponding two fixed directions. The picture becomes quite different and more complicated when the number of directions more than two. The simple generalization of the above-mentioned arguments demands a point traveling in three or more alternating hyperplanes. But in this case the appropriate generalization of the functional G does not annihilate functions from $\mathcal{R}\left(\mathbf{a}^1,\ldots,\mathbf{a}^r\right)$.

There were several attempts to fill this gap in the special case when $r=d$ and $\mathbf{a}^1,\ldots,\mathbf{a}^r$ are the coordinate vectors. Unfortunately, all these attempts failed (see, for example, the attempts in [**47**, **64**] and the refutations in [**10**, **155**]).

At the end of this subsection we want to draw the readers attention to the following problems. All these problems are open and cannot be solved by the methods presented here.

Let Q be a compact subset of $\mathbb{R}^d$. Consider the approximation of a continuous function defined on Q by functions from $\mathcal{R}\left(\mathbf{a}^1,\ldots,\mathbf{a}^r\right)$. Let $r\geq 3$.

PROBLEM 3. *Characterize those functions from $\mathcal{R}\left(\mathbf{a}^1,\ldots,\mathbf{a}^r\right)$ that are extremal to a given continuous function.*

PROBLEM 4. *Establish explicit formulas for the error in approximating from $\mathcal{R}\left(\mathbf{a}^1,\ldots,\mathbf{a}^r\right)$ and construct a best approximation.*

PROBLEM 5. *Find necessary and sufficient geometrical conditions for the set $\mathcal{R}\left(\mathbf{a}^1,\ldots,\mathbf{a}^r\right)$ to be dense in $C(Q)$.*

It should be remarked that in [**150**], Problem 5 was set up for the sum of r subalgebras of $C(Q)$. Lin and Pinkus [**140**] proved that the set $\mathcal{R}\left(\mathbf{a}^1,\ldots,\mathbf{a}^r\right)$ (r may be very large) is not dense in $C(\mathbb{R}^d)$ in the topology of uniform convergence on compact subsets of $\mathbb{R}^d$. That is, there are compact sets $Q\subset\mathbb{R}^d$ such that $\mathcal{R}\left(\mathbf{a}^1,\ldots,\mathbf{a}^r\right)$ is not dense in $C(Q)$. In the case $r=2$, Theorem 1.27 complements this result, by describing compact sets $Q\subset\mathbb{R}^2$, for which $\mathcal{R}\left(\mathbf{a}^1,\mathbf{a}^2\right)$ is dense in $C(Q)$.

1.4. Sums of continuous ridge functions

In this section, we find geometric means of deciding if any continuous multivariate function can be represented by a sum of two continuous ridge functions.

1.4.1. Exposition of the problem. In this section, we will consider the following representation problem associated with the set $\mathcal{R}\left(\mathbf{a}^1,\ldots,\mathbf{a}^r\right)$.

PROBLEM 6. *Let X be a compact subset of $\mathbb{R}^d$. Give geometrical conditions that are necessary and sufficient for*

$$\mathcal{R}\left(\mathbf{a}^1,\ldots,\mathbf{a}^r\right)=C\left(X\right),$$

where $C\left(X\right)$ is the space of continuous functions on X furnished with the uniform norm.

We solve this problem for $r=2$. Problem 6, like Problems 3–5 from the previous section, is open in the case $r\geq 3$. Geometrical characterization of compact sets $X\subset\mathbb{R}^d$ with the property $\mathcal{R}\left(\mathbf{a}^1,\ldots,\mathbf{a}^r\right)=C\left(X\right)$, $r\geq 3$, seems to be beyond the scope of the methods discussed herein. Nevertheless, recall that this problem in a quite abstract form, which involves regular Borel measures on X, was solved by Sternfeld (see Section 1.1).

In the sequel, we will use the notation

$$H_1 = H_1(X) = \{g_1(\mathbf{a}^1 \cdot \mathbf{x}) : g_1 \in C(\mathbb{R})\},$$
$$H_2 = H_2(X) = \{g_2(\mathbf{a}^2 \cdot \mathbf{x}) : g_2 \in C(\mathbb{R})\}.$$

Note that by this notation, $\mathcal{R}(\mathbf{a}^1, \mathbf{a}^2) = H_1 + H_2$.

At the end of this section, we generalize the obtained result from $H_1 + H_2$ to the set of sums $g_1(h_1(\mathbf{x})) + g_2(h_2(\mathbf{x}))$, where h_1, h_2 are fixed continuous functions on X.

1.4.2. The representation theorem.

THEOREM 1.29. *Let X be a compact subset of $\mathbb{R}^d$. The equality*

$$H_1(X) + H_2(X) = C(X)$$

holds if and only if X contains no closed path and there exists a positive integer n_0 such that the lengths of paths in X are bounded by n_0.

PROOF. *Necessity.* Let $H_1 + H_2 = C(X)$. Consider the linear operator

$$A : H_1 \times H_2 \to C(X), \ A[(g_1, g_2)] = g_1 + g_2,$$

where $g_1 \in H_1, g_2 \in H_2$. The norm on $H_1 \times H_2$ we define as

$$\|(g_1, g_2)\| = \|g_1\| + \|g_2\|.$$

It is obvious that the operator A is continuous with respect to this norm. Besides, since $C(X) = H_1 + H_2$, A is a surjection. Consider the conjugate operator

$$A^* : C(X)^* \to [H_1 \times H_2]^*, \ A^*[G] = (G_1, G_2),$$

where the functionals G_1 and G_2 are defined as follows

$$G_1(g_1) = G(g_1), g_1 \in H_1; \ G_2(g_2) = G(g_2), g_2 \in H_2.$$

An element (G_1, G_2) from $[H_1 \times H_2]^*$ has the norm

$$\|(G_1, G_2)\| = \max\{\|G_1\|, \|G_2\|\}. \tag{1.41}$$

Let now $p = (p_1, \ldots, p_m)$ be any path with different points: $p_i \neq p_j$ for any $i \neq j$, $1 \leq i, \ j \leq m$. We associate with p the following functional over $C(X)$

$$L[f] = \frac{1}{m} \sum_{i=1}^{m} (-1)^{i-1} f(p_i).$$

Since $|L(f)| \leq \|f\|$ and $|L(g)| = \|g\|$ for a continuous function $g(\mathbf{x})$ such that $g(p_i) = 1$, for odd indices i, $g(p_j) = -1$, for even indices j and $-1 < g(\mathbf{x}) < 1$ elsewhere, we obtain that $\|L\| = 1$. Let $A^*[L] = (L_1, L_2)$. One can easily verify that

$$\|L_i\| \leq \frac{2}{m}, i = 1, 2.$$

Therefore, from (1.41) we obtain that

$$\|A^*[L]\| \leq \frac{2}{m}. \tag{1.42}$$

Since A is a surjection, there exists $\delta > 0$ such that

$$\|A^*[G]\| \geq \delta \|G\| \ \text{ for any functional } \ G \in C(X)^*$$

Hence

$$\|A^* [L]\| \geq \delta. \tag{1.43}$$

Now from (1.42) and (1.43) we conclude that

$$m \leq \frac{2}{\delta}.$$

This means that for a path with different points, n_0 can be chosen as $\left[\frac{2}{\delta}\right] + 1$.

Let now $p = (p_1, \ldots, p_m)$ be a path with at least two coinciding points. Then we can form a closed path with different points. This may be done by the following way: let i and j be indices such that $p_i = p_j$ and $j - i$ takes its minimal value. Note that in this case all the points $p_i, p_{i+1}, \ldots, p_{j-1}$ are distinct. Now if $j - i$ is an even number, then the path $(p_i, p_{i+1}, \ldots, p_{j-1})$, and if $j - i$ is an odd number, then the path $(p_{i+1}, \ldots, p_{j-1})$ is a closed path with different points. It remains to show that X can not possess closed paths with different points. Indeed, if $q = (q_1, \ldots, q_{2k})$ is a path of this type, then the functional L, associated with q, annihilates all functions from $H_1 + H_2$. On the other hand, $L[f] = 1$ for a continuous function f on X satisfying the conditions $f(t) = 1$ if $t \in \{q_1, q_3, \ldots, q_{2k-1}\}$; $f(t) = -1$ if $t \in \{q_2, q_4, \ldots, q_{2k}\}$; $f(t) \in (-1; 1)$ if $t \in X \backslash q$. This implies on the contrary to our assumption that $H_1 + H_2 \neq C(X)$. The necessity has been proved.

Sufficiency. Let X contains no closed path and the lengths of all paths are bounded by some positive integer n_0. We may suppose that any path has different points. Indeed, in other case we can form a closed path, which contradicts our assumption.

For $i = 1, 2$, let X_i be the quotient space of X obtained by identifying the points a and b whenever $g(a) = g(b)$ for each g in H_i. Let π_i be the natural projection of X onto X_i. For a point $t \in X$ set $T_1 = \pi_1^{-1}(\pi_1 t), T_2 = \pi_2^{-1}(\pi_2 T_1), \ldots$. By $O(t)$ denote the orbit of X containing t. Since the length of any path in X is not more than n_0, we conclude that $O(t) = T_{n_0}$. Since X is compact, the sets $T_1, T_2, \ldots, T_{n_0}$, hence $O(t)$, are compact. By Theorem 1.27, $\overline{H_1 + H_2} = C(X)$.

Now let us show that $H_1 + H_2$ is closed in $C(X)$. Set

$$H_3 = H_1 \cap H_2.$$

Let X_3 and π_3 be the associated quotient space and projection. Fix some $a \in X_3$. Show, within conditions of our theorem, that if $t \in \pi_3^{-1}(a)$, then $O(t) = \pi_3^{-1}(a)$. The inclusion $O(t) \subset \pi_3^{-1}(a)$ is obvious. Suppose that there exists a point $t_1 \in \pi_3^{-1}(a)$ such that $t_1 \notin O(t)$. Then $O(t) \cap O(t_1) = \emptyset$. By $X|O$ denote the quotient space generated by orbits of X. $X|O$ is a normal topological space with its natural quotient topology. Hence we can construct a continuous function $u \in C(X|O)$ such that $u(O(t)) = 0$, $u(O(t_1)) = 1$. The function $v(x) = u(O(x))$, $x \in X$, is continuous on X and belongs to H_3 as a function being constant on each orbit. But, since $O(t) \subset \pi_3^{-1}(a)$ and $O(t_1) \subset \pi_3^{-1}(a)$, the function $v(x)$ can not take different values on $O(t)$ and $O(t_1)$. This contradiction means that there is not a point $t_1 \in \pi_3^{-1}(a)$ such that $t_1 \notin O(t)$. Thus,

$$O(t) = \pi_3^{-1}(a) \tag{1.44}$$

for any $a \in X_3$ and $t \in \pi_3^{-1}(a)$.

Now prove that there exists a positive real number c such that

$$\sup_{z \in X_3} \underset{\pi_3^{-1}(z)}{var} f \leq c \sup_{y \in X_2} \underset{\pi_2^{-1}(y)}{var} f \tag{1.45}$$

for all f in H_1. Note that for $Y \subset X$, $\underset{Y}{var} f$ is the variation of f on the set Y. That is,

$$\underset{Y}{var} f = \sup_{x,y \in Y} |f(x) - f(y)|.$$

Due to (1.44), inequality (1.45) can be written in the following form

$$\sup_{t \in X} \underset{O(t)}{var} f \leq c \sup_{t \in X} \underset{\pi_2^{-1}(\pi_2(t))}{var} f \tag{1.46}$$

for all $f \in H_1$.

Let $t \in X$ and t_1, t_2 be arbitrary points of $O(t)$. Then there is a path $(b_1, b_2, \ldots, b_m)$ with $b_1 = t_1$ and $b_m = t_2$. Besides, by the condition, $m \leq n_0$. Let first $\mathbf{a}^2 \cdot b_1 = \mathbf{a}^2 \cdot b_2$, $\mathbf{a}^1 \cdot b_2 = \mathbf{a}^1 \cdot b_3, \ldots, \mathbf{a}^2 \cdot b_{m-1} = \mathbf{a}^2 \cdot b_m$. Then for any function $f \in H_1$

$$|f(t_1) - f(t_2)| = |f(b_1) - f(b_2) + \cdots - f(b_m)| \leq$$

$$\leq |f(b_1) - f(b_2)| + \cdots + |f(b_{m-1}) - f(b_m)| \leq \frac{n_o}{2} \sup_{t \in X} \underset{\pi_2^{-1}(\pi_2(t))}{var} f. \tag{1.47}$$

It is not difficult to verify that inequality (1.47) holds in all other possible cases of the path $(b_1, \ldots, b_m)$. Now from (1.47) we obtain (1.46), hence (1.45), where $c = \frac{n_0}{2}$. In [**150**], Marshall and O'Farrell proved the following result (see [**150**, Proposition 4]): Let A_1 and A_2 be closed subalgebras of $C(X)$ that contain the constants. Let (X_1, π_1), (X_2, π_2) and (X_3, π_3) be the quotient spaces and projections associated with the algebras A_1, A_2 and $A_3 = A_1 \cap A_2$ respectively. Then $A_1 + A_2$ is closed in $C(X)$ if and only if there exists a positive real number c such that

$$\sup_{z \in X_3} \underset{\pi_3^{-1}(z)}{var} f \leq c \sup_{y \in X_2} \underset{\pi_2^{-1}(y)}{var} f$$

for all f in A_1.

By this proposition, (1.45) implies that $H_1 + H_2$ is closed in $C(X)$. Thus we finally obtain that $H_1 + H_2 = C(X)$. □

Paths with respect to two directions are explicit objects and give geometric means of deciding if $H_1 + H_2 = C(X)$. Let us show this in the example of the bivariate ridge functions $g_1 = x_1 + x_2$ and $g_2 = x_1 - x_2$. If X is the union of two parallel line segments in $\mathbb{R}^2$, not parallel to any of the lines $x_1 + x_2 = 0$ and $x_1 - x_2 = 0$, then Theorem 1.29 holds. If X is any bounded part of the graph of the function $x_2 = \arcsin(\sin x_1)$, then Theorem 1.29 also holds. Let now X be the set

$$\{(0,0), (1,-1), (0,-2), (-1\tfrac{1}{2}, -\tfrac{1}{2}), (0,1), (\tfrac{3}{4}, \tfrac{1}{4}), (0, -\tfrac{1}{2}), \\ (-\tfrac{3}{8}, -\tfrac{1}{8}), (0, \tfrac{1}{4}), (\tfrac{3}{16}, \tfrac{1}{16}), \ldots\}.$$

In this case, there is no positive integer bounding lengths of all paths. Thus Theorem 1.29 fails. Note that since orbits of all paths are closed, Theorem 1.27 from the previous section shows $H_1 + H_2$ is dense in $C(X)$.

If X is any set with interior points, then both Theorem 1.27 and Theorem 1.29 fail, since any such set contains the vertices of some parallelogram with sides parallel to the directions $\mathbf{a}^1$ and $\mathbf{a}^2$, that is a closed path.

Theorem 1.29 admits a direct generalization to the representation by sums $g_1(h_1(\mathbf{x})) + g_2(h_2(\mathbf{x}))$, where $h_1(\mathbf{x})$ and $h_2(\mathbf{x})$ are fixed continuous functions on

X. This generalization needs consideration of new objects – paths with respect to two continuous functions.

DEFINITION 1.30. Let X be a compact set in $\mathbb{R}^d$ and $h_1, h_2 \in C(X)$. A finite ordered subset $(p_1, p_2, \dots, p_m)$ of X with $p_i \neq p_{i+1}$ $(i = 1, \dots, m-1)$, and either $h_1(p_1) = h_1(p_2)$, $h_2(p_2) = h_2(p_3)$, $h_1(p_3) = h_1(p_4), \dots,$ or $h_2(p_1) = h_2(p_2)$, $h_1(p_2) = h_1(p_3)$, $h_2(p_3) = h_2(p_4), \dots,$ is called a path with respect to the functions h_1 and h_2 or, shortly, an h_1-h_2 path.

THEOREM 1.31. *Let X be a compact subset of $\mathbb{R}^d$. All functions $f \in C(X)$ admit a representation*

$$f(\mathbf{x}) = g_1(h_1(\mathbf{x})) + g_2(h_2(\mathbf{x})), \quad g_1, g_2 \in C(\mathbb{R})$$

if and only if the set X contains no closed h_1-h_2 path and there exists a positive integer n_0 such that the lengths of h_1-h_2 paths in X are bounded by n_0.

The proof can be carried out by the same arguments as above.

It should be noted that Theorem 1.31 was first proved by Khavinson in his monograph [**117**]. Khavinson's proof (see [**117**, p.87]) used theorems of Sternfeld [**190**] and Medvedev [**117**, Theorem 2.2], whereas our proof, which generalizes the ideas of Khavinson, was based on the above proposition of Marshall and O'Farrell.

1.5. On the proximinality of ridge functions

In this section, using two results of Garkavi, Medvedev and Khavinson [**61**], we give sufficient conditions for proximinality of sums of two ridge functions with bounded and continuous summands in the spaces of bounded and continuous multivariate functions, respectively. In the first case, we give an example which shows that the corresponding sufficient condition cannot be made weaker for certain subsets of $\mathbb{R}^n$. In the second case, we obtain also a necessary condition for proximinality. All the results are furnished with plenty of examples. The results, examples and following discussions naturally lead us to a conjecture on the proximinality of the considered class of ridge functions.

1.5.1. Problem statement. Let E be a normed linear space and F be its subspace. We say that F is proximinal in E if for any element $e \in E$ there exists at least one element $f_0 \in F$ such that

$$\|e - f_0\| = \inf_{f \in F} \|e - f\|.$$

In this case, the element f_0 is said to be extremal to e.

We are interested in the problem of proximinality of the set of linear combinations of ridge functions in the spaces of bounded and continuous functions respectively. This problem will be considered in the simplest case when the class of approximating functions is the set

$$\mathcal{R} = \mathcal{R}(\mathbf{a}^1, \mathbf{a}^2) = \left\{ g_1(\mathbf{a}^1 \cdot \mathbf{x}) + g_2(\mathbf{a}^2 \cdot \mathbf{x}) : g_i : \mathbb{R} \to \mathbb{R}, i = 1, 2 \right\}.$$

Here $\mathbf{a}^1$and $\mathbf{a}^2$ are fixed directions and we vary over g_i. It is clear that this is a linear space. Consider the following three subspaces of $\mathcal{R}$. The first is obtained by taking only bounded sums $g_1(\mathbf{a}^1 \cdot \mathbf{x}) + g_2(\mathbf{a}^2 \cdot \mathbf{x})$ over some set X in $\mathbb{R}^n$. We denote this subspace by $\mathcal{R}_a(X)$. The second and the third are subspaces of $\mathcal{R}$ with bounded and continuous summands $g_i\left(\mathbf{a}^i \cdot \mathbf{x}\right)$, $i = 1, 2$, on X respectively. These subspaces will be denoted by $\mathcal{R}_b(X)$ and $\mathcal{R}_c(X)$. In the case of $\mathcal{R}_c(X)$, the set X is considered to be compact.

Let $B(X)$ and $C(X)$ be the spaces of bounded and continuous multivariate functions over X respectively. What conditions must one impose on X in order that the sets $\mathcal{R}_a(X)$ and $\mathcal{R}_b(X)$ be proximinal in $B(X)$ and the set $\mathcal{R}_c(X)$ be proximinal in $C(X)$? We are also interested in necessary conditions for proximinality. It follows from one result of Garkavi, Medvedev and Khavinson (see [**61**, Theorem 1]) that $\mathcal{R}_a(X)$ is proximinal in $B(X)$ for all subsets X of $\mathbb{R}^n$. There is also an answer (see [**61**, Theorem 2]) for proximinality of $\mathcal{R}_b(X)$ in $B(X)$. This will be discussed in the following subsection. Is the set $\mathcal{R}_b(X)$ always proximinal in $B(X)$? There is an an example of a set $X \subset \mathbb{R}^n$ and a bounded function f on X for which there does not exist an extremal element in $\mathcal{R}_b(X)$.

In the final subsection, we will obtain sufficient conditions for the existence of extremal elements from $\mathcal{R}_c(X)$ to an arbitrary function $f \in C(X)$. Based on one result of Marshall and O'Farrell [**150**], we will also give a necessary condition for proximinality of $\mathcal{R}_c(X)$ in $C(X)$. All the theorems, following discussions and examples of the paper will lead us naturally to a conjecture on the proximinality of the subspaces $\mathcal{R}_b(X)$ and $\mathcal{R}_c(X)$ in the spaces $B(X)$ and $C(X)$, respectively.

The reader may also be interested in the more general case with the set $\mathcal{R} = \mathcal{R}\left(\mathbf{a}^1, \ldots, \mathbf{a}^r\right)$. In this case, the corresponding sets $\mathcal{R}_a(X)$, $\mathcal{R}_b(X)$ and $\mathcal{R}_c(X)$ are defined similarly. Using the results of [**61**], one can obtain sufficient (but not necessary) conditions for proximinality of these sets. This needs, besides paths, the consideration of some additional and more complicated relations between points of X. Here we will not consider the case $r \geq 3$, since our main purpose is to draw the reader's attention to the arisen problems of proximinality in the simplest case of approximation. For the existing open problems connected with the set $\mathcal{R}\left(\mathbf{a}^1, \ldots, \mathbf{a}^r\right)$, where $r \geq 3$, see [**82**] and [**171**].

1.5.2. Proximinality of $\mathcal{R}_b(X)$ in $B(X)$. Let $\mathbf{a}^1$ and $\mathbf{a}^2$ be two different directions in $\mathbb{R}^n$. In the sequel, we will use paths with respect to the directions $\mathbf{a}^1$ and $\mathbf{a}^2$. Recall that a length of a path is the number of its points and can be equal to ∞ if the path is infinite. A singleton is a path of the unit length. We say that a path $\left(\mathbf{x}^1, \ldots, \mathbf{x}^m\right)$ belonging to some subset X of $\mathbb{R}^n$ is irreducible if there is not another path $\left(\mathbf{y}^1, \ldots, \mathbf{y}^l\right) \subset X$ with $\mathbf{y}^1 = \mathbf{x}^1$, $\mathbf{y}^l = \mathbf{x}^m$ and $l < m$.

The following theorem follows from [**61**, Theorem 2].

THEOREM 1.32. *Let $X \subset \mathbb{R}^n$ and the lengths of all irreducible paths in X be uniformly bounded by some positive integer. Then each function in $B(X)$ has an extremal element in $\mathcal{R}_b(X)$.*

There are a large number of sets in $\mathbb{R}^n$ satisfying the hypothesis of this theorem. For example, if a set X has a cross section according to one of the directions $\mathbf{a}^1$ or $\mathbf{a}^2$, then the set X satisfies the hypothesis of Theorem 1.32. By a cross section according to the direction $\mathbf{a}^1$ we mean any set $X_{\mathbf{a}^1} = \{x \in X : \ \mathbf{a}^1 \cdot \mathbf{x} = c\}$, $c \in \mathbb{R}$, with the property: for any $\mathbf{y} \in X$ there exists a point $\mathbf{y}^1 \in X_{\mathbf{a}^1}$ such that $\mathbf{a}^2 \cdot \mathbf{y} = \mathbf{a}^2 \cdot \mathbf{y}^1$. By the similar way, one can define a cross section according to the direction $\mathbf{a}^2$. For more on cross sections in problems of proximinality of sums of univariate functions see [**60, 116**]. Regarding Theorem 1.32 one may ask if the condition of the theorem is necessary for proximinality of $\mathcal{R}_b(X)$ in $B(X)$. While we do not know a complete answer to this question, we are going to give an example of a set X for which Theorem 1.32 fails. Let $\mathbf{a}^1 = (1;-1)$, $\mathbf{a}^2 = (1;1)$. Consider the set

$$\begin{aligned} X &= \{(2;\frac{2}{3}),(\frac{2}{3};-\frac{2}{3}),(0;0),(1;1),(1+\frac{1}{2};1-\frac{1}{2}),(1+\frac{1}{2}+\frac{1}{4};1-\frac{1}{2}+\frac{1}{4}),\\ &\quad (1+\frac{1}{2}+\frac{1}{4}+\frac{1}{8};1-\frac{1}{2}+\frac{1}{4}-\frac{1}{8}),\dots\}. \end{aligned}$$

In what follows, the elements of X in the given order will be denoted by $\mathbf{x}^0,\mathbf{x}^1,\mathbf{x}^2,\dots$. It is clear that X is a path of the infinite length and $\mathbf{x}^n \to \mathbf{x}^0$ as $n\to\infty$. Let $\sum_{n=1}^{\infty} c_n$ be any divergent series with the terms $c_n > 0$ and $c_n \to 0$ as $n\to\infty$. Besides let f_0 be a function vanishing at the points $\mathbf{x}^0,\mathbf{x}^2,\mathbf{x}^4,\dots$, and taking values $c_1,c_2,c_3,\dots$ at the points $\mathbf{x}^1,\mathbf{x}^3,\ \mathbf{x}^5,\dots$, respectively. It is obvious that f_0 is continuous on X. The set X is compact and satisfies all the conditions of Theorem 1.32. By that theorem, $\overline{\mathcal{R}_c(X)} = C(X)$. Therefore, for any continuous function on X, thus for f_0,

$$\inf_{g\in\mathcal{R}_c(X)} \|f_0 - g\|_{C(X)} = 0. \tag{1.48}$$

Since $\mathcal{R}_c(X) \subset \mathcal{R}_b(X)$, we obtain from (1.48) that

$$\inf_{g\in\mathcal{R}_b(X)} \|f_0 - g\|_{B(X)} = 0. \tag{1.49}$$

Suppose that f_0 has an extremal element $g_1^0\left(\mathbf{a}^1\cdot\mathbf{x}\right) + g_2^0\left(\mathbf{a}^2\cdot\mathbf{x}\right)$ in $\mathcal{R}_b(X)$. By the definition of $\mathcal{R}_b(X)$, the ridge functions $g_i^0, i=1,2$, are bounded on X. From (1.49) it follows that $f_0 = g_1^0\left(\mathbf{a}^1\cdot\mathbf{x}\right) + g_2^0\left(\mathbf{a}^2\cdot\mathbf{x}\right)$. Since $\mathbf{a}^1\cdot\mathbf{x}^{2n} = \mathbf{a}^1\cdot\mathbf{x}^{2n+1}$ and $\mathbf{a}^2\cdot\mathbf{x}^{2n+1} = \mathbf{a}^2\cdot\mathbf{x}^{2n+2}$, for $n=0,1,\dots$, we can write that

$$\sum_{n=0}^{k} c_{n+1} = \sum_{n=0}^{k}\left[f(\mathbf{x}^{2n+1}) - f(\mathbf{x}^{2n})\right]$$

$$= \sum_{n=0}^{k}\left[g_2^0(\mathbf{x}^{2n+1}) - g_2^0(\mathbf{x}^{2n})\right] = g_2^0(\mathbf{a}^2\cdot\mathbf{x}^{2k+1}) - g_2^0(\mathbf{a}^2\cdot\mathbf{x}^0). \tag{1.50}$$

Since $\sum_{n=1}^{\infty} c_n = \infty$, we deduce from (1.50) that the function $g_2^0\left(\mathbf{a}^2\cdot\mathbf{x}\right)$ is not bounded on X. This contradiction means that the function f_0 does not have an extremal element in $\mathcal{R}_b(X)$. Therefore, the space $\mathcal{R}_b(X)$ is not proximinal in $B(X)$.

1.5.3. Proximinality of $\mathcal{R}_c(X)$ in $C(X)$. In this subsection, we give a sufficient condition and also a necessary condition for proximinality of $\mathcal{R}_c(X)$ in $C(X)$.

THEOREM 1.33. *Let the system of linearly independent vectors $\mathbf{a}^1$ and $\mathbf{a}^2$ have a complement to a basis $\{\mathbf{a}^1,\dots,\mathbf{a}^n\}$ in $\mathbb{R}^n$ with the property: for any point $\mathbf{x}^0\in X$ and any positive real number δ there exist a number $\delta_0\in(0,\delta]$ and a point $\mathbf{x}^\sigma$ in the set*

$$\sigma = \{\mathbf{x}\in X : \mathbf{a}^2\cdot\mathbf{x}^0 - \delta_0 \le \mathbf{a}^2\cdot\mathbf{x} \le \mathbf{a}^2\cdot\mathbf{x}^0 + \delta_0\},$$

such that the system

$$
\left\{
\begin{array}{c}
\mathbf{a}^2 \cdot \mathbf{x}' = \mathbf{a}^2 \cdot \mathbf{x}^\sigma \\
\mathbf{a}^1 \cdot \mathbf{x}' = \mathbf{a}^1 \cdot \mathbf{x} \\
\sum_{i=3}^n \left|\mathbf{a}^i \cdot \mathbf{x}' - \mathbf{a}^i \cdot \mathbf{x}\right| < \delta
\end{array}
\right.
\tag{1.51}
$$

has a solution $\mathbf{x}' \in \sigma$ for all points $\mathbf{x} \in \sigma$. Then the space $\mathcal{R}_c(X)$ is proximinal in $C(X)$.

PROOF. Introduce the following mappings and sets:

$$\pi_i : X \to \mathbb{R},\ \pi_i(\mathbf{x}) = \mathbf{a}^i \cdot \mathbf{x},\ Y_i = \pi_i(X),\ i = 1, \dots, n.$$

Since the system of vectors $\{\mathbf{a}^1, \dots, \mathbf{a}^n\}$ is linearly independent, the mapping $\pi = (\pi_1, \dots \pi_n)$ is an injection from X into the Cartesian product $Y_1 \times \cdots \times Y_n$. Besides, π is linear and continuous. By the open mapping theorem, the inverse mapping π^{-1} is continuous from $Y = \pi(X)$ onto X. Let f be a continuous function on X. Then the composition $f \circ \pi^{-1}(y_1, \dots y_n)$ will be continuous on Y, where $y_i = \pi_i(\mathbf{x})$, $i = 1, \dots, n$, are the coordinate functions. Consider the approximation of the function $f \circ \pi^{-1}$ by elements from

$$G_0 = \{g_1(y_1) + g_2(y_2) : \ g_i \in C(Y_i),\ i = 1, 2\}$$

over the compact set Y. Then one may observe that the function f has an extremal element in $\mathcal{R}_c(X)$ if and only if the function $f \circ \pi^{-1}$ has an extremal element in G_0. Thus the problem of proximinality of $\mathcal{R}_c(X)$ in $C(X)$ is reduced to the problem of proximinality of G_0 in $C(Y)$.

Let $T, T_1, \dots, T_{m+1}$ be metric compact spaces and $T \subset T_1 \times \cdots \times T_{m+1}$. For $i = 1, \dots, m$, let φ_i be the continuous mappings from T onto T_i. In **[61]**, Garkavi et al. obtained sufficient conditions for proximinality of the set

$$C_0 = \{\sum_{i=1}^n g_i \circ \varphi_i : \ g_i \in C(T_i),\ i = 1, \dots m\}$$

in the space $C(T)$ of continuous functions on T. Since $Y \subset Y_1 \times Y_2 \times Z_3$, where $Z_3 = Y_3 \times \cdots \times Y_n$, we can use this result in our case, for the approximation of the function $f \circ \pi^{-1}$ by elements from G_0. By this theorem, the set G_0 is proximinal in $C(Y)$ if for any $y_2^0 \in Y_2$ and $\delta > 0$ there exists a number $\delta_0 \in (0,\ \delta)$ such that the set $\sigma(y_2^0, \delta_0) = [y_2^0 - \delta_0, y_2^0 + \delta_0] \cap Y_2$ has $(2, \delta)$ maximal cross section. The last means that there exists a point $y_2^\sigma \in \sigma(y_2^0, \delta_0)$ with the property: for any point $(y_1, y_2, z_3) \in Y$, with the second coordinate y_2 from the set $\sigma(y_2^0, \delta_0)$, there exists a point $(y_1', y_2^\sigma, z_3') \in Y$ such that $y_1 = y_1'$ and $\rho(z_3, z_3') < \delta$, where ρ is a metrics in Z_3. Since these conditions are equivalent to the conditions of Theorem 1.33, the space G_0 is proximinal in the space $C(Y)$. Then by the above conclusion, the space $\mathcal{R}_c(X)$ is proximinal in $C(X)$. □

Let us give some simple examples of compact sets satisfying the hypothesis of Theorem 1.33. For the sake of brevity, we restrict ourselves to the case $n = 3$.

(a) Assume X is a closed ball in $\mathbb{R}^3$ and $\mathbf{a}^1$, $\mathbf{a}^2$ are orthogonal directions. Then Theorem 1.33 holds. Note that in this case, we can take $\delta_0 = \delta$ and $\mathbf{a}^3$ as an orthogonal vector to both the vectors $\mathbf{a}^1$ and $\mathbf{a}^2$.

(b) Let X be the unite cube, $\mathbf{a}^1 = (1;1;0)$, $a^2 = (1;-1;0)$. Then Theorem 1.33 also holds. In this case, we can take $\delta_0 = \delta$ and $\mathbf{a}^3 = (0;0;1)$. Note that the unit cube does not satisfy the hypothesis of the theorem for many directions (take, for example, $\mathbf{a}^1 = (1;2;0)$ and $\mathbf{a}^2 = (2;-1;0)$).

In the following example, one can not always chose δ_0 as equal to δ.

(c) Let $X = \{(x_1, x_2, x_3) : (x_1, x_2) \in Q,\ 0 \leq x_3 \leq 1\}$, where Q is the union of two triangles $A_1B_1C_1$ and $A_2B_2C_2$ with the vertices $A_1 = (0;0)$, $B_1 = (1;2)$, $C_1 = (2;0)$, $A_2 = (1\frac{1}{2};1)$, $B_2 = (2\frac{1}{2};-1)$, $C_2 = (3\frac{1}{2};1)$. Let $\mathbf{a}^1 = (0;1;0)$ and $\mathbf{a}^2 = (1;0;0)$. Then it is easy to see that Theorem 1.33 holds (the vector $\mathbf{a}^3$ can be chosen as $(0;0;1)$). In this case, δ_0 can not be always chosen as equal to δ. Take, for example, $\mathbf{x}^0 = (1\frac{3}{4};0;0)$ and $\delta = 1\frac{3}{4}$. If $\delta_0 = \delta$, then the second equation of the system (1.51) has not a solution for a point $(1;2;0)$ or a point $(2\frac{1}{2};-1;0)$. But if we take δ_0 not more than $\frac{1}{4}$, then for $\mathbf{x}^\sigma = \mathbf{x}^0$ the system has a solution. Note that the last inequality $\left|\mathbf{a}^3 \cdot \mathbf{x}' - \mathbf{a}^3 \cdot \mathbf{x}\right| < \delta$ of the system can be satisfied with the equality $\mathbf{a}^3 \cdot \mathbf{x}' = \mathbf{a}^3 \cdot \mathbf{x}$ if $\mathbf{a}^3 = (0;0;1)$.

It should be remarked that the results of [**61**] tell nothing about necessary conditions for proximinality of the spaces considered there. To fill this gap in our case, we want to give a necessary condition for proximinality of $\mathcal{R}_c(X)$ in $C(X)$. First, let us introduce some notation. By $\mathcal{R}_c^i$, $i = 1, 2$, we will denote the set of continuous ridge functions $g\left(\mathbf{a}^i \cdot \mathbf{x}\right)$ on the given compact set $X \subset \mathbb{R}^n$. Note that $\mathcal{R}_c = \mathcal{R}_c^1 + \mathcal{R}_c^2$. Besides, let $\mathcal{R}_c^3 = \mathcal{R}_c^1 \cap \mathcal{R}_c^2$. For $i = 1, 2, 3$, let X_i be the quotient space obtained by identifying points y_1 and y_2 in X whenever $f(y_1) = f(y_2)$ for each f in $\mathcal{R}_c^i$. By π_i denote the natural projection of X onto X_i, $i = 1, 2, 3$. Note that we have already dealt with the quotient spaces X_1, X_2 and the projections π_1, π_2 in the previous section. Recall that the relation on X, defined by setting $y_1 \approx y_2$ if y_1 and y_2 belong to some path, is an equivalence relation and the equivalence classes are called orbits. By $O(t)$ denote the orbit of X containing t. For $Y \subset X$, let $var_Y f$ be the variation of a function f on the set Y. That is,

$$\underset{Y}{var} f = \sup_{x,y \in Y} |f(x) - f(y)|.$$

The following theorem is valid.

THEOREM 1.34. *Suppose that the space $\mathcal{R}_c(X)$ is proximinal in $C(X)$. Then there exists a positive real number c such that*

$$\sup_{t \in X} \underset{O(t)}{var} f \leq c \sup_{t \in X} \underset{\pi_2^{-1}(\pi_2(t))}{var} f \tag{1.52}$$

for all f in $\mathcal{R}_c^1$.

PROOF. The proof is based on the following result of Marshall and O'Farrell (see [**150**, Proposition 4]): Let A_1 and A_2 be closed subalgebras of $C(X)$ that contain the constants. Let (X_1, π_1), (X_2, π_2) and (X_3, π_3) be the quotient spaces and projections associated with the algebras A_1, A_2 and $A_3 = A_1 \cap A_2$ respectively. Then $A_1 + A_2$ is closed in $C(X)$ if and only if there exists a positive real number c

such that

$$\sup_{z\in X_3}\ \underset{\pi_3^{-1}(z)}{var} f \le c \sup_{y\in X_2}\ \underset{\pi_2^{-1}(y)}{var} f \tag{1.53}$$

for all f in A_1.

If $\mathcal{R}_c(X)$ is proximinal in $C(X)$, then it is necessarily closed and therefore, by the above proposition, (1.53) holds for the algebras $A_1^i = \mathcal{R}_c^i$, $i = 1, 2, 3$. The right-hand side of (1.53) is equal to the right-hand side of (1.52). Let t be some point in X and $z = \pi_3(t)$. Since each function $f \in \mathcal{R}_c^3$ is constant on the orbit of t (note that f is both of the form $g_1(\mathbf{a}^1\cdot\mathbf{x})$ and of the form $g_2(\mathbf{a}^2\cdot\mathbf{x})$), $O(t) \subset \pi_3^{-1}(z)$. Hence,

$$\sup_{t\in X}\underset{O(t)}{var} f \le c \sup_{z\in X_3}\ \underset{\pi_3^{-1}(z)}{var} f \tag{1.54}$$

From (1.53) and (1.54) we obtain (1.52). □

Note that the inequality (1.53) provides not worse but less practicable necessary condition for proximinality than the inequality (1.52) does. On the other hand, there are many cases in which both the inequalities are equivalent. For example, assume the lengths of irreducible paths of X are bounded by some positive integer n_0. In this case, it can be shown that the inequality (1.53), hence (1.52), holds with the constant $c = \frac{n_0}{2}$ and moreover $O(t) = \pi_3^{-1}(z)$ for all $t \in X$, where $z = \pi_3(t)$ (see the proof of [**82**, Theorem 5]). Therefore, the inequalities (1.52) and (1.53) are equivalent for the considered class of sets X. The last argument shows that all the compact sets $X \subset \mathbb{R}^n$ over which $\mathcal{R}_c(X)$ is not proximinal in $C(X)$ should be sought in the class of sets having irreducible paths consisting of sufficiently many points. For example, let $I = [0;1]^2$ be the unit square, $\mathbf{a}^1 = (1;1)$, $\mathbf{a}^2 = (1;\frac{1}{2})$. Consider the path

$$l_k = \{(1;0),(0;1),(\frac{1}{2};0),(0;\frac{1}{2}),(\frac{1}{4};0),\ldots,(0;\frac{1}{2^k})\}.$$

It is clear that l_k is an irreducible path with the length $2k+2$, where k may be very large. Let g_k be a continuous univariate function on $\mathbb{R}$ satisfying the conditions: $g_k(\frac{1}{2^{k-i}}) = i$, $i = 0,\ldots,k$, $g_k(t) = 0$ if $t < \frac{1}{2^k}$, $i-1 \le g_k(t) \le i$ if $t \in (\frac{1}{2^{k-i+1}}, \frac{1}{2^{k-i}})$, $i = 1,\ldots,k$, and $g_k(t) = k$ if $t > 1$. Then it can be easily verified that

$$\sup_{t\in X}\ \underset{\pi_2^{-1}(\pi_2(t))}{var} g_k(\mathbf{a}^1\cdot\mathbf{x}) \le 1. \tag{1.55}$$

Since $\max_{\mathbf{x}\in I} g_k(\mathbf{a}^1\cdot\mathbf{x}) = k$, $\min_{\mathbf{x}\in I} g_k(\mathbf{a}^1\cdot\mathbf{x}) = 0$ and $var_{\mathbf{x}\in O(t_1)} g_k(\mathbf{a}^1\cdot\mathbf{x}) = k$ for $t_1 = (1;0)$, we obtain that

$$\sup_{t\in X}\underset{O(t)}{var} g_k(\mathbf{a}^1\cdot\mathbf{x}) = k. \tag{1.56}$$

Since k may be very large, from (1.55) and (1.56) it follows that the inequality (1.52) cannot hold for the function $g_k(\mathbf{a}^1\cdot\mathbf{x}) \in \mathcal{R}_c^1$. Thus the space $\mathcal{R}_c(I)$ with the directions $\mathbf{a}^1 = (1;1)$ and $\mathbf{a}^2 = (1;\frac{1}{2})$ is not proximinal in $C(I)$.

It should be remarked that if a compact set $X \subset \mathbb{R}^n$ satisfies the hypothesis of Theorem 1.33, then the length of all irreducible paths are uniformly bounded (see the proof of Theorem 1.33 and lemma in [**61**]). We have already seen that if the last condition does not hold, then the proximinality of both $\mathcal{R}_c(X)$ in $C(X)$ and $\mathcal{R}_b(X)$ in $B(X)$ fail for some sets X. In addition to the examples given above and in the previous subsection, one can easily construct many other examples of such sets. All these examples, Theorems 1.32, 1.33, 1.34 and the subsequent remarks justify the statement of the following conjecture:

CONJECTURE. *Let X be some subset of $\mathbb{R}^n$. The space $\mathcal{R}_b(X)$ is proximinal in $B(X)$ and the space $\mathcal{R}_c(X)$ is proximinal in $C(X)$ (in this case, X is considered to be compact) if and only if the lengths of all irreducible paths of X are uniformly bounded.*

REMARK 1.35. Medvedev's result (see [**117**, p.58]), which later came to our attention, in particular, says that the set $R_c(X)$ is closed in $C(X)$ if and only if the lengths of all irreducible paths of X are uniformly bounded. Thus, in the case of $C(X)$, the necessity of the above conjecture was proved by Medvedev.

REMARK 1.36. Note that there are situations in which a continuous function (a specific function on a specially constructed set) has an extremal element in $\mathcal{R}_b(X)$, but not in $\mathcal{R}_c(X)$ (see [**117**, p.73]). One subsection of [**117**] (see p.68 there) was devoted to the proximinality of sums of two univariate functions with continuous and bounded summands in the spaces of continuous and bounded bivariate functions, respectively. If $X \subset \mathbb{R}^2$ and $\mathbf{a}^1, \mathbf{a}^2$ be linearly independent directions in $\mathbb{R}^2$, then the linear transformation $y_1 = \mathbf{a}^1 \cdot \mathbf{x}$, $y_2 = \mathbf{a}^2 \cdot \mathbf{x}$ reduces the problems of proximinality of $\mathcal{R}_b(X)$ in $B(X)$ and $\mathcal{R}_c(X)$ in $C(X)$ to the problems considered in that subsection. But in general, when $X \subset \mathbb{R}^n$ and $n > 2$, they cannot be reduced to those in [**117**].

1.6. On the approximation by weighted ridge functions

In this section, we characterize the best L_2-approximation to a multivariate function by linear combinations of ridge functions multiplied by some fixed weight functions. In the special case, when the weight functions are constants, we obtain explicit formulas for both the best approximation and approximation error.

1.6.1. Problem statement. Ridge approximation in L_2 started to be actively studied in the late 90's by K.I. Oskolkov [**167**], V.E. Maiorov [**143**], A. Pinkus [**171**], V.N. Temlyakov [**197**], P. Petrushev [**170**] and other researchers.

Let D be the unit disk in $\mathbb{R}^2$. In [**141**], Logan and Shepp along with other results gave a closed-form expression for the best L_2-approximation to a function $f \in L_2(D)$ from the set $\mathcal{R}\left(\mathbf{a}^1, \dots, \mathbf{a}^r\right)$. Their solution requires that the directions $\mathbf{a}^1, \dots, \mathbf{a}^r$ be equally-spaced and involves finite sums of convolutions with explicit kernels. In the n-dimensional case, we obtained an expression of simpler form for the best L_2-approximation to square-integrable multivariate functions over a certain domain, provided that $r = n$ and the directions $\mathbf{a}^1, \dots, \mathbf{a}^r$ are linearly independent (see [**81**]).

In this section, we consider the approximation by functions from the following more general set

$$\mathcal{R}\left(\mathbf{a}^1, \dots, \mathbf{a}^r;\ w_1, \dots, w_r\right) = \left\{ \sum_{i=1}^{r} w_i(\mathbf{x}) g_i\left(\mathbf{a}^i \cdot \mathbf{x}\right) : g_i : \mathbb{R} \to \mathbb{R},\ i = 1, \dots, r \right\},$$

where $w_1,\dots,w_r$ are fixed multivariate functions. We characterize the best L_2-approximation from this set in the case $r \le n$. Then, in the special case when the weight functions $w_1,\dots,w_r$ are constants, we will prove two theorems on explicit formulas for the best approximation and the approximation error, respectively. At present, we do not yet know how to approach these problems in other possible cases of r.

1.6.2. Characterization of the best approximation. Let X be a subset of $\mathbb{R}^n$ with a finite Lebesgue measure. Consider the approximation of a function $f(\mathbf{x}) = f(x_1,\dots,x_n)$ in $L_2(X)$ by functions from $\mathcal{R}\left(\mathbf{a}^1,\dots,\mathbf{a}^r;\ w_1,\dots,w_r\right)$, where $r \le n$. We suppose that the functions $w_i(\mathbf{x})$ and the products $w_i(\mathbf{x}) \cdot g_i\left(\mathbf{a}^i \cdot \mathbf{x}\right)$, $i = 1,\dots,r$, belong to the space $L_2(X)$. Besides, we assume that the vectors $\mathbf{a}^1,\dots,\mathbf{a}^r$ are linearly independent. We say that a function $g_w^0 = \sum_{i=1}^{r} w_i(\mathbf{x}) g_i^0\left(\mathbf{a}^i \cdot \mathbf{x}\right)$ in $\mathcal{R}\left(\mathbf{a}^1,\dots,\mathbf{a}^r;\ w_1,\dots,w_r\right)$ is the best approximation (or extremal) to f if

$$\left\|f - g_w^0\right\|_{L_2(X)} = \inf_{g \in \mathcal{R}(\mathbf{a}^1,\dots,\mathbf{a}^r;\ w_1,\dots,w_r)} \|f - g\|_{L_2(X)}.$$

Let the system of vectors $\{\mathbf{a}^1,\dots,\mathbf{a}^r,\mathbf{a}^{r+1},\dots,\mathbf{a}^n\}$ be a completion of the system $\{\mathbf{a}^1,\dots,\mathbf{a}^r\}$ to a basis in $\mathbb{R}^n$. Let $J : X \to \mathbb{R}^n$ be the linear transformation given by the formulas

$$y_i = \mathbf{a}^i \cdot \mathbf{x}, \quad i = 1,\dots,n. \tag{1.57}$$

Since the vectors $\mathbf{a}^i$, $i = 1,\dots,n$, are linearly independent, it is an injection. The Jacobian $\det J$ of this transformation is a constant different from zero.

Let the formulas

$$x_i = \mathbf{b}^i \cdot \mathbf{y}, \quad i = 1,\dots,n,$$

stand for the solution of linear equations (1.57) with respect to x_i, $i = 1,\dots,n$.

Introduce the notation

$$Y = J(X)$$

and

$$Y_i = \left\{y_i \in \mathbb{R} :\ y_i = \mathbf{a}^i \cdot \mathbf{x},\ \ \mathbf{x} \in X\right\},\ i = 1,\dots,n.$$

For any function $u \in L_2(X)$, put

$$u^* = u^*(\mathbf{y}) \overset{def}{=} u\left(\mathbf{b}^1 \cdot \mathbf{y},\dots,\mathbf{b}^n \cdot \mathbf{y}\right).$$

It is obvious that $u^* \in L_2(Y)$. Besides,

$$\int_Y u^*(\mathbf{y})\,d\mathbf{y} = |\det J| \cdot \int_X u(\mathbf{x})\,d\mathbf{x} \tag{1.58}$$

and

$$\|u^*\|_{L_2(Y)} = |\det J|^{1/2} \cdot \|u\|_{L_2(X)}. \tag{1.59}$$

Set

$$L_2^i = \{w_i^*(\mathbf{y}) g(y_i) \in L_2(Y)\},\ i = 1,\dots,r.$$

We need the following auxiliary lemmas.

LEMMA 1.37. *Let $f(\mathbf{x}) \in L_2(X)$. A function* $\sum\limits_{i=1}^{r} w_i(\mathbf{x}) g_i^0 (\mathbf{a}^i \cdot \mathbf{x})$ *is extremal to the function $f(\mathbf{x})$ if and only if* $\sum\limits_{i=1}^{r} w_i^*(\mathbf{y}) g_i^0 (y_i)$ *is extremal from the space $L_2^1 \oplus \cdots \oplus L_2^r$ to the function $f^*(\mathbf{y})$.*

Due to (1.59) the proof of this lemma is obvious.

LEMMA 1.38. *Let $f(\mathbf{x}) \in L_2(X)$. A function* $\sum\limits_{i=1}^{r} w_i(\mathbf{x}) g_i^0 (\mathbf{a}^i \cdot \mathbf{x})$ *is extremal to the function $f(\mathbf{x})$ if and only if*

$$\int\limits_X \left(f(\mathbf{x}) - \sum_{i=1}^{r} w_i(\mathbf{x}) g_i^0 (\mathbf{a}^i \cdot \mathbf{x}) \right) w_j(\mathbf{x}) h(\mathbf{a}^j \cdot \mathbf{x})\, d\mathbf{x} = 0$$

for any ridge function $h(\mathbf{a}^j \cdot \mathbf{x})$ such that $w_j(x) h(\mathbf{a}^j \cdot \mathbf{x}) \in L_2(X)$, $j = 1, \dots, r$.

LEMMA 1.39. *The following formula is valid for the error of approximation to a function $f(\mathbf{x})$ in $L_2(X)$ from $\mathcal{R}(\mathbf{a}^1, \dots, \mathbf{a}^r;\ w_1, \dots, w_r)$:*

$$E(f) = \left(\|f(\mathbf{x})\|^2_{L_2(X)} - \left\| \sum_{i=1}^{r} w_i(\mathbf{x}) g_i^0 (\mathbf{a}^i \cdot \mathbf{x}) \right\|^2_{L_2(X)} \right)^{\frac{1}{2}},$$

where $\sum\limits_{i=1}^{r} w_i(\mathbf{x}) g_i^0 (\mathbf{a}^i \cdot \mathbf{x})$ *is the best approximation to $f(\mathbf{x})$.*

Lemmas 1.38 and 1.39 follow from the well-known facts of functional analysis that the best approximation of an element x in a Hilbert space H from a linear subspace Z of H must be the image of x via the orthogonal projection onto Z and the sum of squares of norms of orthogonal vectors is equal to the square of the norm of their sum.

We say that Y is an *r-set* if it can be represented as $Y_1 \times \cdots \times Y_r \times Y_0$, where Y_0 is some set from the space $\mathbb{R}^{n-r}$. In a special case, Y_0 may be equal to $Y_{r+1} \times \cdots \times Y_n$, but it is not necessary. By $Y^{(i)}$, we denote the Cartesian product of the sets $Y_1, \dots, Y_r, Y_0$ except for Y_i, $i = 1, \dots, r$. That is, $Y^{(i)} = Y_1 \times \cdots \times Y_{i-1} \times Y_{i+1} \times \cdots \times Y_r \times Y_0$, $i = 1, \dots, r$.

THEOREM 1.40. *Let Y be an r-set. A function* $\sum\limits_{i=1}^{r} w_i(\mathbf{x}) g_i^0 (\mathbf{a}^i \cdot \mathbf{x})$ *is the best approximation to $f(\mathbf{x})$ if and only if*

$$g_j^0(y_j) = \frac{1}{\int\limits_{Y^{(j)}} w_j^{*2}(\mathbf{y}) d\mathbf{y}^{(j)}} \int\limits_{Y^{(j)}} \left(f^*(\mathbf{y}) - \sum_{\substack{i=1 \\ i \neq j}}^{r} w_i^*(\mathbf{y}) g_i^0(y_i) \right) w_j^*(\mathbf{y}) d\mathbf{y}^{(j)}, \tag{1.60}$$

for $j = 1, \dots, r$.

PROOF. *Necessity.* Let a function $\sum_{i=1}^{r} w_i(\mathbf{x})g_i^0\left(\mathbf{a}^i\cdot\mathbf{x}\right)$ be extremal to f. Then by Lemma 1.37, the function $\sum_{i=1}^{r} w_i^*(\mathbf{y})g_i^0(y_i)$ in $L_2^1\oplus\cdots\oplus L_2^r$ is extremal to f^*. By Lemma 1.38 and equality (1.58),

$$\int_Y f^*(\mathbf{y})\, w_j^*(\mathbf{y})h(y_j)\,d\mathbf{y} = \int_Y w_j^*(\mathbf{y})h(y_j)\sum_{i=1}^{r} w_i^*(\mathbf{y})g_i^0(y_i)\,d\mathbf{y} \tag{1.61}$$

for any product $w_j^*(\mathbf{y})h(y_j)$ in L_2^j, $j=1,\ldots,r$. Applying Fubini's theorem to the integrals in (1.61), we obtain that

$$\int_{Y_j} h(y_j)\left[\int_{Y^{(j)}} f^*(\mathbf{y})\, w_j^*(\mathbf{y})d\mathbf{y}^{(j)}\right]dy_j$$
$$= \int_{Y_j} h(y_j)\left[\int_{Y^{(j)}} w_j^*(\mathbf{y})\sum_{i=1}^{r} w_i^*(\mathbf{y})g_i^0(y_i)\,d\mathbf{y}^{(j)}\right]dy_j.$$

Since $h(y_j)$ is an arbitrary function such that $w_j^*(\mathbf{y})h(y_j)\in L_2^j$,

$$\int_{Y^{(j)}} f^*(\mathbf{y})\, w_j^*(\mathbf{y})d\mathbf{y}^{(j)} = \int_{Y^{(j)}} w_j^*(\mathbf{y})\sum_{i=1}^{r} w_i^*(\mathbf{y})g_i^0(y_i)\,d\mathbf{y}^{(j)},\quad j=1,\ldots,r.$$

Therefore,

$$\int_{Y^{(j)}} w_j^{*2}(\mathbf{y})g_j^0(y_j)\,d\mathbf{y}^{(j)} = \int_{Y^{(j)}}\left(f^*(\mathbf{y})-\sum_{\substack{i=1\\ i\neq j}}^{r} w_i^*(\mathbf{y})g_i^0(y_i)\right)w_j^*(\mathbf{y})d\mathbf{y}^{(j)},$$

for $j=1,\ldots,r$. Now, since $y_j\notin Y^{(j)}$, we obtain (1.60).

Sufficiency. Note that all the equalities in the proof of the necessity can be obtained in the reverse order. Thus, (1.61) can be obtained from (1.60). Then by (1.58) and Lemma 1.38, we finally conclude that the function $\sum_{i=1}^{r} w_i(\mathbf{x})g_i^0\left(\mathbf{a}^i\cdot\mathbf{x}\right)$ is extremal to $f(\mathbf{x})$. □

In the following, $|Q|$ will denote the Lebesgue measure of a measurable set Q. The following corollary is obvious.

COROLLARY 1.41. *Let Y be an r-set. A function $\sum_{i=1}^{r} g_i^0\left(\mathbf{a}^i\cdot\mathbf{x}\right)$ in $\mathcal{R}\left(\mathbf{a}^1,\ldots,\mathbf{a}^r\right)$ is the best approximation to $f(\mathbf{x})$ if and only if*

$$g_j^0(y_j) = \frac{1}{\left|Y^{(j)}\right|}\int_{Y^{(j)}}\left(f^*(\mathbf{y})-\sum_{\substack{i=1\\ i\neq j}}^{r} g_i^0(y_i)\right)d\mathbf{y}^{(j)},\quad j=1,\ldots,r.$$

In [**81**], this corollary was proven for the case $r=n$.

1.6.3. Formulas for the best approximation and approximation error. In this subsection, we establish explicit formulas for both the best approximation and approximation error, provided that the weight functions are constants. In this case, since we vary over g_i, the set $\mathcal{R}\left(\mathbf{a}^1,\dots,\mathbf{a}^r;\ w_1,\dots,w_r\right)$ coincides with $\mathcal{R}\left(\mathbf{a}^1,\dots,\mathbf{a}^r\right)$. Thus, without loss of generality, we may assume that $w_i(\mathbf{x})=1$ for $i=1,\dots,r$.

For brevity of the further exposition, introduce the notation

$$A=\int\limits_Y f^*(\mathbf{y})\,d\mathbf{y} \text{ and } f_i^*=f_i^*(y_i)=\int\limits_{Y^{(i)}} f^*(\mathbf{y})\,d\mathbf{y}^{(i)},\ i=1,\dots,r.$$

The following theorem is a generalization of the main result of [**81**] from the case $r=n$ to the cases $r<n$.

THEOREM 1.42. *Let Y be an r-set. Set the functions*

$$g_1^0(y_1)=\frac{1}{\left|Y^{(1)}\right|}f_1^*-(r-1)\frac{A}{|Y|}$$

and

$$g_j^0(y_j)=\frac{1}{\left|Y^{(j)}\right|}f_j^*,\ j=2,\dots,r.$$

Then the function $\sum\limits_{i=1}^{r} g_i^0\left(\mathbf{a}^i\cdot\mathbf{x}\right)$ is the best approximation from $\mathcal{R}\left(\mathbf{a}^1,\dots,\mathbf{a}^r\right)$ to $f(\mathbf{x})$.

The proof is simple. It is sufficient to verify that the functions $g_j^0(y_j)$, $j=1,\dots,r$, satisfy the conditions of Corollary 1.41. This becomes obvious if note that

$$\sum_{\substack{i=1\\ i\neq j}}^{r}\frac{1}{\left|Y^{(j)}\right|}\frac{1}{\left|Y^{(i)}\right|}\int\limits_{Y^{(j)}}\left[\int\limits_{Y^{(i)}} f^*(\mathbf{y})\,d\mathbf{y}^{(i)}\right]d\mathbf{y}^{(j)}=(r-1)\frac{1}{|Y|}\int\limits_Y f^*(\mathbf{y})\,d\mathbf{y}$$

for $j=1,\dots,r$.

THEOREM 1.43. *Let Y be an r-set. Then the error of approximation to a function $f(x)$ from the set $\mathcal{R}\left(\mathbf{a}^1,\dots,\mathbf{a}^r\right)$ can be calculated by the formula*

$$E(f)=|\det J|^{-1/2}\left(\|f^*\|^2_{L_2(Y)}-\sum_{i=1}^{r}\frac{1}{\left|Y^{(i)}\right|^2}\|f_i^*\|^2_{L_2(Y)}+(r-1)\frac{A^2}{|Y|}\right)^{1/2}.$$

PROOF. From Eq. (1.59), Lemma 1.39 and Theorem 1.42, it follows that

$$E(f)=|\det J|^{-1/2}\left(\|f^*\|^2_{L_2(Y)}-I\right)^{1/2}, \tag{1.62}$$

where

$$I=\left\|\sum_{i=1}^{r}\frac{1}{\left|Y^{(i)}\right|}f_i^*-(r-1)\frac{A}{|Y|}\right\|^2_{L_2(Y)}.$$

The integral I can be written as a sum of the following four integrals:

$$I_1 = \sum_{i=1}^{r} \frac{1}{\left|Y^{(i)}\right|^2} \|f_i^*\|_{L_2(Y)}^2, \; I_2 = \sum_{i=1}^{r}\sum_{\substack{j=1\\ j\neq i}}^{r} \frac{1}{\left|Y^{(i)}\right|} \frac{1}{\left|Y^{(j)}\right|} \int\limits_Y f_i^* f_j^* d\mathbf{y},$$

$$I_3 = -2(r-1)\frac{1}{|Y|} A \sum_{i=1}^{r} \frac{1}{\left|Y^{(i)}\right|} \int\limits_Y f_i^* d\mathbf{y}, \; I_4 = (r-1)^2 \frac{A^2}{|Y|}.$$

It is not difficult to verify that

$$\int\limits_Y f_i^* f_j^* d\mathbf{y} = \left| Y_0 \times \prod_{\substack{k=1\\ k\neq i,j}}^{r} Y_k \right| A^2, \text{ for } i,j = 1,\dots,r,\ i\neq j, \tag{1.63}$$

and

$$\int\limits_Y f_i^* d\mathbf{y} = \left| Y_0 \times \prod_{\substack{k=1\\ k\neq i}}^{r} Y_k \right| A, \text{ for } i = 1,\dots,r. \tag{1.64}$$

Considering (1.63) and (1.64) in the expressions of I_2 and I_3 respectively, we obtain that

$$I_2 = r(r-1)\frac{A^2}{|Y|} \text{ and } I_3 = -2r(r-1)\frac{A^2}{|Y|}.$$

Therefore,

$$I = I_1 + I_2 + I_3 + I_4 = \sum_{i=1}^{r} \frac{1}{\left|Y^{(i)}\right|^2} \|f_i^*\|_{L_2(Y)}^2 - (r-1)\frac{A^2}{|Y|}.$$

Now the last equality together with (1.62) complete the proof. □

EXAMPLE 1.44. Consider the following set

$$X = \{\mathbf{x} \in \mathbb{R}^4 : y_i = y_i(\mathbf{x}) \in [0;1],\ i = 1,\dots,4\},$$

where

$$\begin{cases} y_1 = x_1 + x_2 + x_3 - x_4 \\ y_2 = x_1 + x_2 - x_3 + x_4 \\ y_3 = x_1 - x_2 + x_3 + x_4 \\ y_4 = -x_1 + x_2 + x_3 + x_4 \end{cases} \tag{1.65}$$

Let the function

$$f = 8x_1x_2x_3x_4 - \sum_{i=1}^{4} x_i^4 + 2\sum_{i=1}^{3}\sum_{j=i+1}^{4} x_i^2 x_j^2$$

be given on X. Consider the approximation of this function by functions from $\mathcal{R}\left(\mathbf{a}^1, \mathbf{a}^2, \mathbf{a}^3\right)$, where $\mathbf{a}^1 = (1;1;1;-1)$, $\mathbf{a}^2 = (1;1;-1;1)$, $\mathbf{a}^3 = (1;-1;1;1)$. By adding $\mathbf{a}^4 = (-1;1;1;1)$, we complete the system of vectors $\mathbf{a}^1, \mathbf{a}^2, \mathbf{a}^3$ to the basis

$\{\mathbf{a}^1, \mathbf{a}^2, \mathbf{a}^3, \mathbf{a}^4\}$ in $\mathbb{R}^4$. The linear transformation J defined by (1.65) maps the set X onto the set $Y = [0;1]^4$. The inverse transformation is given by the formulas

$$\begin{cases} x_1 = \frac{1}{4}y_1 + \frac{1}{4}y_2 + \frac{1}{4}y_3 - \frac{1}{4}y_4 \\ x_2 = \frac{1}{4}y_1 + \frac{1}{4}y_2 - \frac{1}{4}y_3 + \frac{1}{4}y_4 \\ x_3 = \frac{1}{4}y_1 - \frac{1}{4}y_2 + \frac{1}{4}y_3 + \frac{1}{4}y_4 \\ x_4 = -\frac{1}{4}y_1 + \frac{1}{4}y_2 + \frac{1}{4}y_3 + \frac{1}{4}y_4 \end{cases}$$

It can be easily verified that $f^* = y_1y_2y_3y_4$ and Y is a 3-set with $Y_i = [0;1]$, $i = 1, 2, 3$. Besides, $Y_0 = [0;1]$. After easy calculations we obtain that $A = \frac{1}{16}$; $f_i^* = \frac{1}{8}y_i$ for $i = 1, 2, 3$; $\det J = -16$; $\|f^*\|^2_{L_2(Y)} = \frac{1}{81}$; $\|f_i^*\|^2_{L_2(Y)} = \frac{1}{192}$, $i = 1, 2, 3$. Now from Theorems 1.42 and 1.43 it follows that the function $\frac{1}{8}\sum_{i=1}^3 \left(\mathbf{a}^i \cdot \mathbf{x}\right) - \frac{1}{8}$ is the best approximation from $\mathcal{R}\left(\mathbf{a}^1, \mathbf{a}^2, \mathbf{a}^3\right)$ to f and $E(f) = \frac{1}{576}\sqrt{2}\sqrt{47}$.

REMARK 1.45. Most of the material in this chapter is to be found in [**81**–**86**, **91**, **96**].

CHAPTER 2

The smoothness problem in ridge function representation

This chapter discusses the following open problem raised in Buhmann and Pinkus [**20**], and Pinkus [**175**, p. 14]. Assume we are given a function $f(\mathbf{x}) = f(x_1, \ldots, x_n)$ of the form

$$f(\mathbf{x}) = \sum_{i=1}^{k} f_i(\mathbf{a}^i \cdot \mathbf{x}), \tag{2.1}$$

where the $\mathbf{a}^i$, $i = 1, \ldots, k$, are pairwise linearly independent vectors (directions) in $\mathbb{R}^n$, f_i are arbitrarily behaved univariate functions and $\mathbf{a}^i \cdot \mathbf{x}$ are standard inner products. Assume, in addition, that f is of a certain smoothness class, that is, $f \in C^s(\mathbb{R}^n)$, where $s \geq 0$ (with the convention that $C^0(\mathbb{R}^n) = C(\mathbb{R}^n)$). Is it true that there will always exist $g_i \in C^s(\mathbb{R})$ such that

$$f(\mathbf{x}) = \sum_{i=1}^{k} g_i(\mathbf{a}^i \cdot \mathbf{x}) \ ? \tag{2.2}$$

In this chapter, we solve this problem up to some multivariate polynomial. In the special case $n = 2$, we see that this multivariate polynomial can be written as a sum of polynomial ridge functions with the given directions $\mathbf{a}^i$. In addition, we find various conditions on the directions $\mathbf{a}^i$ guaranteeing a positive solution to the problem. We also consider the question on constructing g_i using the information about the known functions f_i.

Most of the material of this chapter may be found in [**2–5, 173**].

2.1. A solution to the problem up to a multivariate polynomial

In this section, we solve the above problem up to a multivariate polynomial. That is, we show that if (2.1) holds for $f \in C^s(\mathbb{R}^n)$ and arbitrarily behaved f_i, then there exist $g_i \in C^s(\mathbb{R})$ such that

$$f(\mathbf{x}) = \sum_{i=1}^{k} g_i(\mathbf{a}^i \cdot \mathbf{x}) + P(\mathbf{x}),$$

where $P(\mathbf{x})$ is a polynomial of degree at most $k-1$. In the special case $n = 2$, we see that this multivariate polynomial can be written as a sum of polynomial ridge functions with the given directions $\mathbf{a}^i$ and thus (2.2) holds with $g_i \in C^s(\mathbb{R})$.

2.1.1. A brief overview of some results. We start this subsection with the simple observation that for $k = 1$ and $k = 2$ the smoothness problem is easily solved. Indeed for $k = 1$ by choosing $\mathbf{c} \in \mathbb{R}^n$ satisfying $\mathbf{a}^1 \cdot \mathbf{c} = 1$, we have that $f_1(t) = f(t\mathbf{c})$ is in $C^s(\mathbb{R})$. The same argument can be carried out for the case $k = 2$. In this case, since the vectors $\mathbf{a}^1$ and $\mathbf{a}^2$ are linearly independent, there exists a vector $\mathbf{c} \in \mathbb{R}^n$ satisfying $\mathbf{a}^1 \cdot \mathbf{c} = 1$ and $\mathbf{a}^2 \cdot \mathbf{c} = 0$. Therefore, we obtain that the function $f_1(t) = f(t\mathbf{c}) - f_2(0)$ is in the class $C^s(\mathbb{R})$. Similarly, one can verify that $f_2 \in C^s(\mathbb{R})$.

The above cases with one and two ridge functions in (2.1) show that the functions f_i inherit smoothness properties of the given f. The picture is absolutely different if the number of directions $k \geq 3$. For $k = 3$, there are ultimately smooth functions which decompose into sums of very badly behaved ridge functions. This phenomena comes from the classical Cauchy Functional Equation (CFE). This equation,

$$h(x+y) = h(x) + h(y), \quad h : \mathbb{R} \to \mathbb{R}, \tag{2.3}$$

looks very simple and has a class of simple solutions $h(x) = cx$, $c \in \mathbb{R}$. However, it easily follows from Hamel basis theory that CFE also has a large class of wild solutions. These solutions are called "wild" because they are extremely pathological. They are, for example, not continuous at a point, not monotone on an interval, not bounded on any set of positive measure (see, e.g., [**1**]). Let h_1 be any wild solution of the equation (2.3). Then the zero function can be represented as

$$0 = h_1(x) + h_1(y) - h_1(x+y). \tag{2.4}$$

Note that the functions involved in (2.4) are bivariate ridge functions with the directions $(1,0)$, $(0,1)$ and $(1,1)$, respectively. This example shows that for $k \geq 3$ the functions f_i in (2.1) may not inherit smoothness properties of the function f, which in the case of (2.4) is the identically zero function. Thus the above problem arises naturally.

However, it was shown by some authors that, additional conditions on f_i or the directions $\mathbf{a}^i$ guarantee smoothness of the representation (2.1). It was first proved by Buhmann and Pinkus [**20**] that if in (2.1) $f \in C^s(\mathbb{R}^n)$, $s \geq k-1$ and $f_i \in L^1_{loc}(\mathbb{R})$ for each i, then $f_i \in C^s(\mathbb{R})$ for $i = 1, \dots, k$. Later Pinkus [**173**] found a strong relationship between CFE and the problem of smoothness in ridge function representation. He generalized extensively the previous result of Buhmann and Pinkus [**20**]. He showed that the solution is quite simple and natural if the functions f_i are taken from a certain class $\mathcal{B}$ of real-valued functions defined on $\mathbb{R}$. $\mathcal{B}$ includes, for example, the set of continuous functions, the set of bounded functions, the set of Lebesgue measurable functions (for the precise definition of $\mathcal{B}$ see the next subsection). The result of Pinkus [**173**] states that if in (2.1) $f \in C^s(\mathbb{R}^n)$ and each $f_i \in \mathcal{B}$, then necessarily $f_i \in C^s(\mathbb{R})$ for $i = 1, \dots, k$.

Note that severe restrictions on the directions $\mathbf{a}^i$ also guarantee smoothness of the representation (2.1). For example, in (2.1) the inclusions $f_i \in C^s(\mathbb{R})$, $i = 1, \dots, k$, are automatically valid if the directions $\mathbf{a}^i$ are linearly independent and if these directions are not linearly independent, then there exists $f \in C^s(\mathbb{R}^n)$ of the form (2.1) such that the $f_i \notin C^s(\mathbb{R})$, $i = 1, \dots, k$ (see [**125**]). Indeed, if the directions $\mathbf{a}^i$ are linearly independent, then for each $i = 1, \dots, k$, we can choose a vector $\mathbf{b}^i$ such that $\mathbf{b}^i \cdot \mathbf{a}^i = 1$, but at the same time $\mathbf{b}^i \cdot \mathbf{a}^j = 0$, for all $j = 1, \dots, k$,

$j \neq i$. Putting $\mathbf{x} = \mathbf{b}^i t$ in (2.1) yields that

$$f(\mathbf{b}^i t) = f_i(t) + \sum_{j=1, j\neq i}^{k} f_j(0), \; i = 1, \dots, k.$$

This shows that all the functions f_i and f belong to the same smoothness class. If the directions $\mathbf{a}^i$ are not linearly independent, then there exist numbers $\lambda_1, \dots, \lambda_k$ such that $\sum_{i=1}^{k} |\lambda_i| > 0$ and $\sum_{i=1}^{k} \lambda_i \mathbf{a}^i = \mathbf{0}$. Let h be any wild solution of CFE. Then it is not difficult to verify that

$$0 = \sum_{i=1}^{k} h_i(\mathbf{a}^i \cdot \mathbf{x}),$$

where $h_i(t) = h(\lambda_i t)$, $i = 1, \dots, k$. Note that in the last representation, the zero function is an ultimately smooth function, while all the functions h_i are highly nonsmooth.

The above result of Pinkus was a starting point for further research on continuous and smooth sums of ridge functions. Much work in this direction was done by Konyagin and Kuleshov [**125, 126**], and Kuleshov [**131**]. They mainly analyze the continuity of f_i, that is, the question of if and when continuity of f guarantees the continuity of f_i. There are also other results concerning different properties, rather than continuity, of f_i. Most results in [**125, 126, 131**] involve certain subsets (convex open sets, convex bodies, etc.) of $\mathbb{R}^n$ instead of only $\mathbb{R}^n$ itself.

In [**3**], Aliev and Ismailov gave a partial solution to the smoothness problem. Their solution comprises the cases in which $s \geq 2$ and $k-1$ directions of the given k directions are linearly independent.

Kuleshov [**130**] generalized Aliev and Ismailov's result [**3**, Theorem 2.3] to all possible cases of s. That is, he proved that if a function $f \in C^s(\mathbb{R}^n)$, where $s \geq 0$, is of the form (2.1) and $(k-1)$-tuple of the given set of k directions $\mathbf{a}^i$ forms a linearly independent system, then there exist $g_i \in C^s(\mathbb{R})$, $i = 1, \dots, k$, such that (2.2) holds (see [**130**, Theorem 3]). In Section 2.2 we give a new constructive proof of Kuleshov's result.

2.1.2. A result of A. Pinkus.

In [**173**], A. Pinkus considered the smoothness problem in ridge function representation. For a given function $f : \mathbb{R}^n \to \mathbb{R}$, he posed and partially answered the following question. If f belongs to some smoothness class and (2.1) holds, what can we say about the smoothness of the functions f_i? He proved that for a large class of representing functions f_i, these f_i are smooth. That is, if apriori we assume that in the representation (2.1) the functions f_i is of a certain class of "reasonably well behaved functions", then they have the same degree of smoothness as the function f. As the mentioned class of "reasonably well behaved functions" one may take, e.g., the set of functions that are continuous at a point, the set of Lebesgue measurable functions, etc. All these classes arise from the class $\mathcal{B}$ considered by Pinkus [**173**] and the classical theory of CFE. In [**173**], $\mathcal{B}$ denotes any linear space of real-valued functions u defined on $\mathbb{R}$, closed under translation, such that if there is a function $v \in C(\mathbb{R})$ for which $u - v$ satisfies CFE, then $u - v$ is necessarily linear, i.e. $u(x) - v(x) = cx$, for some constant $c \in \mathbb{R}$. Such a definition of $\mathcal{B}$ is required in the proof of the following theorem.

THEOREM 2.1 (Pinkus [**173**]). *Assume $f \in C^s(\mathbb{R}^n)$ is of the form (2.1). Assume, in addition, that each $f_i \in \mathcal{B}$. Then necessarily $f_i \in C^s(\mathbb{R})$ for $i = 1, \dots, k$.*

PROOF. We prove this theorem by induction on k. The result is valid when $k = 1$. Indeed, taking any direction $\mathbf{c}$ such that $\mathbf{a}^1 \cdot \mathbf{c} = 1$ and putting $x = \mathbf{c}t$ in (2.1), we obtain that $f_1(t) = f(\mathbf{c}t) \in C^s(\mathbb{R})$. Assume that the result is valid for $k-1$. Let us show that it is valid for k.

Chose any vector $\mathbf{e} \in \mathbb{R}^n$ satisfying $\mathbf{e} \cdot \mathbf{a}^k = 0$ and $\mathbf{e} \cdot \mathbf{a}^i = b_i \neq 0$, for $i = 1, \dots, k-1$. Clearly, there exists a vector with this property. The property of $\mathbf{e}$ enables us to write that

$$f(\mathbf{x} + \mathbf{e}t) - f(\mathbf{x}) = \sum_{i=1}^{k-1} f_i(\mathbf{a}^i \cdot \mathbf{x} + b_i t) - f_i(\mathbf{a}^i \cdot \mathbf{x}).$$

Thus

$$F(\mathbf{x}) := f(\mathbf{x} + \mathbf{e}t) - f(\mathbf{x}) = \sum_{i=1}^{k-1} h_i(\mathbf{a}^i \cdot \mathbf{x}),$$

where

$$h_i(y) = f_i(y + b_i t) - f_i(y),\ i = 1, \dots, k-1.$$

Since $f_i \in \mathcal{B}$ and $\mathcal{B}$ is translation invariant, $h_i \in \mathcal{B}$. In addition, since $F \in C^s(\mathbb{R}^n)$, it follows by our induction assumption that $h_i \in C^s(\mathbb{R})$. Note that this inclusion is valid for all $t \in \mathbb{R}$.

In [**21**], de Bruijn proved that if for any $c \in \mathbb{R}$ the difference $u(y+c) - u(y)$ (u is any real function on $\mathbb{R}$) belongs to the class $C^s(\mathbb{R})$, then u is necessarily of the form $u = v + r$, where $v \in C^s(\mathbb{R})$ and r satisfies CFE. Thus each function f_i is of the form $f_i = v_i + r_i$, where $v_i \in C^s(\mathbb{R})$ and r_i satisfies CFE. By our assumption, each f_i is in $\mathcal{B}$, and from the definition of $\mathcal{B}$ it follows that $r_i = f_i - v_i$ is a linear function. Thus $f_i = v_i + r_i$, where both $v_i, r_i \in C^s(\mathbb{R})$, implying that $f_i \in C^s(\mathbb{R})$. This is valid for $i = 1, \dots, k-1$, and hence also for $i = k$. □

REMARK 2.2. In de Bruijn [**21, 22**], there are delineated various classes of real-valued functions $\mathcal{D}$ with the property that if $\Delta_t f = f(\cdot + t) - f(\cdot) \in \mathcal{D}$ for all $t \in \mathbb{R}$, then $f - s \in \mathcal{D}$, for some s satisfying CFE (for such classes see the next subsection). Some of these classes are $C^\infty(\mathbb{R})$ functions; analytic functions; algebraic polynomials; trigonometric polynomials. Theorem 2.1 can be suitably restated for any of these classes.

2.1.3. Polynomial functions of k-th order. Given $h_1, \dots, h_k \in \mathbb{R}$, we define inductively the difference operator $\Delta_{h_1 \dots h_k}$ as follows

$$\begin{aligned} \Delta_{h_1} f(x) \ &: \ = f(x + h_1) - f(x), \\ \Delta_{h_1 \dots h_k} f \ &: \ = \Delta_{h_k}(\Delta_{h_1 \dots h_{k-1}} f),\ f : \mathbb{R} \to \mathbb{R}. \end{aligned}$$

If $h_1 = \dots = h_k = h$, then we write briefly $\Delta_h^k f$ instead of $\Delta_{\underbrace{h \dots h}_{n \text{ times}}} f$. For various properties of difference operators see [**129**, Section 15.1].

DEFINITION 2.3 (see [**129**]). A function $f : \mathbb{R} \to \mathbb{R}$ is called a polynomial function of order k ($k \in \mathbb{N}$) if for every $x \in \mathbb{R}$ and $h \in \mathbb{R}$ we have

$$\Delta_h^{k+1} f(x) = 0.$$

It can be shown that if $\Delta_h^{k+1} f = 0$ for any $h \in \mathbb{R}$, then $\Delta_{h_1 \dots h_{k+1}} f = 0$ for any $h_1, \dots, h_{k+1} \in \mathbb{R}$ (see [**129**, Theorem 15.3.3]). A polynomial of degree at most k is a polynomial function of order k (see [**129**, Theorem 15.9.4]). The polynomial functions generalize ordinary polynomials, and reduce to the latter under mild regularity assumptions. For example, if a polynomial function is continuous at one point, or bounded on a set of positive measure, then it continuous at all points (see [**34, 132**]), and therefore is a polynomial of degree k (see [**129**, Theorem 15.9.4]).

Basic results concerning polynomial functions are due to S. Mazur-W. Orlicz [**153**], McKiernan [**154**], Djoković [**48**]. The following theorem, which we will use in the sequel, yield implicitly the general construction of polynomial functions.

THEOREM 2.4 (see Theorems 15.9.1 and 15.9.2 in [**129**]). *A function $f : \mathbb{R} \to \mathbb{R}$ is a polynomial function of order k if and only if it admits a representation*

$$f = f_0 + f_1 + \cdots + f_k,$$

where f_0 is a constant and $f_j : \mathbb{R} \to \mathbb{R}$, $j = 1, \dots, k$, are diagonalizations of j-additive symmetric functions $F_j : \mathbb{R}^j \to \mathbb{R}$, i.e.,

$$f_j(x) = F_j(x, \dots, x).$$

Note that a function $F_p : \mathbb{R}^p \to \mathbb{R}$ is called p-additive if for every j, $1 \le j \le p$, and for every $x_1, \dots, x_p, y_j \in \mathbb{R}$

$$F(x_1, \dots, x_j + y_j, \dots, x_p) = F(x_1, \dots, x_p) + F(x_1, \dots, x_{j-1}, y_j, x_{j+1}, \dots, x_p),$$

i.e., F is additive in each of its variables x_j (see [**129**, p.363]). A simple example of a p-additive function is given by the product

$$f_1(x_1) \times \cdots \times f_p(x_p),$$

where the univariate functions f_j, $j = 1, \dots, p$, are additive.

Following de Bruijn, we say that a class $\mathcal{D}$ of real functions has the *difference property* if any function $f : \mathbb{R} \to \mathbb{R}$ such that $\triangle_h f \in \mathcal{D}$ for all $h \in \mathbb{R}$, admits a decomposition $f = g + S$, where $g \in \mathcal{D}$ and S satisfies the Cauchy Functional Equation (2.3). Several classes with the difference property are investigated in de Bruijn [**21, 22**]. Some of these classes are:

1) $C(\mathbb{R})$, continuous functions;
2) $C^s(\mathbb{R})$, functions with continuous derivatives up to order s;
3) $C^\infty(\mathbb{R})$, infinitely differentiable functions;
4) analytic functions;
5) functions which are absolutely continuous on any finite interval;
6) functions having bounded variation over any finite interval;
7) algebraic polynomials;
8) trigonometric polynomials;
9) Riemann integrable functions.

A natural generalization of classes with the difference property are classes of functions with the difference property of k-th order.

DEFINITION 2.5 (see [**58**]). A class $\mathcal{F}$ is said to have the difference property of k-th order if any function $f : \mathbb{R} \to \mathbb{R}$ such that $\triangle_h^k f \in \mathcal{F}$ for all $h \in \mathbb{R}$, admits a decomposition $f = g + H$, where $g \in \mathcal{F}$ and H is a polynomial function of k-th order.

It is not difficult to see that the class $\mathcal{F}$ has the difference property of first order if and only if it has the difference property in de Bruijn's sense. There arises a natural question: which of the above classes have difference properties of higher orders? Gajda [**58**] considered this question in its general form, for functions defined on a locally compact Abelian group and showed that for any $k \in \mathbb{N}$, continuous functions have the difference property of k-th order (see [**58**, Theorem 4]). The proof of this result is based on several lemmas, in particular, on the following lemma, which we will also use in the sequel.

LEMMA 2.6 (see Lemma 5 in [**58**]). *For each $k \in \mathbb{N}$ the class of all continuous functions defined on $\mathbb{R}$ has the difference property of k-th order.*

In fact, Gajda [**58**] proved this lemma for Banach space valued functions, but the simplest case with the space $\mathbb{R}$ has all the difficulties. Unfortunately, the proof of the lemma has an essential gap. The author of [**58**] tried to reduce the proof to mod 1 periodic functions, but made a mistake in proving the continuity of the difference $\Delta_{h_1 \dots h_{k-1}}(f - f^*)$. Here $f^* : \mathbb{R} \to \mathbb{R}$ is a mod 1 periodic function defined on the interval $[0, 1)$ as $f^*(x) = f(x)$ and extended to the whole $\mathbb{R}$ with the period 1. That is, $f^*(x) = f(x)$ for $x \in [0, 1)$ and $f^*(x + 1) = f^*(x)$ for $x \in \mathbb{R}$. In the proof, the author of [**58**] takes a point $x \in [m, m + 1)$ and writes that

$$\Delta_{h_1 \dots h_{k-1}}(f - f^*)(x) = \Delta_{h_1 \dots h_{k-1}}(f(x) - f(x - m)),$$

which is not valid. Even though $f^*(x) = f(x - m)$ for any $x \in [m, m + 1)$, the differences $\Delta_{h_1 \dots h_{k-1}} f^*(x)$ and $\Delta_{h_1 \dots h_{k-1}} f(x - m)$ are completely different, since the latter may involve values of f at points outside $[0, 1)$, which have no relationship with the definition of f^*.

In the next section, we give a new proof for Lemma 2.6 (see Theorem 2.10 below). We hope that our proof is free from mathematical errors and thus the above lemma itself is valid.

2.1.4. Some auxiliary results on polynomial functions. In this subsection, we do further research on polynomial functions and prove some auxiliary results.

LEMMA 2.7. *If $f : \mathbb{R} \to \mathbb{R}$ is a polynomial function of order k, then for any $p \in \mathbb{N}$ and any fixed $\xi_1, \dots, \xi_p \in \mathbb{R}$, the function*

$$g(x_1, \dots, x_p) = f(\xi_1 x_1 + \dots + \xi_p x_p),$$

considered on the p dimensional space $\mathbb{Q}^p$ of rational vectors, is an ordinary polynomial of degree at most k.

PROOF. By Theorem 2.4,

$$f = \sum_{m=0}^{k} f_m, \tag{2.5}$$

where f_0 is a constant and $f_m : \mathbb{R} \to \mathbb{R}$, $1, \dots, m$, are diagonalizations of m-additive symmetric functions $F_m : \mathbb{R}^m \to \mathbb{R}$, i.e.,

$$f_m(x) = F_m(x, \dots, x).$$

For a m-additive function F_m the equality

$$F_m(\xi_1,\dots,\xi_{i-1},r\xi_i,\xi_{i+1},\dots,\xi_m) = rF_m(\xi_1,\dots,\xi_m)$$

holds for all $i=1,\dots,m$ and any $r\in\mathbb{Q}$, $\xi_i\in\mathbb{R}$, $i=1,\dots,m$ (see [**129**, Theorem 13.4.1]). Using this, it is not difficult to verify that for any $(x_1,\dots,x_p)\in\mathbb{Q}^p$,

$$\begin{aligned} f_m(\xi_1 x_1+\cdots+\xi_p x_p) &= F_m(\xi_1 x_1+\cdots+\xi_p x_p,\dots,\xi_1 x_1+\cdots+\xi_p x_p) \\ &= \sum_{\substack{0\le s_i\le m,\ \overline{i=1,p} \\ s_1+\cdots+s_p=m}} A_{s_1\dots s_p} F_m(\underbrace{\xi_1,\dots,\xi_1}_{s_1},\dots,\underbrace{\xi_p,\dots,\xi_p}_{s_p}) x_1^{s_1}\dots x_p^{s_p}. \end{aligned}$$

Here $A_{s_1\dots s_p}$ are some coefficients, namely $A_{s_1\dots s_p}=m!/(s_1!\dots s_p!)$. Considering the last formula in (2.5), we conclude that the function $g(x_1,\dots,x_p)$, restricted to $\mathbb{Q}^p$, is a polynomial of degree at most k. □

LEMMA 2.8. *Assume f is a polynomial function of order k. Then there exists a polynomial function H of order $k+1$ such that $H(0)=0$ and*

$$f(x) = H(x+1) - H(x). \tag{2.6}$$

PROOF. Consider the function

$$H(x) := xf(x) + \sum_{i=1}^{k}(-1)^i \frac{x(x+1)\dots(x+i)}{(i+1)!}\Delta_1^i f(x). \tag{2.7}$$

Clearly, $H(0)=0$. We are going to prove that H is a polynomial function of order $k+1$ and satisfies (2.6).

Let us first show that for any polynomial function g of order m the function $G_1(x) = xg(x)$ is a polynomial function of order $m+1$. Indeed, for any $h_1,\dots,h_{m+2}\in\mathbb{R}$ we can write that

$$\Delta_{h_1\dots h_{m+2}} G_1(x) = (x+h_1+\cdots+h_{m+2})\Delta_{h_1\dots h_{m+2}} g(x)$$

$$+\sum_{i=1}^{m+2} h_i \Delta_{h_1\dots h_{i-1}h_{i+1}\dots h_{m+2}} g(x). \tag{2.8}$$

The last formula is verified directly by using the known product property of differences, that is, the equality

$$\Delta_h(g_1 g_2) = g_1\Delta_h g_2 + g_2\Delta_h g_1 + \Delta_h g_1 \Delta_h g_2. \tag{2.9}$$

Now since g is a polynomial function of order m, all summands in (2.8) is equal to zero; hence we obtain that $G_1(x)$ is a polynomial function of order $m+1$. By induction, we can prove that the function $G_p(x) = x^p g(x)$ is a polynomial function of order $m+p$. Since $\Delta_1^i f(x)$ in (2.7) is a polynomial function of order $k-i$, it follows that all summands in (2.7) are polynomial functions of order $k+1$. Therefore, $H(x)$ is a polynomial function of order $k+1$.

Now let us prove (2.6). Considering the property (2.9) in (2.7) we can write that

$$\Delta_1 H(x) = [f(x) + (x+1)\Delta_1 f(x)] \tag{2.10}$$

$$+\sum_{i=1}^{k}(-1)^i\left[\frac{(x+1)\dots(x+i+1)}{(i+1)!}\Delta_1^{i+1}f(x) + \Delta_1\left(\frac{x(x+1)\dots(x+i)}{(i+1)!}\right)\Delta_1^i f(x)\right].$$

Note that in (2.10)

$$\Delta_1\left(\frac{x(x+1)\dots(x+i)}{(i+1)!}\right)=\frac{(x+1)\dots(x+i)}{i!}.$$

Considering this and the assumption $\Delta_1^{k+1}f(x)=0$, it follows from (2.10) that

$$\Delta_1 H(x)=f(x),$$

that is, (2.6) holds. □

The next lemma is due to Gajda [**58**].

LEMMA 2.9 (see Corollary 1 in [**58**]). *Let $f:\mathbb{R}\to\mathbb{R}$ be a $\bmod 1$ periodic function such that, for any $h_1,\dots,h_k\in\mathbb{R}$, $\Delta_{h_1\dots h_k}f$ is continuous. Then there exist a continuous function $g:\mathbb{R}\to\mathbb{R}$ and a polynomial function H of k-th order such that $f=g+H$.*

The following theorem generalizes de Bruijn's theorem (see [**21**, Theorem 1.1]) on the difference property of continuous functions and shows that Gajda's above lemma (see Lemma 2.6 is valid. Note that the main result of [**58**] also uses this theorem.

THEOREM 2.10. *Assume for any $h_1,\dots,h_k\in\mathbb{R}$, the difference $\Delta_{h_1\dots h_k}f(x)$ is a continuous function of the variable x. Then there exist a function $g\in C(\mathbb{R})$ and a polynomial function H of k-th order with the property $H(0)=0$ such that*

$$f=g+H.$$

PROOF. We prove this theorem by induction. For $k=1$, the theorem is the result of de Bruijn: if f is such that, for each h, $\Delta_h f(x)$ is a continuous function of x, then it can be written in the form $g+H$, where g is continuous and H is additive (that is, satisfies the Cauchy Functional Equation). Assume that the theorem is valid for $k-1$. Let us prove it for k. Without loss of generality we may assume that $f(0)=f(1)$. Otherwise, we can prove the theorem for $f_0(x)=f(x)-[f(1)-f(0)]\,x$ and then automatically obtain its validity for f.

Consider the function

$$F_1(x)=f(x+1)-f(x),\ x\in\mathbb{R}. \tag{2.11}$$

Since for any $h_1,\dots,h_k\in\mathbb{R}$, the difference $\Delta_{h_1\dots h_k}f(x)$ is a continuous function of x and $\Delta_{h_1\dots h_{k-1}}F_1=\Delta_{h_1\dots h_{k-1}1}f$, the difference $\Delta_{h_1\dots h_{k-1}}F_1(x)$ will be a continuous function of x, as well. By assumption, there exist a function $g_1\in C(\mathbb{R})$ and a polynomial function H_1 of $(k-1)$-th order with the property $H_1(0)=0$ such that

$$F_1=g_1+H_1. \tag{2.12}$$

It follows from Lemma 2.8 that there exists a polynomial function H_2 of order k such that $H_2(0)=0$ and

$$H_1(x)=H_2(x+1)-H_2(x). \tag{2.13}$$

Substituting (2.13) in (2.12) we obtain that

$$F_1(x)=g_1(x)+H_2(x+1)-H_2(x). \tag{2.14}$$

It follows from (2.11) and (2.14) that

$$g_1(x)=[f(x+1)-H_2(x+1)]-[f(x)-H_2(x)]. \tag{2.15}$$

Consider the function

$$F_2 = f - H_2. \tag{2.16}$$

Since H_2 is a polynomial function of order k and for any $h_1, \ldots, h_k \in \mathbb{R}$ the difference $\Delta_{h_1 \ldots h_k} f(x)$ is a continuous function of x, we obtain that $\Delta_{h_1 \ldots h_k} F_2(x)$ is also a continuous function of x. In addition, since $f(0) = f(1)$ and $H_2(0) = H_2(1) = 0$, it follows from (2.16) that $F_2(0) = F_2(1)$. We will use these properties of F_2 below.

Let us write (2.15) in the form

$$g_1(x) = F_2(x+1) - F_2(x), \tag{2.17}$$

and define the following mod 1 periodic function

$$\begin{aligned} F^*(x) &= F_2(x) \text{ for } x \in [0,1), \\ F^*(x+1) &= F^*(x) \text{ for } x \in \mathbb{R}. \end{aligned}$$

Consider the function

$$F = F_2 - F^*. \tag{2.18}$$

Let us show that $F \in C(\mathbb{R})$. Indeed since $F(x) = 0$ for $x \in [0,1)$, F is continuous on $(0,1)$. Consider now the interval $[1,2)$. For any $x \in [1,2)$ by the definition of F^* and (2.17) we can write that

$$F(x) = F_2(x) - F_2(x-1) = g_1(x-1). \tag{2.19}$$

Since $g_1 \in C(\mathbb{R})$, it follows from (2.19) that F is continuous on $(1,2)$. Note that by (2.17) $g_1(0) = 0$; hence $F(1) = g_1(0) = 0$. Since $F \equiv 0$ on $[0,1)$, $F(1) = 0$ and $F \in C(1,2)$, we obtain that F is continuous on $(0,2)$. Consider the interval $[2,3)$. For any $x \in [2,3)$ we can write that

$$F(x) = F_2(x) - F_2(x-2) = g_1(x-1) + g_1(x-2). \tag{2.20}$$

Since $g_1 \in C(\mathbb{R})$, F is continuous on $(2,3)$. Note that by (2.19) $\lim_{x \to 2-} F(x) = g_1(1)$ and by (2.20) $F(2) = g_1(1)$. We obtain from these arguments that F is continuous on $(0,3)$. In the same way, we can prove that F is continuous on $(0,m)$ for any $m \in \mathbb{N}$.

Similar arguments can be used to prove the continuity of F on $(-m,0)$ for any $m \in \mathbb{N}$. We show it for the first interval $[-1,0)$. For any $x \in [-1,0)$ by the definition of F^* and (2.17) we can write that

$$F(x) = F_2(x) - F_2(x+1) = -g_1(x).$$

Since $g_1 \in C(\mathbb{R})$, it follows that F is continuous on $(-1,0)$. Besides, $\lim_{x \to 0-} F(x) = -g_1(0) = 0$. This shows that F is continuous on $(-1,1)$, since $F \equiv 0$ on $[0,1)$. Combining all the above arguments we conclude that $F \in C(\mathbb{R})$.

Since $F \in C(\mathbb{R})$ and $\Delta_{h_1 \ldots h_k} F_2(x)$ is a continuous function of x, we obtain from (2.18) that $\Delta_{h_1 \ldots h_k} F^*(x)$ is also a continuous function of x. By Lemma 2.9, there exist a function $g_2 \in C(\mathbb{R})$ and a polynomial function H_3 of order k such that

$$F^* = g_2 + H_3. \tag{2.21}$$

It follows from (2.16), (2.18) and (2.21) that

$$f = F + g_2 + H_2 + H_3. \tag{2.22}$$

Introduce the notation

$$\begin{aligned} H(x) &= H_2(x) + H_3(x) - H_3(0), \\ g(x) &= F(x) + g_2(x) + H_3(0). \end{aligned}$$

Obviously, $g \in C(\mathbb{R})$ and $H(0) = 0$. It follows from (2.22) and the above notation that

$$f = g + H.$$

This completes the proof of the theorem. □

2.1.5. Main results. We start this subsection with the following lemma.

LEMMA 2.11. *Assume we are given pairwise linearly independent vectors $\mathbf{a}^i$, $i = 1, \dots, k$, and a function $f \in C(\mathbb{R}^n)$ of the form (2.1) with arbitrarily behaved univariate functions f_i. Then for any $h_1, \dots, h_{k-1} \in \mathbb{R}$, and all indices $i = 1, \dots, k$, $\Delta_{h_1 \dots h_{k-1}} f_i \in C(\mathbb{R})$.*

PROOF. We prove this lemma for the function f_k. It can be proven for the other functions f_i in the same way. Let $h_1, \dots, h_{k-1} \in \mathbb{R}$ be given. Since the vectors $\mathbf{a}^i$ are pairwise linearly independent, for each $j = 1, \dots, k-1$, there is a vector $\mathbf{b}^j$ such that $\mathbf{b}^j \cdot \mathbf{a}^j = 0$ and $\mathbf{b}^j \cdot \mathbf{a}^k \neq 0$. It is not difficult to see that for any $\lambda \in \mathbb{R}$, $\Delta_{\lambda \mathbf{b}^j} f_j(\mathbf{a}^j \cdot \mathbf{x}) = 0$. Therefore, for any $\lambda_1, \dots, \lambda_{k-1} \in \mathbb{R}$, we obtain from (2.1) that

$$\Delta_{\lambda_1 \mathbf{b}^1 \dots \lambda_{k-1} \mathbf{b}^{k-1}} f(\mathbf{x}) = \Delta_{\lambda_1 \mathbf{b}^1 \dots \lambda_{k-1} \mathbf{b}^{k-1}} f_k(\mathbf{a}^k \cdot \mathbf{x}). \tag{2.23}$$

Note that in multivariate setting the difference operator $\Delta_{\mathbf{h}^1 \dots \mathbf{h}^k} f(\mathbf{x})$ is defined similarly as in the previous section. If in (2.23) we take

$$\begin{aligned} \mathbf{x} &= \frac{\mathbf{a}^k}{\|\mathbf{a}^k\|^2} t, \ t \in \mathbb{R}, \\ \lambda_j &= \frac{h_j}{\mathbf{a}^k \cdot \mathbf{b}^j}, \ j = 1, \dots, k-1, \end{aligned}$$

we will obtain that $\Delta_{h_1 \dots h_{k-1}} f_k \in C(\mathbb{R})$. □

The following theorem is valid.

THEOREM 2.12. *Assume a function $f \in C(\mathbb{R}^n)$ is of the form (2.1). Then there exist continuous functions $g_i : \mathbb{R} \to \mathbb{R}$, $i = 1, \dots, k$, and a polynomial $P(\mathbf{x})$ of degree at most $k-1$ such that*

$$f(\mathbf{x}) = \sum_{i=1}^{k} g_i(\mathbf{a}^i \cdot \mathbf{x}) + P(\mathbf{x}). \tag{2.24}$$

PROOF. By Lemma 2.11 and Theorem 2.10, for each $i = 1, \dots, k$, there exists a function $g_i \in C(\mathbb{R})$ and a polynomial function H_i of $(k-1)$-th order with the property $H_i(0) = 0$ such that

$$f_i = g_i + H_i. \tag{2.25}$$

Consider the function

$$F(\mathbf{x}) = f(\mathbf{x}) - \sum_{i=1}^{k} g_i(\mathbf{a}^i \cdot \mathbf{x}). \tag{2.26}$$

It follows from (2.1), (2.25) and (2.26) that

$$F(\mathbf{x}) = \sum_{i=1}^{k} H_i(\mathbf{a}^i \cdot \mathbf{x}). \tag{2.27}$$

Denote the restrictions of the multivariate functions $H_i(\mathbf{a}^i \cdot \mathbf{x})$ to the space $\mathbb{Q}^n$ by $P_i(\mathbf{x})$, respectively. By Lemma 2.2, the functions $P_i(\mathbf{x})$ are ordinary polynomials of degree at most $k-1$. Since the space $\mathbb{Q}^n$ is dense in $\mathbb{R}^n$, and the functions $F(\mathbf{x})$, $P_i(\mathbf{x})$, $i = 1, \dots, k$, are continuous on $\mathbb{R}^n$, and the equality

$$F(\mathbf{x}) = \sum_{i=1}^{k} P_i(\mathbf{x}), \tag{2.28}$$

holds for all $\mathbf{x} \in \mathbb{Q}^n$, we obtain that (2.28) holds also for all $\mathbf{x} \in \mathbb{R}^n$. Now (2.24) follows from (2.26) and (2.28) by putting $P = \sum_{i=1}^{k} P_i$. □

Now we generalize Theorem 2.12 from $C(\mathbb{R}^n)$ to any space $C^s(\mathbb{R}^n)$ of s-th order continuously differentiable functions.

THEOREM 2.13. *Assume $f \in C^s(\mathbb{R}^n)$ is of the form* (2.1). *Then there exist functions $g_i \in C^s(\mathbb{R})$, $i = 1, \dots, k$, and a polynomial $P(\mathbf{x})$ of degree at most $k-1$ such that* (2.24) *holds.*

The proof is based on Theorems 2.1 and 2.12. On the one hand, it follows from Theorem 2.12 that the s-th order continuously differentiable function $f - P$ can be expressed as $\sum_{i=1}^{k} g_i$ with continuous g_i. On the other hand, since the class $\mathcal{B}$ in Theorem 2.1, in particular, can be taken as $C(\mathbb{R})$, it follows that $g_i \in C^s(\mathbb{R})$.

Note that Theorem 2.13 solves the problem posed in Buhmann and Pinkus [**20**] and Pinkus [**175**, p.14] up to a polynomial. The following theorem shows that in the two dimensional setting $n = 2$ it solves the problem completely.

THEOREM 2.14. *Assume a function $f \in C^s(\mathbb{R}^2)$ is of the form*

$$f(x, y) = \sum_{i=1}^{k} f_i(a_i x + b_i y),$$

where (a_i, b_i) are pairwise linearly independent vectors in $\mathbb{R}^2$ and f_i are arbitrary univariate functions. Then there exist functions $g_i \in C^s(\mathbb{R})$, $i = 1, \dots, k$, such that

$$f(x, y) = \sum_{i=1}^{k} g_i(a_i x + b_i y). \tag{2.29}$$

The proof of this theorem is not difficult. First we apply Theorem 2.13 and obtain that

$$f(x, y) = \sum_{i=1}^{k} \overline{g}_i(a_i x + b_i y) + P(x, y), \tag{2.30}$$

where $\overline{g}_i \in C^s(\mathbb{R})$ and $P(x, y)$ is a bivariate polynomial of degree at most $k-1$. Then we use the known fact that a bivariate polynomial $P(x, y)$ of degree $k-1$

is decomposed into a sum of ridge polynomials with any given k pairwise linearly independent directions (a_i, b_i), $i = 1, \dots, k$ (see e.g. [**141**]). That is,

$$P(x, y) = \sum_{i=1}^{k} p_i(a_i x + b_i y),$$

where p_i are univariate polynomials of degree at most $k - 1$. Considering this in (2.30) gives the desired representation (2.29).

REMARK 2.15. Theorem 2.13 can be restated also for the classes $C^\infty(\mathbb{R})$ of infinitely differentiable functions and $D(\mathbb{R})$ of analytic functions. That is, if under the conditions of Theorem 2.13, we have $f \in C^\infty(\mathbb{R}^n)$ (or $f \in D(\mathbb{R}^n)$), then this function can be represented also in the form (2.24) with $g_i \in C^\infty(\mathbb{R})$ (or $g_i \in D(\mathbb{R})$). This follows, similarly to the case $C^s(\mathbb{R})$ above, from Theorem 2.12 and Remark 2.2. These arguments are also valid for Theorem 2.14.

2.2. A solution to the smoothness problem under certain conditions

Assume we are given a function $f \in C^s(\mathbb{R}^n)$ of the form (2.1). In this section, we discuss various conditions on the directions $\mathbf{a}^i$ guaranteeing the validity of (2.2) with $g_i \in C^s(\mathbb{R})$.

2.2.1. Directions with only rational components. The following theorems, in particular, show that if directions of ridge functions have only rational coordinates then no polynomial term appears in Theorems 2.12 and 2.13.

THEOREM 2.16. *Assume a function $f \in C(\mathbb{R}^n)$ is of the form* (2.1) *and there is a nonsingular linear transformation $T : \mathbb{R}^n \to \mathbb{R}^n$ such that $T\mathbf{a}^i \in \mathbb{Q}^n$, $i = 1, \dots, k$. Then there exist continuous functions $g_i : \mathbb{R} \to \mathbb{R}$, $i = 1, \dots, k$, such that* (2.2) *holds.*

PROOF. Applying the coordinate change $\mathbf{x} \to \mathbf{y}$, given by the formula $\mathbf{x} = T\mathbf{y}$, to both sides of (2.1) we obtain that

$$\tilde{f}(\mathbf{y}) = \sum_{i=1}^{k} f_i(\mathbf{b}^i \cdot \mathbf{y}),$$

where $\tilde{f}(\mathbf{y}) = f(T\mathbf{y})$ and $\mathbf{b}^i = T\mathbf{a}^i$, $i = 1, \dots, k$. Let us repeat the proof of Theorem 2.12 for the function $\tilde{f}$. Since the vectors $\mathbf{b}^i$, $i = 1, \dots, k$, have rational coordinates, it is not difficult to see that the restrictions of the functions H_i to $\mathbb{Q}$ are univariate polynomials. Indeed, for each $\mathbf{b}^i$ we can choose a vector $\mathbf{c}^i$ with rational coordinates such that $\mathbf{b}^i \cdot \mathbf{c}^i = 1$. If in the equality $H_i(\mathbf{b}^i \cdot \mathbf{x}) = P_i(\mathbf{x})$, $\mathbf{x} \in \mathbb{Q}^n$, we take $\mathbf{x} = \mathbf{c}^i t$ with $t \in \mathbb{Q}$, we obtain that $H_i(t) = P_i(\mathbf{c}^i t)$ for all $t \in \mathbb{Q}$. Now since P_i is a multivariate polynomial on $\mathbb{Q}^n$, H_i is a univariate polynomial on $\mathbb{Q}$. Denote this univariate polynomial by L_i. Thus the formula

$$P_i(\mathbf{x}) = L_i(\mathbf{b}^i \cdot \mathbf{x}) \tag{2.31}$$

holds for each $i = 1, \ldots, k$, and all $\mathbf{x} \in \mathbb{Q}^n$. Since $\mathbb{Q}^n$ is dense in $\mathbb{R}^n$, we see that (2.31) holds, in fact, for all $\mathbf{x} \in \mathbb{R}^n$. Thus the polynomial $P(\mathbf{x})$ in (2.24) can be expressed as $\sum_{i=1}^{k} L_i(\mathbf{b}^i \cdot \mathbf{x})$. Considering this in Theorem 2.12, we obtain that

$$\tilde{f}(\mathbf{y}) = \sum_{i=1}^{k} g_i(\mathbf{b}^i \cdot \mathbf{y}), \tag{2.32}$$

where g_i are continuous functions. Using the inverse transformation $\mathbf{y} = T^{-1}\mathbf{x}$ in (2.32) we arrive at (2.2). □

THEOREM 2.17. *Assume a function $f \in C^s(\mathbb{R}^n)$ is of the form (2.1) and there is a nonsingular linear transformation $T : \mathbb{R}^n \to \mathbb{R}^n$ such that $T\mathbf{a}^i \in \mathbb{Q}^n$, $i = 1, \ldots, k$. Then there exist functions $g_i \in C^s(\mathbb{R})$, $i = 1, \ldots, k$, such that (2.2) holds.*

The proof of this theorem easily follows from Theorem 2.16 and Theorem 2.13.

2.2.2. Linear independence of $k-1$ directions. We already know that if the given directions $\mathbf{a}^i$ form a linearly independent set, then the smoothness problem has a positive solution (see Section 2.1). What can we say if all $\mathbf{a}^i$ are not linearly independent? In the sequel, we show that if $k-1$ of the directions $\mathbf{a}^i$, $i = 1, \ldots, k$, are linearly independent, then in (2.1) f_i can be replaced with $g_i \in C^s(\mathbb{R})$. We will also estimate the modulus of continuity of g_i in terms of the modulus of continuity of a function generated from f under a linear transformation.

Let $F : \mathbb{R}^n \to \mathbb{R}$, $n \geq 1$, be any function and $\Omega \subset \mathbb{R}^n$. The function

$$\omega(F; \delta; \Omega) = \sup \{|F(\mathbf{x}) - F(\mathbf{y})| : \mathbf{x}, \mathbf{y} \in \Omega,\ |\mathbf{x} - \mathbf{y}| \leq \delta\},\ 0 \leq \delta \leq diam\Omega,$$

is called the modulus of continuity of the function $F(\mathbf{x}) = F(x_1, \ldots, x_n)$ on the set Ω. We will also use the notation $\omega_{\mathbb{Q}}(F; \delta; \Omega)$, which stands for the function $\omega(F; \delta; \Omega \cap \mathbb{Q}^n)$. Here $\mathbb{Q}$ denotes the set of rational numbers. Note that $\omega_{\mathbb{Q}}(F; \delta; \Omega)$ makes sense if the set $\Omega \cap \mathbb{Q}^n$ is not empty. Clearly, $\omega_{\mathbb{Q}}(F; \delta; \Omega) \leq \omega(F; \delta; \Omega)$. The equality $\omega_{\mathbb{Q}}(F; \delta; \Omega) = \omega(F; \delta; \Omega)$ holds for continuous F and certain sets Ω. For example, it holds if for any $\mathbf{x}, \mathbf{y} \in \Omega$ with $|\mathbf{x} - \mathbf{y}| \leq \delta$ there exist sequences $\{\mathbf{x}_m\}, \{\mathbf{y}_m\} \subset \Omega \cap \mathbb{Q}^n$ such that $\mathbf{x}_m \to \mathbf{x}$, $\mathbf{y}_m \to \mathbf{y}$ and $|\mathbf{x}_m - \mathbf{y}_m| \leq \delta$, for all m. There are many sets Ω, which satisfy this property.

The following lemma is valid.

LEMMA 2.18. *Assume a function $G \in C(\mathbb{R}^n)$ has the form*

$$G(x_1, \ldots, x_n) = \sum_{i=1}^{n} g(x_i) - g(x_1 + \cdots + x_n), \tag{2.33}$$

where g is an arbitrarily behaved function. Then the following inequality holds

$$\omega_{\mathbb{Q}}(g; \delta; [-M, M]) \leq 2\delta\, |g(1) - g(0)| + 3\omega\left(G; \delta; [-M, M]^n\right), \tag{2.34}$$

where $\delta \in \left(0, \frac{1}{2}\right) \cap \mathbb{Q}$ and $M \geq 1$.

PROOF. Consider the function $f(t) = g(t) - g(0)$ and write (2.33) in the form

$$F(x_1, \ldots, x_n) = \sum_{i=1}^{n} f(x_i) - f(x_1 + \cdots + x_n), \tag{2.35}$$

where

$$F(x_1, \ldots, x_n) = G(x_1, \ldots, x_n) - (n-1)g(0).$$

Note that the functions f and g, as well as the functions F and G, have a common modulus of continuity. Thus we prove the lemma if we prove it for the pair $\langle F, f\rangle$.

Since $f(0) = 0$, it follows from (2.35) that

$$F(x_1, 0, \ldots, 0) = F(0, x_2, 0, \ldots, 0) = \cdots = F(0, 0, \ldots, x_n) = 0. \tag{2.36}$$

For the sake of brevity, introduce the notation $\mathcal{F}(x_1, x_2) = F(x_1, x_2, 0, \ldots, 0)$. Obviously, for any real number x,

$$\begin{aligned}
\mathcal{F}(x, x) &= 2f(x) - f(2x);\\
\mathcal{F}(x, 2x) &= f(x) + f(2x) - f(3x);\\
&\ldots\\
\mathcal{F}(x, (k-1)x) &= f(x) + f((k-1)x) - f(kx).
\end{aligned}$$

We obtain from the above equalities that

$$\begin{aligned}
f(2x) &= 2f(x) - \mathcal{F}(x, x),\\
f(3x) &= 3f(x) - \mathcal{F}(x, x) - \mathcal{F}(x, 2x),\\
&\ldots\\
f(kx) &= kf(x) - \mathcal{F}(x, x) - \mathcal{F}(x, 2x) - \cdots - \mathcal{F}(x, (k-1)x).
\end{aligned}$$

Thus for any nonnegative integer k,

$$f(x) = \frac{1}{k} f(kx) + \frac{1}{k}\left[\mathcal{F}(x, x) + \mathcal{F}(x, 2x) + \cdots + \mathcal{F}(x, (k-1)x)\right]. \tag{2.37}$$

Consider now the simple fraction $\frac{p}{m} \in (0, \frac{1}{2})$ and set $m_0 = \left[\frac{m}{p}\right]$. Here $[r]$ denotes the whole number part of r. Clearly, $m_0 \geq 2$ and the remainder $p_1 = m - m_0 p < p$. Taking $x = \frac{p}{m}$ and $k = m_0$ in (2.37) gives us the following equality

$$\begin{aligned}
f\left(\frac{p}{m}\right) &= \frac{1}{m_0} f\left(1 - \frac{p_1}{m}\right)\\
&+ \frac{1}{m_0}\left[\mathcal{F}\left(\frac{p}{m}, \frac{p}{m}\right) + \mathcal{F}\left(\frac{p}{m}, \frac{2p}{m}\right) + \cdots + \mathcal{F}\left(\frac{p}{m}, (m_0 - 1)\frac{p}{m}\right)\right].
\end{aligned} \tag{2.38}$$

On the other hand, since

$$\mathcal{F}\left(\frac{p_1}{m}, 1 - \frac{p_1}{m}\right) = f\left(\frac{p_1}{m}\right) + f\left(1 - \frac{p_1}{m}\right) - f(1),$$

it follows from (2.38) that

$$\begin{aligned}
f\left(\frac{p}{m}\right) &= \frac{f(1)}{m_0}\\
&+ \frac{1}{m_0}\left[\mathcal{F}\left(\frac{p}{m}, \frac{p}{m}\right) + \cdots + \mathcal{F}\left(\frac{p}{m}, (m_0 - 1)\frac{p}{m}\right) + \mathcal{F}\left(\frac{p_1}{m}, 1 - \frac{p_1}{m}\right)\right]\\
&- \frac{1}{m_0} f\left(\frac{p_1}{m}\right).
\end{aligned} \tag{2.39}$$

Put $m_1 = \left[\frac{m}{p_1}\right]$, $p_2 = m - m_1 p_1$. Clearly, $0 \leq p_2 < p_1$. Similar to (2.39), we can write that

$$\begin{aligned}
f\left(\frac{p_1}{m}\right) &= \frac{f(1)}{m_1}\\
&+ \frac{1}{m_1}\left[\mathcal{F}\left(\frac{p_1}{m}, \frac{p_1}{m}\right) + \cdots + \mathcal{F}\left(\frac{p_1}{m}, (m_1 - 1)\frac{p_1}{m}\right) + \mathcal{F}\left(\frac{p_2}{m}, 1 - \frac{p_2}{m}\right)\right]
\end{aligned}$$

$$-\frac{1}{m_1} f\left(\frac{p_2}{m}\right). \tag{2.40}$$

Let us make a convention that (2.39) is the 1-st and (2.40) is the 2-nd formula. One can continue this process by defining the chain of pairs (m_2, p_3), (m_3, p_4) until the pair (m_{k-1}, p_k) with $p_k = 0$ and writing out the corresponding formulas for each pair. For example, the last k-th formula will be of the form

$$f\left(\frac{p_{k-1}}{m}\right) = \frac{f(1)}{m_{k-1}}$$

$$+\frac{1}{m_{k-1}}\left[\mathcal{F}\left(\frac{p_{k-1}}{m}, \frac{p_{k-1}}{m}\right) + \cdots + \mathcal{F}\left(\frac{p_{k-1}}{m}, (m_{k-1}-1)\frac{p_{k-1}}{m}\right) + \mathcal{F}\left(\frac{p_k}{m}, 1-\frac{p_k}{m}\right)\right]$$

$$-\frac{1}{m_{k-1}} f\left(\frac{p_k}{m}\right). \tag{2.41}$$

Note that in (2.41), $f\left(\frac{p_k}{m}\right) = 0$ and $\mathcal{F}\left(\frac{p_k}{m}, 1-\frac{p_k}{m}\right) = 0$. Considering now the k-th formula in the $(k-1)$-th formula, then the obtained formula in the $(k-2)$-th formula, and so forth, we will finally arrive at the equality

$$f\left(\frac{p}{m}\right) = f(1)\left[\frac{1}{m_0} - \frac{1}{m_0 m_1} + \cdots + \frac{(-1)^{k-1}}{m_0 m_1 \cdots m_{k-1}}\right]$$

$$+\frac{1}{m_0}\left[\mathcal{F}\left(\frac{p}{m}, \frac{p}{m}\right) + \cdots + \mathcal{F}\left(\frac{p}{m}, (m_0-1)\frac{p}{m}\right) + \mathcal{F}\left(\frac{p_1}{m}, 1-\frac{p_1}{m}\right)\right]$$

$$-\frac{1}{m_0 m_1}\left[\mathcal{F}\left(\frac{p_1}{m}, \frac{p_1}{m}\right) + \cdots + \mathcal{F}\left(\frac{p_1}{m}, (m_1-1)\frac{p_1}{m}\right) + \mathcal{F}\left(\frac{p_2}{m}, 1-\frac{p_2}{m}\right)\right]$$

$$+\cdots+$$

$$\frac{(-1)^{k-1}}{m_0 m_1 \cdots m_{k-1}}\left[\mathcal{F}\left(\frac{p_{k-1}}{m}, \frac{p_{k-1}}{m}\right) + \cdots + \mathcal{F}\left(\frac{p_{k-1}}{m}, (m_{k-1}-1)\frac{p_{k-1}}{m}\right)\right]. \tag{2.42}$$

Taking into account (2.36) and the definition of $\mathcal{F}$, for any point of the form $\left(\frac{p_i}{m}, c\right)$, $i = 0, 1, \ldots, k-1$, $p_0 = p$, $c \in [0,1]$, we can write that

$$\left|\mathcal{F}\left(\frac{p_i}{m}, c\right)\right| = \left|\mathcal{F}\left(\frac{p_i}{m}, c\right) - \mathcal{F}(0, c)\right| \le \omega\left(F; \frac{p_i}{m}; [0,1]^n\right) \le \omega\left(F; \frac{p}{m}; [0,1]^n\right).$$

Applying this inequality to each term $\mathcal{F}\left(\frac{p_i}{m}, \cdot\right)$ in (2.42), we obtain that

$$\left|f\left(\frac{p}{m}\right)\right| \le \left[\frac{1}{m_0} - \frac{1}{m_0 m_1} + \cdots + \frac{(-1)^{k-1}}{m_0 m_1 \cdots m_{k-1}}\right] |f(1)|$$

$$+\left[1 + \frac{1}{m_0} + \cdots + \frac{1}{m_0 \cdots m_{k-2}}\right] \omega\left(F; \frac{p}{m}; [0,1]^n\right). \tag{2.43}$$

Since $m_0 \le m_1 \le \cdots \le m_{k-1}$, it is not difficult to see that in (2.43)

$$\frac{1}{m_0} - \frac{1}{m_0 m_1} + \cdots + \frac{(-1)^{k-1}}{m_0 m_1 \cdots m_{k-1}} \le \frac{1}{m_0}$$

and

$$1 + \frac{1}{m_0} + \cdots + \frac{1}{m_0 \cdots m_{k-2}} \le \frac{m_0}{m_0 - 1}.$$

Considering the above two inequalities in (2.43) we obtain that

$$\left|f\left(\frac{p}{m}\right)\right| \le \frac{|f(1)|}{m_0} + \frac{m_0}{m_0-1}\omega\left(F; \frac{p}{m}; [0,1]^n\right). \tag{2.44}$$

Since $m_0 = \left[\frac{m}{p}\right] \geq 2$, it follows from (2.44) that

$$\left|f\left(\frac{p}{m}\right)\right| \leq \frac{2p\,|f(1)|}{m} + 2\omega\left(F; \frac{p}{m}; [0,1]^n\right). \tag{2.45}$$

Let now $\delta \in \left(0, \frac{1}{2}\right) \cap \mathbb{Q}$ be a rational increment, $M \geq 1$ and $x, x+\delta$ be two points in $[-M, M] \cap \mathbb{Q}$. By (2.45) we can write that

$$\begin{aligned} |f(x+\delta) - f(x)| &\leq |f(\delta)| + |F(x, \delta, 0, \dots, 0)| \\ &\leq 2\delta\,|f(1)| + 3\omega\left(F; \delta; [-M, M]^n\right). \end{aligned} \tag{2.46}$$

Now (2.34) follows from (2.46) and the definitions of f and F. □

REMARK 2.19. The above lemma shows that the restriction of g to the set of rational numbers $\mathbb{Q}$ is uniformly continuous on any interval $[-M, M] \cap \mathbb{Q}$.

To prove the main result of this section we need the following lemma.

LEMMA 2.20. *Assume a function $G \in C(\mathbb{R}^n)$ has the form*

$$G(x_1, \dots, x_n) = \sum_{i=1}^{n} g(x_i) - g(x_1 + \cdots + x_n),$$

where g is an arbitrary function. Then there exists a function $F \in C(\mathbb{R})$ such that

$$G(x_1, \dots, x_n) = \sum_{i=1}^{n} F(x_i) - F(x_1 + \cdots + x_n) \tag{2.47}$$

and the following inequality holds

$$\omega(F; \delta; [-M, M]) \leq 3\omega\left(G; \delta; [-M, M]^n\right), \tag{2.48}$$

where $0 \leq \delta \leq \frac{1}{2}$ and $M \geq 1$.

PROOF. Consider the function

$$u(t) = g(t) - [g(1) - g(0)]\,t.$$

Obviously, $u(1) = u(0)$ and

$$G(x_1, \dots, x_n) = \sum_{i=1}^{n} u(x_i) - u(x_1 + \cdots + x_n). \tag{2.49}$$

By Lemma 2.18, the restriction of u to $\mathbb{Q}$ is continuous and uniformly continuous on every interval $[-M, M] \cap \mathbb{Q}$. Denote this restriction by v.

Let y be any real number and $\{y_k\}_{k=1}^{\infty}$ be any sequence of rational numbers converging to y. We can choose $M > 0$ so that $y_k \in [-M, M]$ for any $k \in \mathbb{N}$. It follows from the uniform continuity of v on $[-M, M] \cap \mathbb{Q}$ that the sequence $\{v(y_k)\}_{k=1}^{\infty}$ is Cauchy. Thus there exits a finite limit $\lim_{k\to\infty} v(y_k)$. It is not difficult to see that this limit does not depend on the choice of $\{y_k\}_{k=1}^{\infty}$.

Let F denote the following extension of v to the set of real numbers.

$$F(y) = \begin{cases} v(y), \ \text{if } y \in \mathbb{Q}; \\ \lim_{k\to\infty} v(y_k), \ \text{if } y \in \mathbb{R}\backslash\mathbb{Q} \text{ and } \{y_k\} \text{ is a sequence in } \mathbb{Q} \text{ tending to } y. \end{cases}$$

In view of the above arguments, F is well defined on the whole real line. Let us prove that for this function (2.47) is valid.

Consider an arbitrary point $(x_1, \dots, x_n) \in \mathbb{R}^n$ and sequences of rationale numbers $\{y_k^i\}_{k=1}^{\infty}, i = 1, \dots, n$, tending to $x_1, \dots, x_n$, respectively. Taking into account (2.49), we can write that

$$G(y_k^1, \dots, y_k^n) = \sum_{i=1}^{n} v(y_k^i) - v(y_k^1 + \cdots + y_k^n), \text{ for all } k = 1, 2, \dots, \tag{2.50}$$

since v is the restriction of u to $\mathbb{Q}$. Tending $k \to \infty$ in both sides of (2.50) we obtain (2.47).

Let us now prove that $F \in C(\mathbb{R})$ and (2.48) holds. Since $v(1) = v(0)$ we obtain from (2.49) and (2.34) that for $\delta \in \left(0, \frac{1}{2}\right) \cap \mathbb{Q}$, $M \geq 1$ and any numbers $a, b \in [-M, M] \cap \mathbb{Q}$, $|a - b| \leq \delta$, the following inequality holds

$$|v(a) - v(b)| \leq 3\omega\left(G; \delta; [-M, M]^n\right). \tag{2.51}$$

Consider any real numbers r_1 and r_2 satisfying $r_1, r_2 \in [-M, M]$, $|r_1 - r_2| \leq \delta$ and take sequences $\{a_k\}_{k=1}^{\infty} \subset [-M, M] \cap \mathbb{Q}$, $\{b_k\}_{k=1}^{\infty} \subset [-M, M] \cap \mathbb{Q}$ with the property $|a_k - b_k| \leq \delta$, $k = 1, 2, \dots$, and tending to r_1 and r_2, respectively. By (2.51),

$$|v(a_k) - v(b_k)| \leq 3\omega\left(G; \delta; [-M, M]^n\right).$$

If we take limits on both sides of the above inequality, we obtain that

$$|F(r_1) - F(r_2)| \leq 3\omega\left(G; \delta; [-M, M]^n\right),$$

which means that F is uniformly continuous on $[-M, M]$ and

$$\omega\left(F; \delta; [-M, M]\right) \leq 3\omega\left(G; \delta; [-M, M]^n\right).$$

Note that in the last inequality δ is a rational number from the interval $\left(0, \frac{1}{2}\right)$. It is well known that the modulus of continuity $\omega(f; \delta; \Omega)$ of a continuous function f is continuous from the right for any compact set $\Omega \subset \mathbb{R}^n$ and it is continuous from the left for certain compact sets Ω, in particular for rectangular sets (see [**124**]). It follows immediately that (2.48) is valid for all $\delta \in [0, \frac{1}{2}]$. $\square$

The following theorem was first obtained by Kuleshov [**130**]. Below, we prove this using completely different ideas. Our proof, which is taken from [**2**], contains a theoretical method for constructing the functions $g_i \in C^s(\mathbb{R})$ in (2.2). Using this method, we will also estimate the modulus of continuity of g_i in terms of the modulus of continuity of f (see Remark 2.22 below).

THEOREM 2.21. *Assume we are given k directions $\mathbf{a}^i$, $i = 1, \dots, k$, in $\mathbb{R}^n \backslash \{\mathbf{0}\}$ and $k - 1$ of them are linearly independent. Assume that a function $f \in C(\mathbb{R}^n)$ is of the form* (2.1). *Then f can be represented also in the form* (2.2) *with $g_i \in C(\mathbb{R})$, $i = 1, \dots, k$.*

PROOF. Without loss of generality, we may assume that the first $k - 1$ vectors $\mathbf{a}^1, \dots, \mathbf{a}^{k-1}$ are linearly independent. Thus there exist numbers $\lambda_1, \dots, \lambda_{k-1} \in \mathbb{R}$ such that $\mathbf{a}^k = \lambda_1 \mathbf{a}^1 + \cdots + \lambda_{k-1} \mathbf{a}^{k-1}$. We may also assume that the first p numbers $\lambda_1, \dots, \lambda_p$, $1 \leq p \leq k - 1$, are nonzero and the remaining λ_js are zero. Indeed, if necessary, we can rearrange the vectors $\mathbf{a}^1, \dots, \mathbf{a}^{k-1}$ so that this assumption holds. Complete the system $\{\mathbf{a}^1, \dots, \mathbf{a}^{k-1}\}$ to a basis $\{\mathbf{a}^1, \dots, \mathbf{a}^{k-1}, \mathbf{b}^k, \dots, \mathbf{b}^n\}$ and consider the linear transformation $\mathbf{y} = A\mathbf{x}$, where $\mathbf{x} = (x_1, \dots, x_n)^T$, $\mathbf{y} = (y_1, \dots, y_n)^T$ and A is the matrix, rows of which are formed by the coordinates of the vectors $\mathbf{a}^1, \dots, \mathbf{a}^{k-1}, \mathbf{b}^k, \dots, \mathbf{b}^n$. Using this transformation, we can write (2.1) in the form

$$f(A^{-1}\mathbf{y}) = f_1(y_1) + \cdots + f_{k-1}(y_{k-1}) + f_k(\lambda_1 y_1 + \cdots + \lambda_p y_p). \tag{2.52}$$

For the brevity of exposition in the sequel, we put $l = k-1$ and use the notation $w = f_{l+1}$, $\Phi(y_1, \dots, y_l) = f(A^{-1}\mathbf{y})$ and $Y_j = (y_1, \dots, y_{j-1}, y_{j+1}, \dots, y_l), j = 1, \dots, l$. Using this notation, we can write (2.52) in the form

$$\Phi(y_1, \dots, y_l) = f_1(y_1) + \cdots + f_l(y_l) + w(\lambda_1 y_1 + \cdots + \lambda_p y_p). \tag{2.53}$$

In (2.53), taking sequentially $Y_1 = 0$, $Y_2 = 0, \dots,$ $Y_l = 0$ we obtain that

$$f_j(y_j) = \Phi(y_1, \dots, y_l)|_{Y_j=\mathbf{0}} - w(\lambda_j y_j) - \sum_{\substack{i=1 \\ i \neq j}}^{l} f_i(0), \ j = 1, \dots, l. \tag{2.54}$$

Substituting (2.54) in (2.53), we obtain the equality

$$\begin{aligned} w(\lambda_1 y_1 + \cdots + \lambda_p y_p) - \textstyle\sum_{j=1}^{p} w(\lambda_j y_j) - (l-p)w(0) = \\ = \Phi(y_1, \dots, y_l) - \textstyle\sum_{j=1}^{l} \Phi(y_1, \dots, y_l)|_{Y_j=\mathbf{0}} + (l-1)\sum_{j=1}^{l} f_j(0). \end{aligned} \tag{2.55}$$

We see that the right hand side of (2.55) depends only on the variables $y_1, y_2, \dots, y_p$. Denote the right hand side of (2.55) by $H(y_1, \dots, y_p)$. That is, set

$$H(y_1, \dots, y_p) \overset{def}{=} \Phi(y_1, \dots, y_l) - \sum_{j=1}^{l} \Phi(y_1, \dots, y_l)|_{Y_j=\mathbf{0}} + (l-1)\sum_{j=1}^{l} f_j(0). \tag{2.56}$$

We will use the following identity, which follows from (2.55) and (2.56)

$$H(y_1, \dots, y_p) = w(\lambda_1 y_1 + \cdots + \lambda_p y_p) - \sum_{j=1}^{p} w(\lambda_j y_j) - (l-p)w(0). \tag{2.57}$$

It follows from (2.56) and the continuity of f that the function H is continuous on $\mathbb{R}^p$. Then, defining the function

$$G(y_1, \dots, y_p) = H(\frac{y_1}{\lambda_1}, \dots, \frac{y_p}{\lambda_p}) + (l-p)w(0) \tag{2.58}$$

and applying Lemma 2.20, we obtain that there exists a function $F \in C(\mathbb{R})$ such that

$$G(y_1, \dots, y_p) = F(y_1 + \cdots + y_p) - \sum_{j=1}^{p} F(y_p). \tag{2.59}$$

It follows from the formulas (2.57)-(2.59) that

$$w(\lambda_1 y_1 + \cdots + \lambda_p y_p) - \sum_{j=1}^{p} w(\lambda_j y_j) = F(\lambda_1 y_1 + \cdots + \lambda_p y_p) - \sum_{j=1}^{p} F(\lambda_j y_j). \tag{2.60}$$

Let us introduce the following functions

$$\begin{aligned} g_j(y_j) &= \Phi(y_1, \dots, y_l)|_{Y_j=\mathbf{0}} - F(\lambda_j y_j) - \textstyle\sum_{\substack{i=1 \\ i \neq j}}^{l} f_i(0), \ j = 1, \dots, p, \\ g_j(y_j) &= \Phi(y_1, \dots, y_l)|_{Y_j=\mathbf{0}} - \textstyle\sum_{\substack{i=1 \\ i \neq j}}^{l} f_i(0) - w(0), \ j = p+1, \dots, l. \end{aligned} \tag{2.61}$$

Note that $g_j \in C(\mathbb{R})$, $j = 1, \dots, l$. Considering (2.55), (2.60) and (2.61) it is not difficult to verify that

$$\Phi(y_1, \dots, y_l) = g_1(y_1) + \cdots + g_l(y_l) + F(\lambda_1 y_1 + \cdots + \lambda_p y_p). \tag{2.62}$$

In (2.62), denoting $F = g_k$, recalling the definition of Φ and going back to the variable $\mathbf{x} = (x_1, \dots, x_n)$ by using again the linear transformation $\mathbf{y} = A\mathbf{x}$, we finally obtain (2.2). $\square$

REMARK 2.22. Using Theorem 2.21 and Lemma 2.20, one can estimate the modulus of continuity of the functions g_i in representation (2.2) in terms of the modulus of continuity of Φ. To show how one can do this, assume $C > 0$, $M \geq 1$, $0 \leq \delta \leq \frac{1}{2\max\{|\lambda_j|\}}$ and introduce the following sets

$$\mathcal{M}_j = \left\{(x_1, \dots, x_l) \in \mathbb{R}^l : x_j \in [-M/|\lambda_j|, M/|\lambda_j|],\ x_i = 0 \text{ for } i \neq j\right\},$$
$$j = 1, \dots, p;$$
$$\mathcal{C}_j = \left\{(x_1, \dots, x_l) \in \mathbb{R}^l : x_j \in [-C, C],\ x_i = 0 \text{ for } i \neq j\right\},\ j = p+1, \dots, l.$$

It can be easily obtained from (2.61) that

$$\begin{aligned} \omega(g_j; \delta; [-M/|\lambda_j|, M/|\lambda_j|]) &\leq \omega(\Phi; \delta; \mathcal{M}_j) \\ &+ \omega(F; \delta_1; [-M, M]),\ j = 1, \dots, p; \end{aligned} \tag{2.63}$$

$$\omega(g_j; \delta; [-C, C]) \leq \omega(\Phi; \delta; \mathcal{C}_j),\ j = p+1, \dots, l, \tag{2.64}$$

where $\delta_1 = \delta \cdot \max\{|\lambda_j|\}$. To estimate $\omega(F; \delta_1; [-M, M])$ in (2.63), we refer to Lemma 2.20. Applying Lemma 2.20 to the function G in (2.58) we obtain that in addition to (2.59) the following inequality holds.

$$\omega(F; \delta_1; [-M, M]) \leq 3\omega(G; \delta_1; [-M, M]^p). \tag{2.65}$$

Note that here $0 \leq \delta_1 \leq \frac{1}{2}$ as in Lemma 2.20. It follows from (2.58) and (2.65) that

$$\omega(F; \delta_1; [-M, M]) \leq 3\omega(H; \delta_2; \mathcal{K}), \tag{2.66}$$

where $\mathcal{K} = [-M/|\lambda_1|, M/|\lambda_1|] \times \cdots \times [-M/|\lambda_p|, M/|\lambda_p|]$ and $\delta_2 = \delta_1 / \min\{|\lambda_j|\}$. Further, (2.66) and (2.56) together yield that

$$\omega(F; \delta_1; [-M, M]) \leq (3l+3)\omega(\Phi; \delta_2; \mathcal{S}), \tag{2.67}$$

where $\mathcal{S} = \left\{(x_1, \dots, x_l) \in \mathbb{R}^l : (x_1, \dots, x_p) \in \mathcal{K},\ x_i = 0 \text{ for } i > p\right\}$. Now it follows from (2.63) and (2.67) that

$$\begin{aligned} &\omega(g_j; \delta; [-M/|\lambda_j|, M/|\lambda_j|]) \\ &\leq \omega(\Phi; \delta; \mathcal{M}_j) + (3l+3)\omega(\Phi; \delta_2; \mathcal{S}),\ j = 1, \dots, p. \end{aligned} \tag{2.68}$$

Formulas (2.64), (2.67) and (2.68) provide us with upper estimates for the modulus of continuity of the functions g_j, $j = 1, \dots, k$, in terms of the modulus of continuity of Φ. Recall that in these estimates $l = k - 1$, $F = g_k$ and λ_j are coefficients in the expression $\mathbf{a}^k = \lambda_1 \mathbf{a}^1 + \cdots + \lambda_p \mathbf{a}^p$.

Theorems 2.1 and 2.21 together give the following result.

THEOREM 2.23. *Assume we are given k directions $\mathbf{a}^i$, $i = 1, \dots, k$, in $\mathbb{R}^n \backslash \{\mathbf{0}\}$ and $k-1$ of them are linearly independent. Assume that a function $f \in C^s(\mathbb{R}^n)$ is of the form (2.1). Then f can be represented also in the form (2.2), where the functions $g_i \in C^s(\mathbb{R})$, $i = 1, \dots, k$.*

Indeed, on the one hand, it follows from Theorem 2.21 that f can be expressed as (2.2) with continuous g_i. On the other hand, since the class $\mathcal{B}$ in Theorem 2.1, in particular, can be taken as $C(\mathbb{R})$, it follows that $g_i \in C^s(\mathbb{R})$.

REMARK 2.24. In addition to the above $C^s(\mathbb{R})$, Theorem 2.21 can be restated also for some other subclasses of the space of continuous functions. These are $C^\infty(\mathbb{R})$ functions; analytic functions; algebraic polynomials; trigonometric polynomials. More precisely, assume $\mathcal{H}(\mathbb{R})$ is any of these subclasses and $\mathcal{H}(\mathbb{R}^n)$ is the n-variable analog of the $\mathcal{H}(\mathbb{R})$. If under the conditions of Theorem 2.21, we have $f \in \mathcal{H}(\mathbb{R}^n)$, then this function can be represented in the form (2.2) with $g_i \in \mathcal{H}(\mathbb{R})$. This follows, similarly to the case $C^s(\mathbb{R})$ above, from Theorem 2.21 and Remark 2.2.

2.3. A constructive analysis of the smoothness problem

Note that Theorems 2.12, 2.13, 2.14, 2.16, 2.17, 2.21, 2.23 are generally existence results. They tell about existence of smooth ridge functions g_i in the corresponding representation formula (2.2) or (2.24). They are uninformative if we want to construct explicitly these functions.

In this section, we give two theorems which do not only address the smoothness problem, but also are useful in constructing the mentioned g_i.

2.3.1. Bivariate case. We start with the constructive analysis of the smoothness problem for bivariate functions. We show that if a bivariate function of a certain smoothness class is represented by a sum of finitely many, arbitrarily behaved ridge functions, then, under suitable conditions, it also can be represented by a sum of ridge functions of the same smoothness class and these ridge functions can be constructed explicitely.

THEOREM 2.25. *Assume (a_i, b_i), $i = 1, \dots, n$ are pairwise linearly independent vectors in $\mathbb{R}^2$. Assume that a function $f \in C^s(\mathbb{R}^2)$ has the form*

$$f(x, y) = \sum_{i=1}^{n} f_i(a_i x + b_i y),$$

where f_i are arbitrary univariate functions and $s \geq n-2$. Then f can be represented also in the form

$$f(x, y) = \sum_{i=1}^{n} g_i(a_i x + b_i y), \tag{2.69}$$

where the functions $g_i \in C^s(\mathbb{R})$, $i = 1, \dots, n$. In (2.69), the functions g_i, $i = 1, \dots, n$, can be constructed by the formulas

$$\begin{aligned} g_p &= \varphi_{p,n-p-1}, \ p = 1, \dots, n-2; \\ g_{n-1} &= h_{1,n-1}; \ g_n = h_{2,n-1}. \end{aligned}$$

Here all the involved functions $\varphi_{p,n-p-1}$, $h_{1,n-1}$ and $h_{2,n-1}$ can be found inductively as follows

$$\begin{aligned} h_{1,1}(t) &= \frac{\partial^{n-2}}{\partial l_1 \cdots \partial l_{n-2}} f^*(t,0), \\ h_{2,1}(t) &= \frac{\partial^{n-2}}{\partial l_1 \cdots \partial l_{n-2}} f^*(0,t) - \frac{\partial^{n-2}}{\partial l_1 \cdots \partial l_{n-2}} f^*(0,0); \\ h_{1,k+1}(t) &= \frac{1}{e_1 \cdot l_k} \int_0^t h_{1,k}(z)dz, \ k = 1, \dots, n-2; \\ h_{2,k+1}(t) &= \frac{1}{e_2 \cdot l_k} \int_0^t h_{2,k}(z)dz, \ k = 1, \dots, n-2; \end{aligned}$$

and

$$\varphi_{p,1}(t) = \frac{\partial^{n-p-2} f^*}{\partial l_{p+1} \cdots \partial l_{n-2}} \left(\frac{\widetilde{a}_p t}{\widetilde{a}_p^2 + \widetilde{b}_p^2}, \frac{\widetilde{b}_p t}{\widetilde{a}_p^2 + \widetilde{b}_p^2} \right) - h_{1,p+1}\left(\frac{\widetilde{a}_p t}{\widetilde{a}_p^2 + \widetilde{b}_p^2} \right)$$

$$-h_{2,p+1}\left(\frac{\widetilde{b}_p t}{\widetilde{a}_p^2 + \widetilde{b}_p^2} \right) - \sum_{j=1}^{p-1} \varphi_{j,p-j+1}\left(\frac{\widetilde{a}_j \widetilde{a}_p + \widetilde{b}_j \widetilde{b}_p}{\widetilde{a}_p^2 + \widetilde{b}_p^2} t \right),$$

$$p = 1, \dots, n-2 \left(\text{for } p = n-2, \ \frac{\partial^{n-p-2} f^*}{\partial l_{p+1} \cdots \partial l_{n-2}} := f^* \right);$$

$$\varphi_{p,k+1}(t) = \frac{1}{(\widetilde{a}_p, \widetilde{b}_p) \cdot l_{k+p}} \int_0^t \varphi_{p,k}(z)dz, \ p = 1, \dots, n-3, \ k = 1, \dots, n-p-2.$$

In the above formulas

$$\widetilde{a}_p = \frac{a_p b_n - a_n b_p}{a_{n-1} b_n - a_n b_{n-1}}; \ \widetilde{b}_p = \frac{a_{n-1} b_p - a_p b_{n-1}}{a_{n-1} b_n - a_n b_{n-1}}, \ p = 1, \dots, n-2,$$

$$l_p = \left(\frac{\widetilde{b}_p}{\sqrt{\widetilde{a}_p^2 + \widetilde{b}_p^2}}, \frac{-\widetilde{a}_p}{\sqrt{\widetilde{a}_p^2 + \widetilde{b}_p^2}} \right), \ p = 1, \dots, n-2.$$

$$f^*(x,y) = f\left(\frac{b_n x - b_{n-1} y}{a_{n-1} b_n - a_n b_{n-1}}, \frac{a_n x - a_{n-1} y}{a_n b_{n-1} - a_{n-1} b_n} \right).$$

PROOF. Since the vectors (a_{n-1}, b_{n-1}) and (a_n, b_n) are linearly independent, there is a nonsingular linear transformation $S : (x,y) \to (x', y')$ such that $S : (a_{n-1}, b_{n-1}) \to (1,0)$ and $S : (a_n, b_n) \to (0,1)$. Thus, without loss of generality we may assume that the vectors (a_{n-1}, b_{n-1}) and (a_n, b_n) coincide with the coordinate vectors $e_1 = (1,0)$ and $e_2 = (0,1)$ respectively. Therefore, to prove the first part of the theorem it is enough to show that if a function $f \in C^s(\mathbb{R}^2)$ is expressed in the form

$$f(x,y) = \sum_{i=1}^{n-2} f_i(a_i x + b_i y) + f_{n-1}(x) + f_n(y),$$

with arbitrary f_i, then there exist functions $g_i \in C^s(\mathbb{R})$, $i = 1, \dots, n$, such that f is also expressed in the form

$$f(x,y) = \sum_{i=1}^{n-2} g_i(a_i x + b_i y) + g_{n-1}(x) + g_n(y). \tag{2.70}$$

By $\Delta_l^{(\delta)}F$ we denote the increment of a function F in a direction $l = (l', l'')$. That is,

$$\Delta_l^{(\delta)}F(x, y) = F(x + l'\delta, y + l''\delta) - F(x, y).$$

We also use the notation $\frac{\partial F}{\partial l}$ which denotes the derivative of F in the direction l.

It is easy to check that the increment of a ridge function $g(ax+by)$ in a direction perpendicular to (a, b) is zero. Let $l_1, \ldots, l_{n-2}$ be unit vectors perpendicular to the vectors $(a_1, b_1), \ldots, (a_{n-2}, b_{n-2})$ correspondingly. Then for any set of numbers $\delta_1, \ldots, \delta_{n-2} \in \mathbb{R}$ we have

$$\Delta_{l_1}^{(\delta_1)} \cdots \Delta_{l_{n-2}}^{(\delta_{n-2})} f(x, y) = \Delta_{l_1}^{(\delta_1)} \cdots \Delta_{l_{n-2}}^{(\delta_{n-2})} [f_{n-1}(x) + f_n(y)] . \tag{2.71}$$

Denote the left hand side of (2.71) by $S(x, y)$. That is, set

$$S(x, y) \stackrel{def}{=} \Delta_{l_1}^{(\delta_1)} \cdots \Delta_{l_{n-2}}^{(\delta_{n-2})} f(x, y).$$

Then from (2.71) it follows that for any real numbers $\delta_{n-1}$and δ_n,

$$\Delta_{e_1}^{(\delta_{n-1})} \Delta_{e_2}^{(\delta_n)} S(x, y) = 0,$$

or in expanded form,

$$S(x + \delta_{n-1}, y + \delta_n) - S(x, y + \delta_n) - S(x + \delta_{n-1}, y) + S(x, y) = 0.$$

Putting in the last equality $\delta_{n-1} = -x$, $\delta_n = -y$, we obtain that

$$S(x, y) = S(x, 0) + S(0, y) - S(0, 0).$$

This means that

$$\Delta_{l_1}^{(\delta_1)} \cdots \Delta_{l_{n-2}}^{(\delta_{n-2})} f(x, y)$$

$$= \Delta_{l_1}^{(\delta_1)} \cdots \Delta_{l_{n-2}}^{(\delta_{n-2})} f(x, 0) + \Delta_{l_1}^{(\delta_1)} \cdots \Delta_{l_{n-2}}^{(\delta_{n-2})} f(0, y) - \Delta_{l_1}^{(\delta_1)} \cdots \Delta_{l_{n-2}}^{(\delta_{n-2})} f(0, 0).$$

By the hypothesis of the theorem, the derivative $\frac{\partial^{n-2}}{\partial l_1 \cdots \partial l_{n-2}} f(x, y)$ exists at any point $(x, y) \in \mathbb{R}^2$. Thus, it follows from the above formula that

$$\frac{\partial^{n-2} f}{\partial l_1 \cdots \partial l_{n-2}}(x, y) = h_{1,1}(x) + h_{2,1}(y), \tag{2.72}$$

where $h_{1,1}(x) = \frac{\partial^{n-2}}{\partial l_1 \cdots \partial l_{n-2}} f(x, 0)$ and $h_{2,1}(y) = \frac{\partial^{n-2}}{\partial l_1 \cdots \partial l_{n-2}} f(0, y) - \frac{\partial^{n-2}}{\partial l_1 \cdots \partial l_{n-2}} f(0, 0)$. Note that $h_{1,1}$ and $h_{2,1}$ belong to the class $C^{s-n+2}(\mathbb{R})$.

By $h_{1,2}$ and $h_{2,2}$ denote the antiderivatives of $h_{1,1}$ and $h_{2,1}$ satisfying the condition $h_{1,2}(0) = h_{2,2}(0) = 0$ and multiplied by the numbers $1/(e_1 \cdot l_1)$ and $1/(e_2 \cdot l_1)$ correspondingly. That is,

$$\begin{aligned} h_{1,2}(x) &= \frac{1}{e_1 \cdot l_1} \int_0^x h_{1,1}(z) dz; \\ h_{2,2}(y) &= \frac{1}{e_2 \cdot l_1} \int_0^y h_{2,1}(z) dz. \end{aligned}$$

Here $e \cdot l$ denotes the scalar product between vectors e and l. Obviously, the function

$$F_1(x, y) = h_{1,2}(x) + h_{2,2}(y)$$

obeys the equality

$$\frac{\partial F_1}{\partial l_1}(x, y) = h_{1,1}(x) + h_{2,1}(y). \tag{2.73}$$

From (2.72) and (2.73) we obtain that

$$\frac{\partial}{\partial l_1}\left[\frac{\partial^{n-3} f}{\partial l_2 \cdots \partial l_{n-2}} - F_1\right] = 0.$$

Hence, for some ridge function $\varphi_{1,1}(a_1x + b_1y)$,

$$\frac{\partial^{n-3} f}{\partial l_2 \cdots \partial l_{n-2}}(x, y) = h_{1,2}(x) + h_{2,2}(y) + \varphi_{1,1}(a_1x + b_1y). \tag{2.74}$$

Here all the functions $h_{2,1}, h_{2,2}(y), \varphi_{1,1} \in C^{s-n+3}(\mathbb{R})$.

Set the following functions

$$\begin{aligned} h_{1,3}(x) &= \frac{1}{e_1 \cdot l_2}\int_0^x h_{1,2}(z)dz; \\ h_{2,3}(y) &= \frac{1}{e_2 \cdot l_2}\int_0^y h_{2,2}(z)dz; \\ \varphi_{1,2}(t) &= \frac{1}{(a_1, b_1) \cdot l_2}\int_0^t \varphi_{1,1}(z)dz. \end{aligned}$$

Note that the function

$$F_2(x, y) = h_{1,3}(x) + h_{2,3}(y) + \varphi_{1,2}(a_1x + b_1y)$$

obeys the equality

$$\frac{\partial F_2}{\partial l_2}(x, y) = h_{1,2}(x) + h_{2,2}(y) + \varphi_{1,1}(a_1x + b_1y). \tag{2.75}$$

From (2.74) and (2.75) it follows that

$$\frac{\partial}{\partial l_2}\left[\frac{\partial^{n-4} f}{\partial l_3 \cdots \partial l_{n-2}} - F_2\right] = 0.$$

The last equality means that for some ridge function $\varphi_{2,1}(a_2x + b_2y)$,

$$\frac{\partial^{n-4} f}{\partial l_3 \cdots \partial l_{n-2}}(x, y) = h_{1,3}(x) + h_{2,3}(y) + \varphi_{1,2}(a_1x + b_1y) + \varphi_{2,1}(a_2x + b_2y). \tag{2.76}$$

Here all the functions $h_{1,3}$, $h_{2,3}$, $\varphi_{1,2}$, $\varphi_{2,1} \in C^{s-n+4}(\mathbb{R})$.

Note that in the left hand sides of (2.72), (2.74) and (2.76) we have the mixed directional derivatives of f and the order of these derivatives is decreased by one in each consecutive step. Continuing the above process, until it reaches the function f, we obtain the desired representation (2.70).

The formulas for g_i are obtained in the process of the above proof. These formulas involve certain functions which can be found inductively as described in the proof. The validity of the formulas for the functions $h_{1,k}$ and $h_{2,k}$, $k = 1, \ldots, n-1$, is obvious. The formulas for $\varphi_{p,1}$ and $\varphi_{p,k+1}$ can be obtained from (2.74), (2.76) and the subsequent (assumed but not written) equations if we put $x = \widetilde{a}_p t/(\widetilde{a}_p^2 + \widetilde{b}_p^2)$ and $y = \widetilde{b}_p t/(\widetilde{a}_p^2 + \widetilde{b}_p^2)$. Note that $(\widetilde{a}_p, \widetilde{b}_p)$, $p = 1, \ldots, n-2$, are the images of vectors (a_p, b_p) under the linear transformation S which takes the vectors (a_{n-1}, b_{n-1}) and (a_n, b_n) to the coordinate vectors $e_1 = (1, 0)$ and $e_2 = (0, 1)$, respectively. Besides, note that for $p = 1, \ldots, n-2$, the vectors l_p are perpendicular to the vectors $(\widetilde{a}_p, \widetilde{b}_p)$, respectively and f^* is the function generated from f by the above linear transformation. □

Theorem 2.25 can be applied to some higher order partial differential equations in two variables, e.g., to the following homogeneous equation

$$\prod_{i=1}^{r}\left(\alpha_i\frac{\partial}{\partial x}+\beta_i\frac{\partial}{\partial y}\right)u(x,y)=0, \tag{2.77}$$

where (α_i,β_i), $i=1,\dots,r$, are pairwise linearly independent vectors in $\mathbb{R}^2$. Clearly, the general solution to this equation are all functions of the form

$$u(x,y)=\sum_{i=1}^{r}v_i(\beta_i x-\alpha_i y), \tag{2.78}$$

where $v_i\in C^r(\mathbb{R})$, $i=1,\dots,r$. Based on Theorem 2.25, for the general solution, one can demand only smoothness of the sum u and dispense with smoothness of the summands v_i. More precisely, the following corollary is valid.

COROLLARY 2.26. *Assume a function $u\in C^r(\mathbb{R}^2)$ is of the form* (2.78) *with arbitrarily behaved v_i. Then u is a solution to Equation* (2.77).

REMARK 2.27. If in Theorem 2.25 $s\geq n-1$, then the functions g_i, $i=1,\dots,n$, can be constructed (up to polynomials) by the method discussed in Buhmann and Pinkus [**20**]. This method is based on the fact that for a direction $\mathbf{c}=(c_1,\dots,c_m)$ orthogonal to a given direction $\mathbf{a}\in\mathbb{R}^m\backslash\{\mathbf{0}\}$, the operator

$$D_{\mathbf{c}}=\sum_{k=1}^{m}c_k\frac{\partial}{\partial x_k}$$

acts on m-variable ridge functions $g(\mathbf{a}\cdot\mathbf{x})$ as follows

$$D_{\mathbf{c}}g(\mathbf{a}\cdot\mathbf{x})=(\mathbf{c}\cdot\mathbf{a})\,g'(\mathbf{a}\cdot\mathbf{x}).$$

Thus, if in our case for fixed $r\in\{1,\dots,n\}$, vectors l_k, $k\in\{1,\dots,n\}$, $k\neq r$, are perpendicular to the vectors (a_k,b_k), then

$$\prod_{\substack{k=1\\k\neq r}}^{n}D_{l_k}f(x,y)=\prod_{\substack{k=1\\k\neq r}}^{n}D_{l_k}\sum_{i=1}^{n}g_i(a_ix+b_iy)$$

$$=\sum_{i=1}^{n}\left(\prod_{\substack{k=1\\k\neq r}}^{n}((a_i,b_i)\cdot l_k)\right)g_i^{(n-1)}(a_ix+b_iy)=\prod_{\substack{k=1\\k\neq r}}^{n}((a_r,b_r)\cdot l_k)\,g_r^{(n-1)}(a_rx+b_ry).$$

Now g_r can be easily constructed from the above formula (up to a polynomial of degree at most $n-2$). Note that this method is not feasible if in Theorem 2.25 the function f is of the class $C^{n-2}(\mathbb{R}^2)$.

2.3.2. Multivariate case. In this subsection, we generalize ideas from the previous subsection to prove constructively that if a multivariate function of a certain smoothness class is represented by a sum of k arbitrarily behaved ridge functions, then, under suitable conditions, it can be represented by a sum of ridge functions of the same smoothness class and some polynomial of a certain degree. The appearance of a polynomial term is mainly related to the fact that in $\mathbb{R}^n$ $(n\geq 3)$ there are many directions orthogonal to a given direction. Such a result was proved nonconstructively in Section 2.1 (see Theorem 2.13), but here under a mild hypothesis on the degree of smoothness, we give a new proof for this theorem, which will provide us with a recipe for constructing the functions g_i in (2.24).

The following theorem is valid.

THEOREM 2.28. *Assume $f \in C^s(\mathbb{R}^n)$ is of the form (2.1). Let $s \geq k - p + 1$, where p is the number of vectors $\mathbf{a}^i$ forming a maximal linearly independent system. Then there exist functions $g_i \in C^s(\mathbb{R})$ and a polynomial $P(\mathbf{x})$ of total degree at most $k - p + 1$ such that (2.24) holds and g_i can be constructed algorithmically.*

PROOF. We start the proof by choosing a maximal linearly independent system in $\{\mathbf{a}^1, \dots, \mathbf{a}^k\}$. The case when the system $\{\mathbf{a}^1, \dots, \mathbf{a}^k\}$ itself is linearly independent is obvious (see Section 2.1). Thus we omit this special case here. Without loss of generality we may assume that the first p vectors $\mathbf{a}^1, \dots, \mathbf{a}^p$, $p < k$, are linearly independent. Thus, the vectors $\mathbf{a}^j$, $j = p+1, \dots, k$, can be expressed as lincar combinations $\lambda_1^j \mathbf{a}^1 + \cdots + \lambda_p^j \mathbf{a}^p$, where $\lambda_1^j, \dots, \lambda_p^j$ are real numbers. In addition, we can always apply a nonsingular linear transformation S of the coordinates such that $S : \mathbf{a}^i \to \mathbf{e}_i$, $i = 1, \dots, p$, where $\mathbf{e}_i$ denotes the i-th unit vector. This reduces the initial representation (2.1) to the following simpler form

$$f(\mathbf{x}) = f_1(x_1) + \cdots + f_p(x_p) + \sum_{i=1}^{m} f_{p+i}(\mathbf{a}^i \cdot \mathbf{x}). \tag{2.79}$$

Note that we keep the notation of (2.1), but here $\mathbf{x} = (x_1, \dots, x_p)$, $\mathbf{a}^i = (\lambda_1^i, \dots, \lambda_p^i) \in \mathbb{R}^p$ and $m = k - p$. Obviously, we prove Theorem 2.28 if we prove it for the representation (2.79). Thus, in the sequel, we prove that if $f \in C^s(\mathbb{R}^n)$ is of the form (2.79) and $s \geq m+1$, then there exist functions $g_i \in C^s(\mathbb{R})$ and a polynomial $P(\mathbf{x})$ of total degree at most $m+1$ such that

$$f(\mathbf{x}) = g_1(x_1) + \cdots + g_p(x_p) + \sum_{i=1}^{m} g_{p+i}(\mathbf{a}^i \cdot \mathbf{x}) + P(\mathbf{x}).$$

In the process of the proof, we also see how these g_i are constructed.

For each $i = 1, \dots, m$, let $\{\mathbf{e}_1^{(i)}, \dots, \mathbf{e}_{p-1}^{(i)}\}$ denote an orthonormal basis in the hyperplane perpendicular to $\mathbf{a}^i$. By $\Delta_{\mathbf{e}}^{(\delta)} F$ we denote the increment of a function F in a direction $\mathbf{e}$ of length δ. That is,

$$\Delta_{\mathbf{e}}^{(\delta)} F(\mathbf{x}) = F(\mathbf{x} + \delta \mathbf{e}) - F(\mathbf{x}).$$

We also use the notation $\frac{\partial F}{\partial \mathbf{e}}$ to denote the derivative of F in a direction $\mathbf{e}$.

It is easy to check that the increment of a ridge function $g(\mathbf{a} \cdot \mathbf{x})$ in any direction perpendicular to $\mathbf{a}$ is zero. For example,

$$\Delta_{\mathbf{e}_j^{(i)}}^{(\delta)} g(\mathbf{a}^i \cdot \mathbf{x}) = 0,$$

for all $i = 1, \dots, m$, $j = 1, \dots, p-1$. Therefore, for any indices $i_1, \dots, i_m \in \{1, \dots, p-1\}$, $q \in \{1, \dots, p\}$ and numbers $\delta_1, \dots, \delta_m, \delta \in \mathbb{R}$ we have the formula

$$\Delta_{\mathbf{e}_{i_1}^{(1)}}^{(\delta_1)} \Delta_{\mathbf{e}_{i_2}^{(2)}}^{(\delta_2)} \cdots \Delta_{\mathbf{e}_{i_m}^{(m)}}^{(\delta_m)} \Delta_{\mathbf{e}_q}^{(\delta)} f(\mathbf{x}) = \Delta_{\mathbf{e}_{i_1}^{(1)}}^{(\delta_1)} \Delta_{\mathbf{e}_{i_2}^{(2)}}^{(\delta_2)} \cdots \Delta_{\mathbf{e}_{i_m}^{(m)}}^{(\delta_m)} \Delta_{\mathbf{e}_q}^{(\delta)} f_q(x_q),$$

where $\mathbf{e}_q$ denotes the q-th unit vector. This means that for each $q = 1, \dots, p$, the mixed directional derivative

$$\frac{\partial^{m+1} f}{\partial \mathbf{e}_{i_1}^{(1)} \cdots \partial \mathbf{e}_{i_m}^{(m)} \partial x_q}(\mathbf{x})$$

depends only on the variable x_q. Denote this derivative by $h_{i_1,\dots,i_m}^{0,q}(x_q)$:

$$h_{i_1,\dots,i_m}^{0,q}(x_q) = \frac{\partial^{m+1} f}{\partial \mathbf{e}_{i_1}^{(1)} \cdots \partial \mathbf{e}_{i_m}^{(m)} \partial x_q}(\mathbf{x}),\ q = 1, \dots, p. \tag{2.80}$$

Since $f \in C^s(\mathbb{R}^p)$, we obtain that $h_{i_1,\dots,i_m}^{0,q} \in C^{s-m-1}(\mathbb{R})$. It follows from (2.80) that

$$d\left(\frac{\partial^m f}{\partial \mathbf{e}_{i_1}^{(1)} \cdots \partial \mathbf{e}_{i_m}^{(m)}}\right) - h_{i_1,\dots,i_m}^{0,1}(x_1)dx_1 + \cdots + h_{i_1,\dots,i_m}^{0,p}(x_p)dx_p. \tag{2.81}$$

We conclude from (2.81) that

$$\frac{\partial^m f}{\partial \mathbf{e}_{i_1}^{(1)} \cdots \partial \mathbf{e}_{i_m}^{(m)}}(\mathbf{x}) = h_{i_1,\dots,i_m}^{1,1}(x_1) + \cdots + h_{i_1,\dots,i_m}^{1,p}(x_p) + c_{i_1,\dots,i_m}, \tag{2.82}$$

where the functions $h_{i_1,\dots,i_m}^{1,q}(x_q)$, $q = 1, \dots, p$, are antiderivatives of $h_{i_1,\dots,i_m}^{0,q}(x_q)$ satisfying the condition $h_{i_1,\dots,i_m}^{1,q}(0) = 0$ and $c_{i_1,\dots,i_m}$ is a constant. Note that $h_{i_1,\dots,i_m}^{1,q} \in C^{s-m}(\mathbb{R})$, $q = 1, \dots, p$. Obviously, for any pair $k, t \in \{1, \dots, p-1\}$,

$$\frac{\partial^{m+1} f}{\partial \mathbf{e}_{i_1}^{(1)} \cdots \partial \mathbf{e}_{i_{m-1}}^{(m-1)} \partial \mathbf{e}_k^{(m)} \partial \mathbf{e}_t^{(m)}} = \frac{\partial^{m+1} f}{\partial \mathbf{e}_{i_1}^{(1)} \cdots \partial \mathbf{e}_{i_{m-1}}^{(m-1)} \partial \mathbf{e}_t^{(m)} \partial \mathbf{e}_k^{(m)}} \tag{2.83}$$

It follows from (2.82) and (2.83) that

$$(\mathbf{e}_q \cdot \mathbf{e}_k^{(m)}) \left(h_{i_1,\dots,i_{m-1},t}^{1,q}\right)' (x_q) = (\mathbf{e}_q \cdot \mathbf{e}_t^{(m)}) \left(h_{i_1,\dots,i_{m-1},k}^{1,q}\right)' (x_q) + c,$$

where c is a constant depending on the parameters $i_1, \dots, i_{m-1}, k, t$ and q. Recall that by construction, $h_{i_1,\dots,i_m}^{1,q}(0) = 0$. Hence

$$(\mathbf{e}_q \cdot \mathbf{e}_k^{(m)}) h_{i_1,\dots,i_{m-1},t}^{1,q}(x_q) = (\mathbf{e}_q \cdot \mathbf{e}_t^{(m)}) h_{i_1,\dots,i_{m-1},k}^{1,q}(x_q) + c x_q. \tag{2.84}$$

Since for each $q = 1, \dots, p$, the vectors $\mathbf{e}_q$ and $\mathbf{a}^m$ are linearly independent, there exists an index $i_m(q) \in \{1, \dots, p-1\}$ such that the vector $\mathbf{e}_{i_m(q)}^{(m)}$ is not orthogonal to $\mathbf{e}_q$. That is, $\mathbf{e}_q \cdot \mathbf{e}_{i_m(q)}^{(m)} \neq 0$. For each $q = 1, \dots, p$, fix the index $i_m(q)$ and define the following functions

$$h_{i_1,\dots,i_{m-1}}^{2,q}(x_q) = \frac{1}{\mathbf{e}_q \cdot \mathbf{e}_{i_m(q)}^{(m)}} \int_0^{x_q} h_{i_1,\dots,i_{m-1},i_m(q)}^{1,q}(z)dz. \tag{2.85}$$

and

$$F_{i_1,\dots,i_{m-1}}(\mathbf{x}) = h_{i_1,\dots,i_{m-1}}^{2,1}(x_1) + \cdots + h_{i_1,\dots,i_{m-1}}^{2,p}(x_p).$$

It is easy to obtain from (2.84) and (2.85) that for any $i_m \in \{1, \dots, p-1\}$,

$$\frac{\partial F_{i_1,\dots,i_{m-1}}}{\partial \mathbf{e}_{i_m}^{(m)}}(\mathbf{x}) = h_{i_1,\dots,i_m}^{1,1}(x_1) + \cdots + h_{i_1,\dots,i_m}^{1,p}(x_p) + P_{i_1,\dots,i_m}^{(1)}, \tag{2.86}$$

where $P^{(1)}_{i_1,\dots,i_m}$ is a polynomial of total degree not greater than 1. It follows from (2.82) and (2.86) that

$$\frac{\partial}{\partial \mathbf{e}^{(m)}_{i_m}}\left[\frac{\partial^{m-1} f}{\partial \mathbf{e}^{(1)}_{i_1}\cdots\partial \mathbf{e}^{(m-1)}_{i_{m-1}}} - F_{i_1,\dots,i_{m-1}}\right](\mathbf{x}) = c_{i_1,\dots,i_m} - P^{(1)}_{i_1,\dots,i_m}(\mathbf{x}). \tag{2.87}$$

Note that the last equality is valid for all vectors $\mathbf{e}^{(m)}_{i_m}$, which form a basis in the hyperplane orthogonal to $\mathbf{a}^m$. Thus from (2.87) we conclude that the following expansion is valid

$$\begin{aligned} \frac{\partial^{m-1} f}{\partial \mathbf{e}^{(1)}_{i_1}\cdots\partial \mathbf{e}^{(m-1)}_{i_{m-1}}}(\mathbf{x}) &= h^{2,1}_{i_1,\dots,i_{m-1}}(x_1)+\cdots+h^{2,p}_{i_1,\dots,i_{m-1}}(x_p) \\ &+\varphi^{2,1}_{i_1,\dots,i_{m-1}}(\mathbf{a}^m\cdot\mathbf{x}) + P^{(2)}_{i_1,\dots,i_{m-1}}(\mathbf{x}). \end{aligned} \tag{2.88}$$

Here all the functions $h^{2,1}_{i_1,\dots,i_{m-1}},\dots,h^{2,p}_{i_1,\dots,i_{m-1}}(x_p),\varphi^{2,1}_{i_1,\dots,i_{m-1}}$ belong to the space $C^{s-m+1}(\mathbb{R})$ and $P^{(2)}_{i_1,\dots,i_{m-1}}$ is a polynomial of total degree not greater than 2.

Since for each $q=1,\dots,p$, the vector $\mathbf{e}_q$ is not collinear to $\mathbf{a}^{m-1}$, there is an index $i_{m-1}(q)\in\{1,\dots,p-1\}$ such that $\mathbf{e}^{(m-1)}_{i_{m-1}(q)}$ is not orthogonal to $\mathbf{e}_q$. Similarly, since $\mathbf{a}^{m-1}$ is not collinear to $\mathbf{a}^m$, there is an index $i_{m-1}(m)\in\{1,\dots,p-1\}$ such that $\mathbf{e}^{(m-1)}_{i_{m-1}(m)}$ is not orthogonal to $\mathbf{a}^m$. Fix the indices $i_{m-1}(q)$, $i_{m-1}(m)$ and consider the following functions

$$h^{3,q}_{i_1,\dots,i_{m-2}}(x_q) = \frac{1}{\mathbf{e}_q\cdot\mathbf{e}^{(m-1)}_{i_{m-1}(q)}}\int_0^{x_q} h^{2,q}_{i_1,\dots,i_{m-2},i_{m-1}(q)}(z)dz,\ q=1,\dots,p, \tag{2.89}$$

$$\varphi^{3,1}_{i_1,\dots,i_{m-2}}(t) = \frac{1}{\mathbf{a}^m\cdot\mathbf{e}^{(m-1)}_{i_{m-1}(m)}}\int_0^t \varphi^{2,1}_{i_1,\dots,i_{m-2},i_{m-1}(m)}(z)dz, \tag{2.90}$$

and

$$F_{i_1,\dots,i_{m-2}}(\mathbf{x}) = h^{3,1}_{i_1,\dots,i_{m-2}}(x_1)+\cdots+h^{3,p}_{i_1,\dots,i_{m-2}}(x_p)+\varphi^{3,1}_{i_1,\dots,i_{m-2}}(\mathbf{a}^m\cdot\mathbf{x}). \tag{2.91}$$

Similar to (2.84), one can easily verify that for any pair $k,t\in\{1,\dots,p-1\}$ and for all $q=1,\dots,p$, the following equalities are valid.

$$\begin{aligned} &(\mathbf{e}_q\cdot\mathbf{e}^{(m-1)}_k)h^{2,q}_{i_1,\dots,i_{m-2},t}(x_q) \\ &= (\mathbf{e}_q\cdot\mathbf{e}^{(m-1)}_t)h^{2,q}_{i_1,\dots,i_{m-2},k}(x_q) + H_q(x_q), \\ &(\mathbf{a}^m\cdot\mathbf{e}^{(m-1)}_k)\varphi^{2,1}_{i_1,\dots,i_{m-2},t}(\mathbf{a}^m\cdot\mathbf{x}) \\ &= (\mathbf{a}^m\cdot\mathbf{e}^{(m-1)}_t)\varphi^{2,1}_{i_1,\dots,i_{m-2},k}(\mathbf{a}^m\cdot\mathbf{x}) + \Phi(\mathbf{x}), \end{aligned} \tag{2.92}$$

where H_q and Φ are univariate and n-variable polynomials of degree not greater than 2. Indeed, applying the Schwarz formula

$$\frac{\partial^{m+1} f}{\partial \mathbf{e}^{(1)}_{i_1}\cdots\partial \mathbf{e}^{(m-2)}_{i_{m-2}}\partial \mathbf{e}^{(m-1)}_k\partial \mathbf{e}^{(m-1)}_t\partial \mathbf{e}^{(m)}_{i_m(q)}} = \frac{\partial^{m+1} f}{\partial \mathbf{e}^{(1)}_{i_1}\cdots\partial \mathbf{e}^{(m-2)}_{i_{m-2}}\partial \mathbf{e}^{(m-1)}_t\partial \mathbf{e}^{(m-1)}_k\partial \mathbf{e}^{(m)}_{i_m(q)}}$$

on the symmetry of derivatives, it follows from (2.82) that for any $k,t\in\{1,\dots,p-1\}$

$$(\mathbf{e}_q\cdot\mathbf{e}^{(m-1)}_k)\left(h^{1,q}_{i_1,\dots,i_{m-2},t,i_m(q)}\right)'(x_q) = (\mathbf{e}_q\cdot\mathbf{e}^{(m-1)}_t)\left(h^{1,q}_{i_1,\dots,i_{m-2},k,i_m(q)}\right)'(x_q)+d,$$

where d is a constant depending on the parameters $i_1, \ldots, i_{m-2}, k, t$ and $i_m(q)$. Since, by construction, $h^{1,q}_{i_1,\ldots,i_m}(0) = 0$, we obtain that

$$(\mathbf{e}_q \cdot \mathbf{e}_k^{(m-1)}) h^{1,q}_{i_1,\ldots,i_{m-2},t,i_m(q)}(x_q) = (\mathbf{e}_q \cdot \mathbf{e}_t^{(m-1)}) h^{1,q}_{i_1,\ldots,i_{m-2},k,i_m(q)}(x_q) + dx_q.$$

The last equality together with (2.85) yield that

$$(\mathbf{e}_q \cdot \mathbf{e}_k^{(m-1)}) \left(h^{2,q}_{i_1,\ldots,i_{m-2},t}\right)'(x_q) = (\mathbf{e}_q \cdot \mathbf{e}_t^{(m-1)}) \left(h^{2,q}_{i_1,\ldots,i_{m-2},k}\right)'(x_q) + dx_q.$$

Therefore, the first equality in (2.92) holds. Considering this and applying the corresponding Schwarz formula to (2.88) we obtain the second equality in (2.92).

Taking into account (2.89), (2.90) and the relations (2.92), we obtain from (2.91) that for any $i_{m-1} \in \{1, \ldots, p-1\}$,

$$\frac{\partial F_{i_1,\ldots,i_{m-2}}}{\partial \mathbf{e}^{(m-1)}_{i_{m-1}}}(\mathbf{x}) = h^{2,1}_{i_1,\ldots,i_{m-1}}(x_1) + \cdots + h^{2,p}_{i_1,\ldots,i_{m-1}}(x_p) \tag{2.93}$$

$$+\varphi^{2,1}_{i_1,\ldots,i_{m-1}}(\mathbf{a}^m \cdot \mathbf{x}) + \widetilde{P}^{(2)}_{i_1,\ldots,i_{m-1}}(\mathbf{x}),$$

where $\widetilde{P}^{(2)}_{i_1,\ldots,i_{m-1}}$ is a polynomial of degree not greater than 2. It follows from (2.88) and (2.93) that

$$\frac{\partial}{\partial \mathbf{e}^{(m-1)}_{i_{m-1}}} \left[\frac{\partial^{m-2} f}{\partial \mathbf{e}^{(1)}_{i_1} \cdots \partial \mathbf{e}^{(m-2)}_{i_{m-2}}} - F_{i_1,\ldots,i_{m-2}} \right](\mathbf{x}) \tag{2.94}$$
$$= P^{(2)}_{i_1,\ldots,i_{m-1}}(\mathbf{x}) - \widetilde{P}^{(2)}_{i_1,\ldots,i_{m-1}}(\mathbf{x}).$$

Note that the last equality is valid for all vectors $\mathbf{e}^{(m-1)}_{i_{m-1}}$, which form a basis in the hyperplane orthogonal to $\mathbf{a}^{m-1}$. Considering this, from (2.94) we derive the following representation

$$\frac{\partial^{m-2} f}{\partial \mathbf{e}^{(1)}_{i_1} \cdots \partial \mathbf{e}^{(m-2)}_{i_{m-2}}}(\mathbf{x}) = h^{3,1}_{i_1,\ldots,i_{m-2}}(x_1) + \cdots + h^{3,p}_{i_1,\ldots,i_{m-2}}(x_p) \tag{2.95}$$
$$+\varphi^{3,1}_{i_1,\ldots,i_{m-2}}(\mathbf{a}^m \cdot \mathbf{x}) + \varphi^{3,2}_{i_1,\ldots,i_{m-2}}(\mathbf{a}^{m-1} \cdot \mathbf{x}) + P^{(3)}_{i_1,\ldots,i_{m-2}}(\mathbf{x}).$$

Here all the functions $h^{3,1}_{i_1,\ldots,i_{m-2}}, \ldots, h^{3,p}_{i_1,\ldots,i_{m-2}}(x_p)$, $\varphi^{3,1}_{i_1,\ldots,i_{m-2}}$, $\varphi^{3,2}_{i_1,\ldots,i_{m-2}}$ belong to the space $C^{s-m+2}(\mathbb{R})$ and $P^{(3)}_{i_1,\ldots,i_{m-2}}$ is a polynomial of total degree not greater than 3.

Note that in the left hand sides of (2.82), (2.88) and (2.95) we have the mixed directional derivatives of f and the order of these derivatives is decreased by one at each consecutive step. Continuing the above process, until it reaches the function f, we obtain the desired representation. Note that the above proof gives a recipe for constructing the smooth ridge functions g_i. Writing out explicit recurrent formulas for g_i, as in Theorem 2.25, is technically cumbersome here and hence is avoided. □

REMARK 2.29. Note that using Theorem 2.28, the degree of polynomial $P(\mathbf{x})$ in Theorem 2.13 can be reduced. Indeed, it follows from (2.27) and (2.28) that the the above polynomial $P(\mathbf{x})$ is of the form (2.1). On the other hand, by Theorem 2.28 there exist functions $g_i^* \in C^s(\mathbb{R})$, $i = 1, \ldots, k$, and a polynomial $G(\mathbf{x})$ of degree at most $k - p + 1$ such that

$$P(\mathbf{x}) = \sum_{i=1}^{k} g_i^*(\mathbf{a}^i \cdot \mathbf{x}) + G(\mathbf{x}).$$

Now considering this in (2.24) we see that our assertion is true.

At the end of this chapter, we want to draw the reader's attention to the following uniqueness question. Assume we are given pairwise linearly independent vectors $\mathbf{a}^i$, $i = 1, \dots, k$, in $\mathbb{R}^n$ and a function $f : \mathbb{R}^n \to \mathbb{R}$ of the form (2.1). How many different ways can f be written as a sum of ridge functions with the directions $\mathbf{a}^i$? Clearly, representation (2.1) is not unique, since we can always add some constants c_i to f_i without changing the resulting sum in (2.1) provided that $\sum_{i=1}^k c_i = 0$. It turns out that under minimal requirements representation (2.1) is unique up to polynomials of degree at most $k-2$. More precisely, if, in addition to (2.1), f also has the form (2.2) and $f_i, g_i \in \mathcal{B}$, $i = 1, \dots, k$, then the functions $f_i - g_i$ are univariate polynomials of degree at most $k-2$. This result is due to Pinkus [**175**, Theorem 3.1]. It follows immediately from this result that in Theorems 2.14, 2.16, 2.17, 2.21, 2.23 the functions g_i is unique up to a univariate polynomial. This is also valid for g_i in Theorems 2.12 and 2.13, but in this case for the proof we must apply a slightly different result of Pinkus [**175**, Corollary 3.2]: Assume a multivariate polynomial f of degree m is of the form (2.1) and $f_i \in \mathcal{B}$ for $i = 1, \dots, k$. Then f_i are univariate polynomials of degree at most $l = \max\{m, k-2\}$.

A different uniqueness problem, in a more general setting, will be analyzed in Chapter 4. In that problem we will look for sets $Q \subset \mathbb{R}^n$ for which representation (2.1), considered on Q, is unique.

CHAPTER 3

Approximation of multivariate functions by sums of univariate functions

It is clear that in the special case, when directions of ridge functions coincide with the coordinate directions, the problem of approximation by linear combinations of these functions turn into the problem of approximation by sums of univariate functions. This is also the simplest case in ridge function approximation. The simplicity of the approximation guarantees its practicability in application areas, where complicated multivariate functions are main obstacles. In mathematics, this type of approximation has arisen, for example, in connection with the classical functional equations [**19**], the numerical solution of certain PDE boundary value problems [**15**], dimension theory [**189, 190**], etc. In this chapter, we obtain some results concerning the problem of best approximation by sums of univariate functions.

Most of the material of this chapter is taken from [**77–79, 87**].

3.1. Characterization of some bivariate function classes by formulas for the error of approximation

This section is devoted to calculation formulas for the error of approximation of bivariate functions by sums of univariate functions. Certain classes of bivariate functions depending on some numerical parameter are constructed and characterized in terms of the approximation error calculation formulas.

3.1.1. Exposition of the problem. The approximation problem considered here is to approximate a continuous and real-valued function of two variables by sums of two continuous functions of one variable. To make the problem precise, let Q be a compact set in the xOy plane. Consider the approximation of a continuous function $f \in C(Q)$ by functions from the manifold $D = \{\varphi(x) + \psi(y)\}$, where $\varphi(x), \psi(y)$ are defined and continuous on the projections of Q into the coordinate axes x and y, respectively. The approximation error is defined as the distance from f to D:

$$E(f) = dist(f, D) = \inf_D \|f - \varphi - \psi\|_{C(Q)} =$$
$$= \inf_D \max_{(x,y)\in Q} |f(x,y) - \varphi(x) - \psi(y)|.$$

A function $\varphi_0(x) + \psi_0(y)$ from D, if it exists, is called an extremal element or a best approximating sum if

$$E(f) = \|f - \varphi_0 - \psi_0\|_{C(Q)}.$$

To show that $E(f)$ depends also on Q, in some cases to avoid confusion, we will write $E(f, Q)$ instead of $E(f)$.

In this section we deal with calculation formulas for $E(f)$. In 1951 Diliberto and Straus published a paper [**47**], in which along with other results they established a formula for $E(f, R)$, where R here and throughout this section is a rectangle with sides parallel to the coordinate axes, containing supremum over all closed lightning bolts. Later the same formula was established by other authors differently, in cases of both rectangle (see [**166**]) and more general sets (see [**114, 151**]). Although the formula was valid for all continuous functions, it was not easily calculable. Some authors started to seek easily calculable formulas for the approximation error for some subsets of continuous functions. Rivlin and Sibner [**176**] proved a result, which allow one to find the exact value of $E(f, R)$ for a function $f(x, y)$ having the continuous and nonnegative derivative $\frac{\partial^2 f}{\partial x \partial y}$. This result in a more general case (for functions of n variables) was proved by Flatto [**52**]. Babaev [**11**] generalized Rivlin and Sibner's result (as well as Flatto's result, see [**13**]). More precisely, he considered the class $M(R)$ of continuous functions $f(x, y)$ with the property

$$\Delta_{h_1,h_1} f = f(x, y) + f(x + h_1, y + h_2) - f(x, y + h_2) - f(x + h_1, y) \geq 0$$

for each rectangle $[x, x + h_1] \times [y, y + h_2] \subset R$, and proved that if $f(x, y)$ belongs to $M(R)$, where $R = [a_1, b_1] \times [a_2, b_2]$, then

$$E(f, R) = \frac{1}{4} \left[f(a_1, a_2) + f(b_1, b_2) - f(a_1, b_2) - f(b_1, a_2) \right].$$

As seen from this formula, to calculate $E(f)$ it is sufficient to find only values of $f(x, y)$ at the vertices of R. One can see that the formula also gives a sufficient condition for membership in the class $M(R)$, i.e. if

$$E(f, S) = \frac{1}{4} \left[f(x_1, y_1) + f(x_2, y_2) - f(x_1, y_2) - f(x_2, y_1) \right],$$

for a given f and for each $S = [x_1, x_2] \times [y_1, y_2] \subset R$, then the function $f(x, y)$ is from $M(R)$.

Our purpose is to construct new classes of continuous functions, which will depend on a numerical parameter, and characterize each class in terms of the approximation error calculation formulas. The mentioned parameter will show which points of R the calculation formula involves. We will also construct a best approximating sum $\varphi_0 + \psi_0$ to a function from constructed classes.

3.1.2. Definition of the main classes. Let throughout this section $R = [a_1, b_1] \times [a_2, b_2]$ be a rectangle and $c \in (a_1, b_1]$. Denote $R_1 = [a_1, c] \times [a_2, b_2]$ and $R_2 = [c, b_1] \times [a_2, b_2]$. It is clear that $R = R_1 \cup R_2$ and if $c = b_1$, then $R = R_1$.

We associate each rectangle $S = [x_1, x_2] \times [y_1, y_2]$ lying in R with the following functional:

$$L(f, S) = \frac{1}{4} \left[f(x_1, y_1) + f(x_2, y_2) - f(x_1, y_2) - f(x_2, y_1) \right].$$

DEFINITION 3.1. We say that a continuous function $f(x, y)$ belongs to the class $V_c(R)$ if

1) $L(f, S) \geq 0$, for each $S \subset R_1$;
2) $L(f, S) \leq 0$, for each $S \subset R_2$;
3) $L(f, S) \geq 0$, for each $S = [a_1, b_1] \times [y_1, y_2], \quad S \subset R$.

It can be shown that for any $c \in (a_1, b_1]$ the class $V_c(R)$ is not empty. Indeed, one can easily verify that the function

$$v_c(x,y) = \begin{cases} w(x,y) - w(c,y), & (x,y) \in R_1 \\ w(c,y) - w(x,y), & (x,y) \in R_2 \end{cases}$$

where $w(x,y) = \left(\frac{x-a_1}{b_1-a_1}\right)^{\frac{1}{n}} \cdot y$ and $n \geq \log_2 \frac{b_1-a_1}{c-a_1}$, satisfies conditions 1)-3) and therefore belongs to $V_c(R)$. The class $V_c(R)$ has the following obvious properties:

a) For given functions $f_1, f_2 \in V_c(R)$ and numbers $\alpha_1, \alpha_2 \geq 0$, $\alpha_1 f_1 + \alpha_2 f_2 \in V_c(R)$. $V_c(R)$ is a closed subset of the space of continuous functions.
b) $V_{b_1}(R) = M(R)$.
c) If f is a common element of $V_{c_1}(R)$ and $V_{c_2}(R)$, $a_1 < c_1 < c_2 \leq b_1$ then $f(x,y) = \varphi(x) + \psi(y)$ on the rectangle $[c_1, c_2] \times [a_2, b_2]$.

The properties a) and b) are clear. The property c) also becomes clear if note that according to the definition of the classes $V_{c_1}(R)$ and $V_{c_2}(R)$, for each rectangle

$$S \subset [c_1, c_2] \times [a_2, b_2]$$

we have

$$L(f,S) \leq 0 \text{ and } L(f,S) \geq 0,$$

respectively. Hence

$$L(f,S) = 0 \text{ for each } S \subset [c_1, c_2] \times [a_2, b_2].$$

Thus it is not difficult to understand that f is of the form $\varphi(x) + \psi(y)$ on the rectangle $[c_1, c_2] \times [a_2, b_2]$.

LEMMA 3.2. *Assume a function $f(x,y)$ has the continuous derivative $\frac{\partial^2 f}{\partial x \partial y}$ on the rectangle R and satisfies the following conditions*

1) $\frac{\partial^2 f}{\partial x \partial y} \geq 0$, *for all* $(x,y) \in R_1$;
2) $\frac{\partial^2 f}{\partial x \partial y} \leq 0$, *for all* $(x,y) \in R_2$;
3) $\frac{df(a_1,y)}{dy} \leq \frac{df(b_1,y)}{dy}$, *for all* $y \in [a_2, b_2]$.

Then $f(x,y)$ belongs to $V_c(R)$.

The proof of this lemma is very simple and can be obtained by integrating both sides of inequalities in conditions 1)-3) through sets $[x_1, x_2] \times [y_1, y_2] \subset R_1$, $[x_1, x_2] \times [y_1, y_2] \subset R_2$ and $[y_1, y_2] \subset [a_2, b_2]$, respectively.

EXAMPLE 3.3. Consider the function $f(x,y) = y \sin \pi x$ on the unit square $K = [0,1] \times [0,1]$ and rectangles $K_1 = \left[0, \frac{1}{2}\right] \times [0,1]$, $K_2 = \left[\frac{1}{2}, 1\right] \times [0,1]$. It is not difficult to verify that this function satisfies all conditions of the lemma and therefore belongs to $V_{\frac{1}{2}}(K)$.

3.1.3. Construction of an extremal element. The following theorem is valid.

THEOREM 3.4. *The approximation error of a function $f(x,y)$ from the class $V_c(R)$ can be calculated by the formula*

$$E(f,R) = L(f,R_1) = \frac{1}{4}\left[f(a_1,a_2) + f(c,b_2) - f(a_1,b_2) - f(c,a_2)\right].$$

Let y_0 be any solution from $[a_2, b_2]$ of the equation

$$L(f, Y) = \frac{1}{2} L(f, R_1), \quad Y = [a_1, c] \times [a_2, y].$$

Then the function $\varphi_0(x) + \psi_0(y)$, where

$$\varphi_0(x) = f(x, y_0),$$

$$\psi_0(y) = \frac{1}{2} [f(a_1, y) + f(c, y) - f(a_1, y_0) - f(c, y_0)]$$

is a best approximating sum from the manifold D to f.

To prove this theorem we need the following lemma.

LEMMA 3.5. *Let $f(x, y)$ be a function from $V_c(R)$ and $X = [a_1, x] \times [y_1, y_2]$ be a rectangle with fixed $y_1, y_2 \in [a_2, b_2]$. Then the function $h(x) = L(f, X)$ has the properties:*

1) *$h(x) \geq 0$, for any $x \in [a_1, b_1]$;*
2) *$\max\limits_{[a_1,b_1]} h(x) = h(c)$ and $\min\limits_{[a_1,b_1]} h(x) = h(a_1) = 0$.*

PROOF. If $X \subset R_1$, then the validity of $h(x) \geq 0$ follows from the definition of $V_c(R)$. If X is from R but not lying in R_1, then denoting $X' = [x, b_1] \times [y_1, y_2]$, $S = X \cup X'$, and using the obvious equality

$$L(f, S) = L(f, X) + L(f, X')$$

we deduce from the definition of $V_c(R)$ that $h(x) \geq 0$.

To prove the second part of the lemma, it is enough to show that $h(x)$ increases on the interval $[a_1, c]$ and decreases on the interval $[c, b_1]$. Indeed, if $a_1 \leq x_1 \leq x_2 \leq c$, then

$$h(x_2) = L(f, X_2) = L(f, X_1) + L(f, X_{12}), \tag{3.1}$$

where $X_1 = [a_1, x_1] \times [y_1, y_2]$, $X_2 = [a_1, x_2] \times [y_1, y_2]$, $X_{12} = [x_1, x_2] \times [y_1, y_2]$. Taking into consideration that $L(f, X_1) = h(x_1)$ and X_{12} lies in R_1 we obtain from (3.1) that $h(x_2) \geq h(x_1)$. If $c \leq x_1 \leq x_2 \leq b_1$, then X_{12} lies in R_2 and we obtain from (3.1) that $h(x_2) \leq h(x_1)$.

PROOF OF THEOREM 3.4 It is obvious that $L(f, R_1) = L(f - \varphi - \psi, R_1)$ for each sum $\varphi(x) + \psi(y)$. Hence

$$L(f, R_1) \leq \|f - \varphi - \psi\|_{C(R_1)} \leq \|f - \varphi - \psi\|_{C(R)}.$$

Since a sum $\varphi(x) + \psi(y)$ is arbitrary, $L(f, R_1) \leq E(f, R)$. To complete the proof it is sufficient to construct a sum $\varphi_0(x) + \psi_0(y)$ for which the equality

$$\|f - \varphi_0 - \psi_0\|_{C(R)} = L(f, R_1) \tag{3.2}$$

holds.

Consider the function

$$g(x, y) = f(x, y) - f(x, a_2) - f(a_1, y) + f(a_1, a_2).$$

This function has the following obvious properties

1) $g(x, a_2) = g(a_1, y) = 0$;
2) $L(f, R_1) = L(g, R_1) = \frac{1}{4} g(c, b_2)$;
3) $E(f, R) = E(g, R)$;
4) The function of one variable $g(c, y)$ increases on the interval $[a_2, b_2]$.

The last property of g allows us to write that

$$0 = g(c, a_2) \le \frac{1}{2} g(c, b_2) \le g(c, b_2).$$

Since $g(x, y)$ is continuous, there exists at least one solution $y = y_0$ of the equation

$$g(c, y) = \frac{1}{2} g(c, b_2)$$

or, in other notation, of the equation

$$L(f, Y) = \frac{1}{2} L(f, R_1), \quad \text{where } Y = [a_1, c] \times [a_2, y],$$

Introduce the functions

$$\begin{aligned} \varphi_1(x) &= g(x, y_0), \\ \psi_1(y) &= \frac{1}{2}\left(g(c, y) - g(c, y_0)\right), \\ G(x, y) &= g(x, y) - \varphi_1(x) - \psi_1(y). \end{aligned}$$

Calculate the norm of $G(x, y)$ on R. Consider the rectangles $R' = [a_1, b_1] \times [y_0, b_2]$ and $R'' = [a_1, b_1] \times [a_2, y_0]$. It is clear that

$$\|G\|_{C(R)} = \max\left\{\|G\|_{C(R')}, \|G\|_{C(R'')}\right\}.$$

First calculate the norm $\|G\|_{C(R')}$:

$$\|G\|_{C(R')} = \max_{(x,y)\in R'} |G(x, y)| = \max_{y\in[y_0,b_2]} \max_{x\in[a_1,b_1]} |G(x, y)|. \tag{3.3}$$

For a fixed point y (we keep it fixed until (3.6)) from the interval $[y_0, b_2]$ we can write that

$$\max_{x\in[a_1,b_1]} G(x, y) = \max_{x\in[a_1,b_1]} \left(g(x, y) - g(x, y_0)\right) - \psi_1(y) \tag{3.4}$$

and

$$\min_{x\in[a_1,b_1]} G(x, y) = \min_{x\in[a_1,b_1]} \left(g(x, y) - g(x, y_0)\right) - \psi_1(y). \tag{3.5}$$

By Lemma 3.5, the function

$$h_1(x) = 4L(f, X) = g(x, y) - g(x, y_0), \quad \text{where } X = [a_1, x] \times [y_0, y],$$

reaches its maximum on $x = c$ and minimum on $x = a_1$:

$$\max_{x\in[a_1,b_1]} h_1(x) = g(c, y) - g(c, y_0)$$

$$\min_{x\in[a_1,b_1]} h_1(x) = g(a_1, y) - g(a_1, y_0) = 0.$$

Considering these facts in (3.4) and (3.5) we obtain that

$$\max_{x\in[a_1,b_1]} G(x, y) = g(c, y) - g(c, y_0) - \psi_1(y) = \frac{1}{2}\left(g(c, y) - g(c, y_0)\right),$$

$$\min_{x\in[a_1,b_1]} G(x, y) = -\psi_1(y) = -\frac{1}{2}\left(g(c, y) - g(c, y_0)\right).$$

Consequently,

$$\max_{x\in[a_1,b_1]} |G(x, y)| = \frac{1}{2}\left(g(c, y) - g(c, y_0)\right). \tag{3.6}$$

Taking (3.6) and the 4-th property of g into account in (3.3) yields

$$\|G\|_{C(R')} = \frac{1}{2}\left(g(c, b_2) - g(c, y_0)\right) = \frac{1}{4}g(c, b_2).$$

Similarly it can be shown that

$$\|G\|_{C(R'')} = \frac{1}{4}g(c, b_2).$$

Hence

$$\|G\|_{C(R)} = \frac{1}{4}g(c, b_2) = L(f, R_1).$$

But by the definition of G,

$$G(x, y) = g(x, y) - \varphi_1(x) - \psi_1(y) = f(x, y) - \varphi_0(x) - \psi_0(y),$$

where

$$\begin{aligned}\varphi_0(x) &= \varphi_1(x) + f(x, a_2) - f(a_1, a_2) + f(a_1, y_0) = f(x, y_0),\\ \psi_0(y) &= \psi_1(y) + f(a_1, y) - f(a_1, y_0) = \\ &= \frac{1}{2}\left(f(a_1, y) + f(c, y) - f(a_1, y_0) - f(c, y_0)\right).\end{aligned}$$

Therefore,

$$\|f - \varphi_0 - \psi_0\|_{C(R)} = L(f, R_1).$$

We proved (3.2) and hence Theorem 3.4. Note that the function $\varphi_0(x) + \psi_0(y)$ is a best approximating sum from the manifold D to f.

REMARK 3.6. In the special case $c = b_1$, Theorem 3.4 turns into Babaev's result from [**11**].

COROLLARY 3.7. *Let a function $f(x, y)$ have the continuous derivative $\frac{\partial^2 f}{\partial x \partial y}$ on the rectangle R and satisfy the following conditions*

1) $\frac{\partial^2 f}{\partial x \partial y} \geq 0$, *for all* $(x, y) \in R_1$;
2) $\frac{\partial^2 f}{\partial x \partial y} \leq 0$, *for all* $(x, y) \in R_2$;
3) $\frac{df(a_1, y)}{dy} \leq \frac{df(b_1, y)}{dy}$, *for all* $y \in [a_2, b_2]$.

Then

$$E(f, R) = L(f, R_1) = \frac{1}{4}\left[f(a_1, a_2) + f(c, b_2) - f(a_1, b_2) - f(c, a_2)\right].$$

The proof of this corollary can be directly obtained from Lemma 3.2 and Theorem 3.4.

REMARK 3.8. Rivlin and Sibner [**176**] proved Corollary 3.7 in the special case $c = b_1$.

EXAMPLE 3.9. As we know (see Example 3.3) the function $f = y \sin \pi x$ belongs to $V_{\frac{1}{2}}(K)$, where $K = [0, 1] \times [0, 1]$. By Theorem 3.4, $E(f, K) = \frac{1}{4}$ and the function $\frac{1}{2} \sin \pi x + \frac{1}{2}y - \frac{1}{4}$ is a best approximating sum.

The following theorem shows that in some cases the approximation error formula in Theorem 3.4 is valid for more general sets than rectangles with sides parallel to the coordinate axes.

THEOREM 3.10. *Let $f(x,y)$ be a function from $V_c(R)$ and $Q \subset R$ be a compact set which contains all vertices of R_1 (points $(a_1,a_2),(a_1,b_2),(c,a_2),(c,b_2)$). Then*

$$E(f,Q) = L(f,R_1) = \frac{1}{4}\left[f(a_1,a_2) + f(c,b_2) - f(a_1,b_2) - f(c,a_2)\right].$$

PROOF. Since $Q \subset R$, $E(f,Q) \leq E(f,R)$. On the other hand by Theorem 3.4, $E(f,R) = L(f,R_1)$. Hence $E(f,Q) \leq L(f,R_1)$. It can be shown, as it has been shown in the proof of Theorem 3.4, that $L(f,R_1) \leq E(f,Q)$. But then automatically $E(f,Q) = L(f,R_1)$. □

EXAMPLE 3.11. Calculate the approximation error of the function $f(x,y) = -(x-2)^{2n}y^m$ (n and m are positive integers) on the domain

$$Q = \left\{(x,y) : 0 \leq x \leq 2, 0 \leq y \leq (x-1)^2 + 1\right\}.$$

It can be easily verified that $f \in V_2(R)$, where $R = [0,4]\times[0,2]$. Besides, Q contains all vertices of $R_1 = [0,2]\times[0,2]$. Consequently, by Theorem 3.10, $E(f,Q) = L(f,R_1) = 2^{2(n-1)+m}$.

3.1.4. Characterization of $V_c(R)$.

The following theorem characterizes the class $V_c(R)$ in terms of the approximation error calculation formulas.

THEOREM 3.12. *The following conditions are necessary and sufficient for a continuous function $f(x,y)$ belong to $V_c(R)$:*

1) *$E(f,S) = L(f,S)$, for each rectangle $S = [x_1,x_2]\times[y_1,y_2], S \subset R_1$;*
2) *$E(f,S) = -L(f,S)$, for each rectangle $S = [x_1,x_2]\times[y_1,y_2], S \subset R_2$;*
3) *$E(f,S) = L(f,S_1)$, for each rectangle $S = [a_1,b_1]\times[y_1,y_2], S \subset R$ and $S_1 = [a_1,c]\times[y_1,y_2]$.*

PROOF. The necessity easily follows from the definition of $V_c(R)$, Babaev's above-mentioned result and Theorem 3.4. The sufficiency is clear if pay attention to the fact that $E(f,S) \geq 0$. □

3.1.5. Classes $V_c^-(R), U(R)$ and $U_c^-(R)$.

By $V_c^-(R)$ we denote the class of functions $f(x,y)$ such that $-f \in V_c(R)$. It is clear that $E(f,R) = -L(f,R_1)$ for each $f \in V_c^-(R)$.

We define $U_c(R), a_1 \leq c < b_1$, as a class of continuous functions $f(x,y)$ with the properties

1) $L(f,S) \leq 0$, for each rectangle $S = [x_1,x_2]\times[y_1,y_2],\ S \subset R_1$;
2) $L(f,S) \geq 0$, for each rectangle $S = [x_1,x_2]\times[y_1,y_2],\ S \subset R_2$;
3) $L(f,S) \geq 0$, for each rectangle $S = [a_1,b_1]\times[y_1,y_2],\ S \subset R$.

Using the same techniques in the proof of Theorem 3.4 it can be shown that the following theorem is valid:

THEOREM 3.13. *The approximation error of a function $f(x,y)$ from the class $U_c(R)$ can be calculated by the formula*

$$E(f,R) = L(f,R_2) = \frac{1}{4}\left[f(c,a_2) + f(b_1,b_2) - f(c,b_2) - f(b_1,a_2)\right].$$

Let y_0 be any solution from $[a_2, b_2]$ of the equation

$$L(f,Y) = \frac{1}{2}L(f,R_2), \qquad Y = [c,b_1] \times [a_2,y] .$$

Then the function $\varphi_0(x) + \psi_0(y)$, where

$$\varphi_0(x) = f(x,y_0), \quad \psi_0(y) = \frac{1}{2}\left[f(c,y) + f(b_1,y) - f(c,y_0) - f(b_1,y_0)\right],$$

is a best approximating sum from the manifold D to f.

By $U_c^-(R)$ denote the class of functions $f(x,y)$ such that $-f \in U_c(R)$. It is clear that $E(f,R) = -L(f,R_2)$ for each $f \in U_c^-(R)$.

REMARK 3.14. The correspondingly modified versions of Theorems 3.10, 3.12 and Corollary 3.7 are valid for the classes $V_c^-(R), U_c(R)$ and $U_c^-(R)$.

EXAMPLE 3.15. Consider the function $f(x,y) = \left(x - \frac{1}{2}\right)^2 y$ on the unit square $K = [0,1] \times [0,1]$. It can be easily verified that $f \in U_{\frac{1}{2}}(K)$. Hence, by Theorem 3.13, $E(f,K) = \frac{1}{16}$ and the function $\frac{1}{2}\left(x - \frac{1}{2}\right)^2 + \frac{1}{8}y - \frac{1}{16}$ is a best approximating function.

3.2. Approximation by sums of univariate functions on certain domains

The purpose of this section is to develop a method for obtaining explicit formulas for the error of approximation of bivariate functions by sums of univariate functions. It should be remarked that formulas of this type were known only for functions defined on a rectangle with sides parallel to the coordinate axes. Our method, based on a maximization process over closed bolts, allows the consideration of functions defined on hexagons, octagons and stairlike polygons with sides parallel to the coordinate axes.

3.2.1. Problem statement. Let Q be a compact set in $\mathbb{R}^2$. Consider the approximation of a continuous function $f \in C(Q)$ by functions from the set $D = \{\varphi(x) + \psi(y)\}$, where $\varphi(x), \psi(y)$ are defined and continuous on the projections of Q into the coordinate axes x and y, respectively. The approximation error is defined as follows

$$E(f,Q) = \inf_{\varphi+\psi\in D} \|f - \varphi - \psi\|_{C(Q)} .$$

Our purpose is to develop a method for obtaining explicit formulas providing precise and easy computation of $E(f,Q)$ for polygons Q with sides parallel to the coordinate axes. This method will be based on the herein developed *closed bolts maximization process* and can be used in alternative proofs of the known results from [**11**], [**77**] and [**176**]. First, we show efficiency of the method in the example of a hexagon with sides parallel to the coordinate axes. Then we formulate an analogous theorem for staircase polygons and two theorems for octagons, which can be proved in a similar way, and touch some aspects of the question about the case of an arbitrary polygon with sides parallel to the coordinate axes. The condition posed on sides of polygons (being parallel to the coordinate axes) is essential for our method. This has several reasons, which get clear through the proof of Theorem 3.16. Here we are able to explain one of these reasons: by [**60**, Theorem 3], a continuous function $f(x,y)$

defined on a polygon with sides parallel to the coordinate axes has an extremal element, the existence of which is required in our method. Now let K be a rectangle (not speaking about polygons) with sides not parallel to the coordinate axes. Does any function $f \in C(K)$ have an extremal element? No one knows (see [**60**]).

In the sequel, all the considered polygons are supposed to have sides parallel to the coordinate axes.

3.2.2. The maximization process. Let H be a closed hexagon. It is clear that H can be uniquely represented in the form

$$H = R_1 \cup R_2, \tag{3.7}$$

where R_1, R_2 are rectangles and there does not exist any rectangle R such that $R_1 \subset R \subset H$ or $R_2 \subset R \subset H$.

We associate each closed bolt $p = \{p_1, p_2, \cdots p_{2n}\}$ with the following functional

$$l(f,p) = \frac{1}{2n}\sum_{k=1}^{2n}(-1)^{k-1}f(p_k).$$

Denote by $M(H)$ the class of bivariate continuous functions f on H satisfying the condition

$$f(x_1, y_1) + f(x_2, y_2) - f(x_1, y_2) - f(x_2, y_1) \geq 0$$

for any rectangle $[x_1, x_2] \times [y_1, y_2] \subset H$.

THEOREM 3.16. *Let H be a hexagon and (3.7) be its representation. Let $f \in M(H)$. Then*

$$E(f, H) = \max\{|l(f,h)|, |l(f,r_1)|, |l(f,r_2)|\}, \tag{3.8}$$

where h, r_1, r_2 are closed bolts formed by vertices of the polygons H, R_1, R_2 respectively.

PROOF. Without loss of generality, we may assume that the rectangles R_1 and R_2 are of the following form

$$R_1 = [a_1, a_2] \times [b_1, b_3], \quad R_2 = [a_1, a_3] \times [b_1, b_2], \quad a_1 < a_2 < a_3, \ b_1 < b_2 < b_3.$$

Introduce the notation

$$\begin{aligned} &f_{11} = f(a_1, b_1), \ f_{12} = -f(a_1, b_2), \ f_{13} = -f(a_1, b_3); \\ &f_{21} = -f(a_2, b_1), \ f_{22} = -f(a_2, b_2), \ f_{23} = f(a_2, b_3); \\ &f_{31} = -f(a_3, b_1), \ f_{32} = f(a_3, b_2). \end{aligned} \tag{3.9}$$

It is clear that

$$\begin{aligned} |l(f,r_1)| &= \frac{1}{4}(f_{11} + f_{13} + f_{23} + f_{21}), \\ |l(f,r_2)| &= \frac{1}{4}(f_{11} + f_{12} + f_{32} + f_{31}), \\ |l(f,h)| &= \frac{1}{6}(f_{11} + f_{13} + f_{23} + f_{22} + f_{32} + f_{31}). \end{aligned} \tag{3.10}$$

Let $p = \{p_1, p_2, \cdots p_{2n}\}$ be any closed bolt. We group the points $p_1, p_2, \cdots p_{2n}$ by putting

$$p_+ = \{p_1, p_3, \cdots p_{2n-1}\}, \ p_- = \{p_2, p_4, \cdots p_{2n}\}.$$

First, assume that $l(f,p) \geq 0$. We apply the following algorithm, which we call *the maximization process over closed bolts*, to p.

Step 1. Consider sequentially the units p_ip_{i+1} $\left(i=\overline{1,2n}, p_{2n+1}=p_1\right)$ with the vertices $p_i\left(x_i,y_i\right)$, $p_{i+1}\left(x_{i+1},y_{i+1}\right)$ having equal abscissae: $x_i=x_{i+1}$. Four cases are possible.

1) $p_i\in p_+$ and $y_{i+1}>y_i$. In this case, replace the unit p_ip_{i+1} by a new unit q_iq_{i+1} with the vertices $q_i=(a_1,y_i)$, $q_{i+1}=(a_1,y_{i+1})$.

2) $p_i\in p_+$ and $y_{i+1}<y_i$. In this case, replace the unit p_ip_{i+1} by a new unit q_iq_{i+1} with the vertices $q_i=\left(a_2,y_i\right)$, $q_{i+1}=\left(a_2,y_{i+1}\right)$ if $b_2<y_i\le b_3$ or with the vertices $q_i=\left(a_3,y_i\right), q_{i+1}=\left(a_3,y_{i+1}\right)$ if $b_1\le y_i\le b_2$.

3) $p_i\in p_-$ and $y_{i+1}<y_i$. In this case, replace p_ip_{i+1} by a new unit q_iq_{i+1} with the vertices $q_i=(a_1,y_i)$, $q_{i+1}=(a_1,y_{i+1})$.

4) $p_i\in p_-$ and $y_{i+1}>y_i$. In this case, replace p_ip_{i+1} by a new unit q_iq_{i+1} with the vertices $q_i=(a_2,y_i)$, $q_{i+1}=(a_2,y_{i+1})$ if $b_2<y_{i+1}\le b_3$ or with the vertices $q_i=(a_3,y_i)$, $q_{i+1}=(a_3,y_{i+1})$ if $b_1\le y_{i+1}\le b_2$.

Since $f\in M(II)$, it is not difficult to verify that

$$\begin{aligned} &f(p_i)-f(p_{i+1})\le f(q_i)-f(q_{i+1}) \quad \text{for cases 1) and 2),}\\ &-f(p_i)+f(p_{i+1})\le -f(q_i)+f(q_{i+1}) \quad \text{for cases 3) and 4)} \end{aligned} \tag{3.11}$$

It is clear that after Step 1 the bolt p will be replaced by the ordered set $q=\{q_1,q_2,\cdots,q_{2n}\}$. We do not say a bolt but an ordered set because of a possibility of coincidence of some successive points q_i,q_{i+1} (this, for example, may happen if the 1-st case takes place for the units $p_{i-1}p_i$ and $p_{i+1}p_{i+2}$). Let us exclude simultaneously successive and coincident points from q. Then we obtain some closed bolt, which we denote by $q'=\{q_1',q_2',\cdots,q_{2m}'\}$. It is not difficult to understand that all points of the bolt q' are located on straight lines $x=a_1$, $x=a_2$, $x=a_3$.

From inequalities (3.11) and the fact that $2m\le 2n$, we deduce that

$$l(f,p)\le l(f,q'). \tag{3.12}$$

Step 2. Consider sequentially units $q_i'q_{i+1}'$ $\left(i=\overline{1,2m}, q_{2m+1}'=q_1'\right)$ with the vertices $q_i'=\left(x_i',y_i'\right)$, $q_{i+1}'\left(x_{i+1}',y_{i+1}'\right)$ having equal ordinates: $y_i'=y_{i+1}'$. The following four cases are possible.

1) $q_i'\in q_+'$ and $x_{i+1}'>x_i'$. In this case, replace the unit $q_i'q_{i+1}'$ by a new unit $p_i'p_{i+1}'$ with the vertices $p_i'=\left(x_i',b_1\right)$, $p_{i+1}'=\left(x_{i+1}',b_1\right)$.

2) $q_i'\in q_+'$ and $x_{i+1}'<x_i'$. In this case, replace the unit $q_i'q_{i+1}'$ by a new unit $p_i'p_{i+1}'$ with the vertices $p_i'=\left(x_i',b_2\right)$, $p_{i+1}'=\left(x_{i+1}',b_2\right)$ if $x_i'=a_3$ and with the vertices $p_i'=\left(x_i',b_3\right), p_{i+1}'=\left(x_{i+1}',b_3\right)$ if $x_i'=a_2$.

3) $q_i'\in q_-'$ and $x_{i+1}'<x_i'$. In this case, replace $q_i'q_{i+1}'$ by a new unit $p_i'p_{i+1}'$ with the vertices $p_i'=\left(x_i',b_1\right)$, $p_{i+1}'=\left(x_{i+1}',b_1\right)$.

4) $q_i'\in q_-'$ and $x_{i+1}'>x_i'$. In this case, replace $q_i'q_{i+1}'$ by a new unit $p_i'p_{i+1}'$ with the vertices $p_i'=\left(x_i',b_2\right)$, $p_{i+1}'=\left(x_{i+1}',b_2\right)$ if $x_{i+1}'=a_3$ and with the vertices $p_i'=\left(x_i',b_3\right)$, $p_{i+1}'=\left(x_{i+1}',b_3\right)$ if $x_{i+1}'=a_2$.

It is easy to see that after Step 2 the bolt q' will be replaced by the bolt $p'=\{p_1',p_2',\cdots p_{2m}'\}$ and

$$l(f,q')\le l(f,p'). \tag{3.13}$$

From (3.12) and (3.13) we obtain that

$$l(f,p)\le l(f,p'). \tag{3.14}$$

It is clear that each point of the set p_+' coincides with one of the points (a_1,b_1), (a_2,b_3), (a_3,b_2) and each point of the set p_-' coincides with one of the points (a_1, b_2), (a_1, b_3), (a_2, b_1), (a_2, b_2), (a_3, b_1). Denote by m_{ij} the number of points of the bolt p' coinciding with the point (a_i, b_j), $i, j = \overline{1,3}$, $i + j \neq 6$. By (3.9), we can write that

$$l(f, p') = \frac{1}{2m} \sum_{\substack{i,j=\overline{1,3} \\ i+j \leq 5}} m_{ij} f_{ij}. \tag{3.15}$$

On the straight line $x = a_i$ or $y = b_i$, $i = \overline{1,3}$, the number of points of the set p'_+ is equal to the number of points of the set p'_-. Hence

$$m_{11} = m_{12} + m_{13} = m_{21} + m_{31};\ m_{23} = m_{22} + m_{21} = m_{13};\ m_{32} = m_{31} = m_{12} + m_{22}.$$

From these equalities we deduce that

$$\begin{gathered} m_{11} = m_{12} + m_{21} + m_{22};\ m_{13} = m_{21} + m_{22}; \\ m_{23} = m_{21} + m_{22};\ m_{31} = m_{12} + m_{22}. \end{gathered} \tag{3.16}$$

Consequently,

$$2m = \sum_{\substack{i,j=\overline{1,3} \\ i+j \leq 5}} m_{ij} = 4m_{12} + 4m_{21} + 6m_{22}. \tag{3.17}$$

Considering (3.16) and (3.17) in (3.15) and taking (3.10) into account, we obtain that

$$\begin{aligned} l(f, p') &= \frac{4m_{12} \left| l(f, r_2) \right| + 4m_{21} \left| l(f, r_1) \right| + 6m_{22} \left| l(f, h) \right|}{4m_{12} + 4m_{21} + 6m_{22}} \\ &\leq \max \left\{ \left| l(f, r_1) \right|, \left| l(f, r_2) \right|, \left| l(f, h) \right| \right\}. \end{aligned}$$

Therefore, due to (3.14),

$$l(f, p) \leq \max \left\{ \left| l(f, r_1) \right|, \left| l(f, r_2) \right|, \left| l(f, h) \right| \right\}. \tag{3.18}$$

Note that in the beginning of the proof the bolt p has been chosen so that $l(f, p) \geq 0$. Let now $p = \{p_1, p_2, \cdots p_{2n}\}$ be any closed bolt such that $l(f, p) \leq 0$. Since $l(f, p'') = -l(f, p) \geq 0$ for the bolt $p'' = \{p_2, p_3, \cdots, p_{2n}, p_1\}$,we obtain from (3.18) that

$$-l(f, p) \leq \max \left\{ \left| l(f, r_1) \right|, \left| l(f, r_2) \right|, \left| l(f, h) \right| \right\}. \tag{3.19}$$

From (3.18) and (3.19) we deduce on the strength of arbitrariness of p that

$$\sup_{p \subset H} \left\{ \left| l(f, p) \right| \right\} = \max \left\{ \left| l(f, r_1) \right|, \left| l(f, r_2) \right|, \left| l(f, h) \right| \right\}, \tag{3.20}$$

where the sup is taken over all closed bolts of the hexagon H.

The hexagon H satisfies the conditions of Theorem 1.33 on the existence of a best approximation. By the duality theorem of Khavinson [**114**, Theorem 2] (see Section 3.3 for such theorems), we obtain that

$$E(f, H) = \sup_{p \subset H} \left\{ \left| l(f, p) \right| \right\}. \tag{3.21}$$

From (3.20) and (3.21) we finally conclude that

$$E(f, H) = \max \left\{ \left| l(f, r_1) \right|, \left| l(f, r_2) \right|, \left| l(f, h) \right| \right\}.$$

□

COROLLARY 3.17. *Let a function $f(x, y)$ have the continuous nonnegative derivative $\dfrac{\partial^2 f}{\partial x \partial y}$ on H. Then the formula (3.8) is valid.*

The proof is very simple and can be obtained by integrating the inequality $\frac{\partial^2 f}{\partial x \partial y} \geq 0$ over an arbitrary rectangle $[x_1, x_2] \times [y_1, y_2] \subset H$ and applying Theorem 3.16.

The method used in the proof of Theorem 3.16 can be generalized to obtain similar results for stairlike polygons. For example, let S be a closed polygon of the following form

$$S = \bigcup_{i=1}^{N-1} P_i,$$

where $N \geq 2$, $P_i = [a_i, a_{i+1}] \times [b_1, b_{N+1-i}]$, $i = \overline{1, N-1}$, $a_1 < a_2 < \cdots < a_N$, $b_1 < b_2 < \cdots < b_N$. Such polygons will be called *stairlike polygons* (see [**79**]).

A closed $2m$-gon F with sides parallel to the coordinate axes is called a maximal $2m$-gon of the polygon S if $F \subset S$ and there is no another $2m$-gon F' such that $F \subset F' \subset S$. Clearly, if F is a maximal $2m$-gon of the polygon S, then $m \leq N$. A closed bolt formed by the vertices of a maximal polygon F is called a maximal bolt of S. By S^B denote the set of all maximal bolts of the stairlike polygon S.

THEOREM 3.18. *Let S be a stairlike polygon. The approximation error of a function $f \in M(S)$ can be computed by the formula*

$$E(f, S) = \max\left\{|r(f, h)|,\ h \in S^B\right\}.$$

For the proof of this theorem see [**79**].

3.2.3. E-bolts. The main idea in the proof of Theorem 3.16 can be successfully used in obtaining formulas of type (3.8) for functions $f(x, y)$ defined on another simple polygons. The following two theorems include cases of some octagons and can be proved in a similar way.

THEOREM 3.19. *Let $a_1 < a_2 < a_3 < a_4$, $b_1 < b_2 < b_3$ and Q be an octagon of the following form*

$$Q = \bigcup_{i=1}^{4} R_i,$$

where $R_1 = [a_1, a_2] \times [b_1, b_2]$, $R_2 = [a_2, a_3] \times [b_1, b_2]$, $R_3 = [a_3, a_4] \times [b_1, b_2]$, $R_4 = [a_2, a_3] \times [b_2, b_3]$. Let $f \in M(Q)$. Then the following formula holds

$$E(f, Q) = \max\left\{|l(f, q)|, |l(f, r_{123})|, |l(f, r_{124})|, |l(f, r_{234})|, |l(f, r_{24})|\right\},$$

where q, r_{123}, r_{124}, r_{234}, r_{24} are closed bolts formed by the vertices of the polygons Q, $R_1 \cup R_2 \cup R_3$, $R_1 \cup R_2 \cup R_4$, $R_2 \cup R_3 \cup R_4$ and $R_2 \cup R_4$, respectively.

THEOREM 3.20. *Let $a_1 < a_2 < a_3 < a_4$, $b_1 < b_2 < b_3$ and Q be an octagon of the following form*

$$Q = \bigcup_{i=1}^{3} R_i,$$

where $R_1 = [a_1, a_4] \times [b_1, b_2]$, $R_2 = [a_1, a_2] \times [b_2, b_3]$, $R_3 = [a_3, a_4] \times [b_2, b_3]$. Let $f \in M(Q)$. Then

$$E(f, Q) = \max\left\{|l(f, r)|, |l(f, r_{12})|, |l(f, r_{13})|\right\},$$

where r, r_{12}, r_{13} *are closed bolts formed by the vertices of the polygons* $R = [a_1, a_4] \times [b_1, b_3]$, $R_1 \cup R_2, R_1 \cup R_3$, *respectively.*

Although the closed bolts maximization process can be applied to bolts of an arbitrary polygon, some combinatorial difficulties arise when grouping values at points of maximized bolts (bolts obtained after the maximization process, see (3.15)-(3.18)). While we do not know a complete answer to this problem, we can describe points of a polygon F with which points of maximized bolts coincide and state a conjecture concerning the approximation error.

Let $F = A_1 A_2 \dots A_{2n}$ be any polygon with sides parallel to the coordinate axes. The vertices A_1, A_2, …, A_{2n} in the given order form a closed bolt, which we denote by r_F. By $[r_F]$ denote the length of r_F. In our case, $[r_F] = 2n$.

DEFINITION 3.21. Let F and S be polygons with sides parallel to the coordinate axes. We say that the closed bolt r_F is an e-bolt (extended bolt) of S if $r_F \subset S$ and there does not exist any polygon F' such that $F \subset F'$, $r_{F'} \subset S$, $[r_{F'}] \leq [r_F]$.

For example, in Theorem 3.20 the octagon Q has 3 e-bolts. They are r, r_{12} and r_{13}. In Theorem 3.19, the octagon Q has 5 e-bolts, which are $q, r_{123}, r_{124}, r_{234}$ and r_{24}. The polygon $S_{2n} = \bigcup_{i=1}^{n-1} R_i$, where $R_i = [a_i, a_{i+1}] \times [b_1, b_{n+1-i}], i = \overline{1, n-1}, a_1 < a_2 < \dots < a_n, b_1 < b_2 < \dots < b_n$ has exactly $2^{n-1} - 1$ e-bolts. It is not difficult to observe that the set of points of a closed bolt obtained after the maximization process is a subset of the set of points of all e-bolts. This condition and Theorems 3.16, 3.18, 3.19, 3.20 justify the statement of the following conjecture:

Let S be any polygon with sides parallel to the coordinate axes and $f \in M(S)$. Then

$$E(f, S) = \max_{h \in S^E} \{|l(f, h)|\},$$

where S^E is a set of all e-bolts of the polygon S.

3.2.4. Error estimates. Theorem 3.16 allows us to consider classes wider than $M(H)$ and establish sharp estimates for the approximation error.

THEOREM 3.22. *Let H be a hexagon and (3.7) be its representation. The following sharp estimates are valid for a function $f(x, y)$ having the continuous derivative $\frac{\partial^2 f}{\partial x \partial y}$ on H:*

$$A \leq E(f, H) \leq BC + \frac{3}{2}\left(B\,|l(g, h)| - |l(f, h)|\right), \tag{3.22}$$

where

$$B = \max_{(x,y)\in H} \left|\frac{\partial^2 f(x, y)}{\partial x \partial y}\right|, \quad g = g(x, y) = x \cdot y,$$

$$A = \max\{|l(f, h)|, \ |l(f, r_1)|, |l(f, r_2)|\}, \ C = \max\{|l(g, h)|, |l(g, r_1)|, \ |l(g, r_2)|\},$$

where h, r_1, r_2 are closed bolts formed by vertices of the polygons H, R_1 and R_2, respectively.

REMARK 3.23. Inequalities similar to (3.22) were established in Babaev [**12**] for the approximation of a function $f(x) = f(x_1, \dots, x_n)$, defined on a parallelepiped with sides parallel to the coordinate axes, by sums $\sum_{i=1}^{n} \varphi_i(x \backslash x_i)$. For the approximation of bivariate functions, Babaev's result contains only rectangular case.

REMARK 3.24. Estimates (3.22) are easily calculable in contrast to those established in [**14**] for continuous functions defined on certain domains, which are different from polygons.

To prove Theorem 3.22 we need the following lemmas.

LEMMA 3.25. *Let X be a normed space, F be a subspace of X. The following inequality is valid for an element $x = x_1 + x_2$ from X:*

$$|E(x_1) - E(x_2)| \leq E(x) \leq E(x_1) + E(x_2),$$

where

$$E(x) = E(x, F) = \inf_{y \in F} \|x - y\|.$$

LEMMA 3.26. *If $f \in M(H)$, then*

$$|l(f, r_i)| \leq \frac{3}{2} |l(f, h)|, i = 1, 2.$$

Lemma 3.25 is obvious. To prove Lemma 3.26, note that for any $f \in M(H)$

$$6 |l(f, h)| = 4 |l(f, r_i)| + 4 |l(f, r_3)|, \quad i = 1, 2,$$

where r_3 is a closed bolt formed by the vertices of the rectangle $R_3 = H \backslash R_i$.

Now let us prove Theorem 3.22.

PROOF. It is not difficult to verify that if $\frac{\partial^2 u}{\partial x \partial y} \geq 0$ on H for some $u(x, y)$, $\frac{\partial^2 u(x,y)}{\partial x \partial y} \in C(H)$, then $u \in M(H)$ (see the proof of Corollary 3.17. Set $f_1 = f + Bg$. Since $\frac{\partial^2 f_1}{\partial x \partial y} \geq 0$ on H, $f_1 \in M(H)$. By Lemma 3.26,

$$|l(f_1, r_i)| \leq \frac{3}{2} |l(f_1, h)|, i = 1, 2. \tag{3.23}$$

Theorem 3.16 implies that

$$E(f_1, H) = \max \{|l(f_1, h)|, |l(f_1, r_1)|, |l(f_1, r_2)|\}. \tag{3.24}$$

We deduce from (3.23) and (3.24) that

$$E(f_1, H) \leq \frac{3}{2} |l(f_1, h)|.$$

First, let the closed bolt h start at the point (a_1, b_1). Then it is clear that

$$E(f_1, H) \leq \frac{3}{2} l(f_1, h). \tag{3.25}$$

By Lemma 3.25,

$$E(f, H) - E(Bg, H) \leq E(f_1, H). \tag{3.26}$$

Inequalities (3.25) and (3.26) yield

$$E(f, H) \leq BE(g, H) + \frac{3}{2} l(f_1, h). \tag{3.27}$$

Since the functional $l(f, h)$ is linear,

$$l(f_1, h) = l(f, h) + Bl(g, h).$$

Considering this expression of $l(f_1, h)$ in (3.27), we obtain that

$$E(f, H) \le BE(g, H) + \frac{3}{2}Bl(g, h) + \frac{3}{2}l(f, h). \tag{3.28}$$

Now consider the function $f_2 = Bg - f$. Obviously, $\frac{\partial^2 f_2}{\partial x \partial y} \ge 0$ on H. It can be shown, in the same way as (3.28) has been obtained, that

$$E(f, H) \le BE(g, H) + \frac{3}{2}Bl(g, h) - \frac{3}{2}l(f, h). \tag{3.29}$$

From (3.28) and (3.29) it follows that

$$E(f, H) \le BE(g, H) + \frac{3}{2}Bl(g, h) - \frac{3}{2}\left|l(f, h)\right|. \tag{3.30}$$

Since $g \in M(H)$ and h starts at the point (a_1, b_1), we have $l(g, h) \ge 0$.

Let now h start at a point such that $l(u, h) \le 0$ for any $u \in M(H)$. Then in a similar way as above we can prove that

$$E(f, H) \le BE(g, H) - \frac{3}{2}Bl(g, h) - \frac{3}{2}\left|l(f, h)\right|, \tag{3.31}$$

where $l(g, h) \le 0$. From (3.30), (3.31) and the fact that $E(g, H) = C$ (in view of Theorem 3.16), it follows that

$$E(f, H) \le BC + \frac{3}{2}\left(B\left|l(g, h)\right| - \left|l(f, h)\right|\right).$$

The upper bound in (3.22) has been established. Note that it is attained by $f = g = xy$.

The proof of the lower bound in (3.22) is simple. One of the obvious properties of the functional $l(f, p)$ is that $|l(f, p)| \le E(f, H)$ for any continuous function f on H and a closed bolt p. Hence,

$$A = \max\left\{\left|l(f, h)\right|, \left|l(f, r_1)\right|, \left|l(f, r_2)\right|\right\} \le E(f, H).$$

Note that by Theorem 3.16 the lower bound in (3.22) is attained by an arbitrary function from $M(H)$. □

REMARK 3.27. Using Theorems 3.19 and 3.20 one can obtain sharp estimates of type (3.22) for bivariate functions defined on the corresponding simple polygons with sides parallel to the coordinate axes.

3.3. On the theorem of M. Golomb

Let $X_1, \dots, X_n$ be compact spaces and $X = X_1 \times \cdots \times X_n$. Consider the approximation of a function $f \in C(X)$ by sums $g_1(x_1) + \cdots + g_n(x_n)$, where $g_i \in C(X_i)$, $i = 1, \dots, n$. In [**64**], M.Golomb obtained a formula for the error of this approximation in terms of measures constructed on special points of X, called "projection cycles". However, his proof had a gap, which was pointed out later by Marshall and O'Farrell [**151**]. But the question if the formula was correct, remained open. The purpose of this section is to prove that Golomb's formula is valid, and moreover it holds in a stronger form.

3.3.1. History of Golomb's formula. Let $X_i, i = 1, \ldots, n$, be compact Hausdorff spaces. Consider the approximation to a continuous function f, defined on $X = X_1 \times \cdots \times X_n$, from the manifold

$$M = \left\{ \sum_{i=1}^{n} g_i(x_i) : g_i \in C(X_i),\ i = 1, \ldots, n \right\}.$$

The approximation error is defined as the distance from f to M:

$$E(f) \stackrel{def}{=} dist(f, M) = \inf_{g \in M} \|f - g\|_{C(X)} \cdot$$

The well-known duality relation says that

$$E(f) = \sup_{\substack{\mu \in M^{\perp} \\ \|\mu\| \leq 1}} \left| \int_X f d\mu \right|, \tag{3.32}$$

where $M^{\perp}$ is the space of regular Borel measures annihilating all functions in M and $\|\mu\|$ stands for the total variation of a measure μ. It should be noted that the sup in (3.32) is attained by some measure μ^* with total variation $\|\mu^*\| = 1$. We are interested in the problem: is it possible to replace in (3.32) the class $M^{\perp}$ by some subclass of it consisting of measures of simple structure? For the case $n = 2$, this problem was first considered by Diliberto and Straus [**47**]. They showed that the measures generated by closed bolts are sufficient for the equality (3.32).

In case of general topological spaces, a lightning bolt is defined similarly to the case $\mathbb{R}^2$. Let $X = X_1 \times X_2$ and π_i be the projections of X onto X_i, $i = 1, 2$. A lightning bolt (or, simply, a bolt) is a finite ordered set $\{a_1, \ldots, a_k\}$ contained in X, such that $a_i \neq a_{i+1}$, for $i = 1, 2, \ldots, k-1$, and either $\pi_1(a_1) = \pi_1(a_2)$, $\pi_2(a_2) = \pi_2(a_3)$, $\pi_1(a_3) = \pi_1(a_4), \ldots$, or $\pi_2(a_1) = \pi_2(a_2)$, $\pi_1(a_2) = \pi_1(a_3)$, $\pi_2(a_3) = \pi_2(a_4), \ldots$ A bolt $\{a_1, \ldots, a_k\}$ is said to be closed if k is an even number and the set $\{a_2, \ldots, a_k, a_1\}$ is also a bolt.

Let $l = \{a_1, \ldots, a_{2k}\}$ be a closed bolt. Consider a measure μ_l having atoms $\pm\frac{1}{2k}$ with alternating signs at the vertices of l. That is,

$$\mu_l = \frac{1}{2k} \sum_{i=1}^{2k} (-1)^{i-1} \delta_{a_i} \quad \text{or} \quad \mu_l = \frac{1}{2k} \sum_{i=1}^{2k} (-1)^{i} \delta_{a_i},$$

where δ_{a_i} is a point mass at a_i. It is clear that $\mu_l \in M^{\perp}$ and $\|\mu_l\| \leq 1$. $\|\mu_l\| = 1$ if and only if the set of vertices of the bolt l having even indices does not intersect with that having odd indices. The following duality relation was first established by Diliberto and Straus [**47**]

$$E(f) = \sup_{l \subset X} \left| \int_X f d\mu_l \right|, \tag{3.33}$$

where $X = X_1 \times X_2$ and the sup is taken over all closed bolts of X. In fact, Diliberto and Straus obtained the formula (3.33) for the case when X is a rectangle in $\mathbb{R}^2$ with sides parallel to the coordinate axis. The same result was independently proved by Smolyak (see [**166**]). Yet another proof of (3.33), in the case when X is a Cartesian product of two compact Hausdorff spaces, was given by Light and Cheney [**137**]. For X's other than a rectangle in $\mathbb{R}^2$, the theorem under some additional assumptions appeared in the works [**76**, **114**, **151**]. But we shall not discuss these works here.

Golomb's paper [**64**] made a start to a systematic study of approximation of multivariate functions by various compositions, including sums of univariate functions. Golomb generalized the notion of a closed bolt to the n-dimensional case and obtained the analogue of formula (3.33) for the error of approximation from the manifold M. The objects introduced in [**64**] were called *projection cycles* and they are defined as sets of the form

$$p = \{b_1, \ldots, b_k;\ c_1, \ldots, c_k\} \subset X, \tag{3.34}$$

with the property that $b_i \neq c_j$, $i, j = 1, \ldots, k$ and for all $\nu = 1, \ldots, n$, the group of the ν-th coordinates of $c_1, \ldots, c_k$ is a permutation of that of the ν-th coordinates of $b_1, \ldots, b_k$. Some points in the b-part $(b_1, \ldots, b_k)$ or c-part $(c_1, \ldots, c_k)$ of p may coincide. The measure associated with p is

$$\mu_p = \frac{1}{2k}\left(\sum_{i=1}^{k} \delta_{b_i} - \sum_{i=1}^{k} \delta_{c_i}\right).$$

It is clear that $\mu_p \in M^{\perp}$ and $\|\mu_p\| = 1$. Besides, if $n = 2$, then a projection cycle is the union of closed bolts after some suitable permutation of its points. Golomb's result states that

$$E(f) = \sup_{p \subset X} \left| \int_X f d\mu_p \right|, \tag{3.35}$$

where $X = X_1 \times \cdots \times X_n$ and the sup is taken over all projection cycles of X. It can be proved that in the case $n = 2$, the formulas (3.33) and (3.35) are equivalent. Unfortunately, the proof of (3.35) had a gap, which was pointed out many years later by Marshall and O'Farrell [**151**]. But the question if the formula (3.35) was correct, remained unsolved (see also the monograph by Khavinson [**117**]). Note that Golomb's result was used and cited in the literature, for example, in works [**113**, **184**].

In the following subsection, we will construct families of normalized measures (that is, measures with the total variation equal to 1) on projection cycles. Each measure μ_p defined above will be a member of some family. We will also consider minimal projection cycles and measures constructed on them. By properties of these measures, we show that Golomb's formula (3.35) is valid in a stronger form.

3.3.2. Measures supported on projection cycles.

Let us give an equivalent definition of a projection cycle. This will be useful in constructing of certain measures having simple structure and capability of approximating arbitrary measures in $M^{\perp}$.

In the sequel, χ_a will denote the characteristic function of a single point set $\{a\} \subset \mathbb{R}$.

DEFINITION 3.28. Let $X = X_1 \times \cdots \times X_n$ and π_i be the projections of X onto the sets X_i, $i = 1, \ldots, n$. We say that a set $p = \{x_1, \ldots, x_m\} \subset X$ is a projection cycle if there exists a vector $\lambda = (\lambda_1, \ldots, \lambda_m)$ with nonzero real coordinates such that

$$\sum_{j=1}^{m} \lambda_j \chi_{\pi_i(x_j)} = 0, \quad i = 1, \ldots, n. \tag{3.36}$$

Let us give some explanatory remarks concerning Definition 3.28. Fix the subscript i. Let the set $\{\pi_i(x_j),\ j = 1, \ldots, m\}$ have s_i different values, which we denote by $\gamma_1^i, \gamma_2^i, \ldots, \gamma_{s_i}^i$. Then (3.36) implies that

$$\sum_j \lambda_j = 0,$$

where the sum is taken over all j such that $\pi_i(x_j) = \gamma_k^i$, $k = 1, \ldots, s_i$. Thus for fixed i, we have s_i homogeneous linear equations in $\lambda_1, \ldots, \lambda_m$. The coefficients of these equations are the integers 0 and 1. By varying i, we obtain $s = \sum_{i=1}^{n} s_i$ such equations. Hence (3.36), in its expanded form, stands for the system of these equations. One can observe that if this system has a solution $(\lambda_1, \ldots, \lambda_m)$ with nonzero real components λ_i, then it also has a solution $(n_1, \ldots, n_m)$ with nonzero integer components n_i, $i = 1, \ldots, m$. This means that in Definition 3.28, we can replace the vector λ by the vector $n = (n_1, \ldots, n_m)$, where $n_i \in \mathbb{Z}\backslash\{0\}$, $i = 1, \ldots, m$. Thus, Definition 3.28 is equivalent to the following definition.

DEFINITION 3.29. A set $p = \{x_1, \ldots, x_m\} \subset X$ is called a projection cycle if there exist nonzero integers $n_1, \ldots, n_m$ such that

$$\sum_{j=1}^{m} n_j \chi_{\pi_i(x_j)} = 0, \quad i = 1, \ldots, n. \tag{3.37}$$

LEMMA 3.30. *Definition 3.29 is equivalent to Golomb's definition of a projection cycle.*

PROOF. Let $p = \{x_1, \ldots, x_m\}$ be a projection cycle with respect to Definition 3.29. By b and c denote the set of all points x_i such that the integers n_i associated with them in (3.37) are positive and negative correspondingly. Write out each point x_i n_i times if $n_i > 0$ and $-n_i$ times if $n_i < 0$. Then the set $\{b; c\}$ is a projection cycle with respect to Golomb's definition. The inverse is also true. Let a set $p_1 = \{b_1, \ldots, b_k;\ c_1, \ldots, c_k\}$ be a projection cycle with respect to Golomb's definition. Here, some points b_i or c_i may be repeated. Let $p = \{x_1, \ldots, x_m\}$ stand for the set p_1, but with no repetition of its points. Let n_i show how many times x_i appear in p_1. We take n_i positive if x_i appears in the b-part of p_1 and negative if it appears in the c-part of p_1. Clearly, the set $\{x_1, \ldots, x_m\}$ is a projection cycle with respect to Definition 3.29, since the integers n_i, $i = 1, \ldots, m$, satisfy (3.37). □

In the sequel, we will use Definition 3.28. A pair $\langle p, \lambda \rangle$, where p is a projection cycle in X and λ is a vector associated with p by (3.36), will be called a "projection cycle-vector pair" of X. To each such pair $\langle p, \lambda \rangle$ with $p = \{x_1, \ldots, x_m\}$ and $\lambda = (\lambda_1, \ldots, \lambda_m)$, we correspond the measure

$$\mu_{p,\lambda} = \frac{1}{\sum_{j=1}^{m} |\lambda_j|} \sum_{j=1}^{m} \lambda_j \delta_{x_j}. \tag{3.38}$$

Clearly, $\mu_{p,\lambda} \in M^{\perp}$ and $\|\mu_{p,\lambda}\| = 1$. We will also deal with measures supported on some certain subsets of projection cycles called *minimal projection cycles*. A projection cycle is said to be minimal if it does not contain any projection cycle as its proper subset. For example, the set $p=\{(0,0,0),\ (0,0,1),\ (0,1,0),\ (1,0,0),\ (1,1,1)\}$ is a minimal projection cycle in $\mathbb{R}^3$, since the vector $\lambda = (2,-1,-1,-1,1)$ satisfies Eq. (3.36) and there is no such vector for any other subset of p. Adding one point $(0,1,1)$ from the right to p, we will also have a projection cycle, but not minimal. Note that in this case, λ can be taken as $(3,-1,-1,-2,2,-1)$.

REMARK 3.31. A minimal projection cycle under the name of a *loop* was introduced and used in the works of Klopotowski, Nadkarni, Rao [**119, 120**].

To prove our main result we need some auxiliary facts.

LEMMA 3.32. (1) *The vector $\lambda = (\lambda_1, \dots, \lambda_m)$ associated with a minimal projection cycle $p = (x_1, \dots, x_m)$ is unique up to multiplication by a constant.*

(2) *If in (1), $\sum_{j=1}^m |\lambda_j| = 1$, then all the numbers λ_j, $j = 1, \dots, m$, are rational.*

PROOF. Let $\lambda^1 = (\lambda_1^1, \dots, \lambda_m^1)$ and $\lambda^2 = (\lambda_1^2, \dots, \lambda_m^2)$ be any two vectors associated with p. That is,

$$\sum_{j=1}^{m} \lambda_j^1 \chi_{\pi_i(x_j)} = 0 \text{ and } \sum_{j=1}^{m} \lambda_j^2 \chi_{\pi_i(x_j)} = 0, \quad i = 1, \dots, n.$$

After multiplying the second equality by $c = \frac{\lambda_1^1}{\lambda_1^2}$ and subtracting from the first, we obtain that

$$\sum_{j=2}^{m} (\lambda_j^1 - c\lambda_j^2) \chi_{\pi_i(x_j)} = 0, \quad i = 1, \dots, n.$$

Now since the cycle p is minimal, $\lambda_j^1 = c\lambda_j^2$, for all $j = 1, \dots, m$.

The second part of the lemma is a consequence of the first part. Indeed, let $n = (n_1, \dots, n_m)$ be a vector with the nonzero integer coordinates associated with p. Then the vector $\lambda' = (\lambda_1', \dots, \lambda_m')$, where $\lambda_j' = \frac{n_j}{\sum_{j=1}^m |n_j|}$, $j = 1, \dots, m$, is also associated with p. All coordinates of λ' are rational and therefore by the first part of the lemma, it is the unique vector satisfying $\sum_{j=1}^m \left|\lambda_j'\right| = 1$. □

By this lemma, a minimal projection cycle p uniquely (up to a sign) defines the measure

$$\mu_p = \sum_{j=1}^{m} \lambda_j \delta_{x_j}, \quad \sum_{j=1}^{m} |\lambda_j| = 1.$$

LEMMA 3.33. *Let μ be a normalized orthogonal measure on a projection cycle $l \subset X$. Then it is a convex combination of normalized orthogonal measures on minimal projection cycles of l. That is,*

$$\mu = \sum_{i=1}^{s} t_i \mu_{l_i}, \quad \sum_{i=1}^{s} t_i = 1, \ t_i > 0,$$

where l_i, $i = 1, \dots, s$, are minimal projection cycles in l.

This lemma follows from the result of Navada (see [**161**, Theorem 2]): Let $S \subset X_1 \times \cdots \times X_n$ be a finite set. Then any extreme point of the convex set of measures μ on S, $\mu \in M^\perp$, $\|\mu\| \le 1$, has its support on a minimal projection cycle contained in S.

REMARK 3.34. In the case $n = 2$, Lemma 3.33 was proved by Medvedev (see [**117**, p.77]).

LEMMA 3.35 (see p.73 in [**117**]). *Let $X = X_1 \times \cdots \times X_n$ and π_i be the projections of X onto the sets X_i, $i = 1, \dots, n$. In order that a measure $\mu \in C(X)^*$ be orthogonal to the subspace M, it is necessary and sufficient that*

$$\mu \circ \pi_i^{-1} = 0, \ i = 1, \dots, n.$$

LEMMA 3.36 (see p.75 in [**117**]). *Let $\mu \in M^\perp$ and $\|\mu\| = 1$. Then there exist a net of measures $\{\mu_\alpha\} \subset M^\perp$ weak* converging in $C(X)^*$ to μ and satisfying the following properties:*

1) $\|\mu_\alpha\| = 1$;
2) *The closed support of each μ_α is a finite set.*

The main result of this section is the following theorem.

THEOREM 3.37. *The error of approximation from the manifold M obeys the equality*

$$E(f) = \sup_{l \subset X} \left| \int_X f d\mu_l \right|,$$

where the sup *is taken over all minimal projection cycles of X.*

PROOF. Let $\widetilde{\mu}$ be a measure with finite support $\{x_1, \dots, x_m\}$ and orthogonal to the space M. Put $\lambda_j = \widetilde{\mu}(x_j)$, $j = 1, \dots m$. By Lemma 3.35, $\widetilde{\mu}(\pi_i^{-1}(\pi_i(x_j))) = 0$, for all $i = 1, \dots, n$, $j = 1, \dots, m$. Fix the indices i and j. Then we have the equation $\sum_k \lambda_k = 0$, where the sum is taken over all indices k such that $\pi_i(x_k) = \pi_i(x_j)$. Varying i and j, we obtain a system of such equations, which concisely can be written as

$$\sum_{k=1}^{m} \lambda_k \chi_{\pi_i(x_k)} = 0, \quad i = 1, \dots, n.$$

This means that the finite support of $\widetilde{\mu}$ forms a projection cycle. Therefore, a net of measures approximating the given measure μ in Lemma 3.36 are all of the form (3.38).

Let now $\mu_{p,\lambda}$ be any measure of the form (3.38). Since $\mu_{p,\lambda} \in M^\perp$ and $\|\mu_{p,\lambda}\| = 1$, we can write

$$\left| \int_X f d\mu_{p,\lambda} \right| = \left| \int_X (f - g) d\mu_{p,\lambda} \right| \le \|f - g\|, \tag{3.39}$$

where g is an arbitrary function in M. It follows from (3.39) that

$$\sup_{\langle p,\lambda\rangle}\left|\int_X f d\mu_{p,\lambda}\right| \leq E(f), \tag{3.40}$$

where the sup is taken over all projection cycle-vector pairs of X.

Consider the general duality relation (3.32). Let μ_0 be a measure attaining the supremum in (3.32) and $\{\mu_{p,\lambda}\}$ be a net of measures of the form (3.38) approximating μ_0 in the weak* topology of $C(X)^*$. We already know that this is possible. For any $\varepsilon > 0$, there exists a measure μ_{p_0,λ_0} in $\{\mu_{p,\lambda}\}$ such that

$$\left|\int_X f d\mu_0 - \int_X f d\mu_{p_0,\lambda_0}\right| < \varepsilon.$$

From the last inequality we obtain that

$$\left|\int_X f d\mu_{p_0,\lambda_0}\right| > \left|\int_X f d\mu_0\right| - \varepsilon = E(f) - \varepsilon.$$

Hence,

$$\sup_{\langle p,\lambda\rangle}\left|\int_X f d\mu_{p,\lambda}\right| \geq E(f). \tag{3.41}$$

From (3.40) and (3.41) it follows that

$$\sup_{\langle p,\lambda\rangle}\left|\int_X f d\mu_{p,\lambda}\right| = E(f). \tag{3.42}$$

By Lemma 3.33,

$$\mu_{p,\lambda} = \sum_{i=1}^{s} t_i \mu_{l_i},$$

where l_i, $i = 1, \ldots, s$, are minimal projection cycles in p and $\sum_{i=1}^{s} t_i = 1$, $t_i > 0$. Let k be an index in the set $\{1, \ldots, s\}$ such that

$$\left|\int_X f d\mu_{l_k}\right| = \max\left\{\left|\int_X f d\mu_{l_i}\right|, \ i = 1, \ldots, s\right\}.$$

Then

$$\left|\int_X f d\mu_{p,\lambda}\right| \leq \left|\int_X f d\mu_{l_k}\right|. \tag{3.43}$$

Now since

$$\left|\int_X f d\mu_l\right| \leq E(f),$$

for any minimal cycle l, from (3.42) and (3.43) we obtain the assertion of the theorem. □

REMARK 3.38. Theorem 3.37 not only proves Golomb's formula, but also improves it. Indeed, based on Lemma 3.30, one can easily observe that the formula (3.35) is equivalent to the formula

$$E(f) = \sup_{\langle p,\lambda\rangle} \left| \int_X f d\mu_{p,\lambda} \right|,$$

where the sup is taken over all projection cycle-vector pairs $\langle p, \lambda\rangle$ of X provided that all the numbers $\lambda_i / \sum_{j=1}^m |\lambda_j|$, $i = 1, \dots, m$, are rational. But by Lemma 3.32, minimal projection cycles enjoy this property.

CHAPTER 4

Generalized ridge functions and linear superpositions

A ridge function $g(\mathbf{a} \cdot \mathbf{x})$ with a direction $\mathbf{a} \in \mathbb{R}^d \backslash \{\mathbf{0}\}$ admits a natural generalization to a multivariate function of the form $g(\alpha_1(x_1) + \cdots + \alpha_d(x_d))$, where $\alpha_i(x_i)$, $i = \overline{1, d}$, are real, presumably well behaved, fixed univariate functions. We know from Chapter 1 that finitely many directions $\mathbf{a}^j$ are not enough for sums $\sum g_j(\mathbf{a}^j \cdot \mathbf{x})$ to approximate multivariate functions. However, we will see in this chapter that sums of the form $\sum g_j(\alpha_1^j(x_1) + \cdots + \alpha_d^j(x_d))$ with finitely many $\alpha_i^j(x_i)$ is capable not only approximating multivariate functions but also precisely representing them. First we study the problem of representation of a function $f : X \to \mathbb{R}$, where X is any set, as a linear superposition $\sum_j g_j(h_j(x))$ with arbitrary but fixed functions $h_j : X \to \mathbb{R}$. Then we apply the obtained result and the famous Kolmogorov superposition theorem to prove representability of an arbitrarily behaved multivariate function in the form of a sum of generalized ridge functions $\sum g_j(\alpha_1^j(x_1) + \cdots + \alpha_d^j(x_d))$. We also study the uniqueness of representation of functions by linear superpositions. Clearly, exact representation by a sum of generalized ridge functions is not always possible. In fact, the possibility of representation depends on the inner functions α_i^j. This gives rise to the problem of finding efficient approximation methods. In the final part of this chapter, we discuss a special algorithmic approximation method called the Diliberto-Straus algorithm. Under reasonable assumptions, we prove the convergence of this algorithm. In the simplest case of approximation by sums of two univariate functions, we see that Diliberto-Straus's original result holds for a large class of convex compact sets.

The material of this chapter is taken from [**9, 84, 94**].

4.1. Representation theorems

In this section, we study some problems of representation of real functions by linear superpositions and linear combinations of generalized ridge functions.

4.1.1. Problem statement and historical notes. Let X be any set and $h_i : X \to \mathbb{R}$, $i = 1, \ldots, r$, be arbitrarily fixed functions. Consider the set

(4.1)
$$\mathcal{B}(X) = \mathcal{B}(h_1, \ldots, h_r; X) = \left\{ \sum_{i=1}^{r} g_i(h_i(x)),\ x \in X,\ g_i : \mathbb{R} \to \mathbb{R},\ i = 1, \ldots, r \right\}$$

Members of this set will be called linear superpositions with respect to the functions $h_1, \ldots, h_r$ (see [**202**]). For a detailed study of linear superpositions and their approximation-theoretic properties we refer the reader to the monograph by

Khavinson [**117**]. Note that sums of generalized ridge functions $\sum g_j(\alpha_1^j(x_1) + \cdots + \alpha_d^j(x_d))$ with fixed $\alpha_i^j(x_i)$ are a special case of linear superpositions. In Section 1.2, we considered linear superpositions defined on a subset of the d-dimensional Euclidean space, while here X is a set of arbitrary nature. As in Section 1.2, we are interested in the question: what conditions on X guarantee that each function on X will be in the set $\mathcal{B}(X)$? The simplest case $X \subset \mathbb{R}^d$, $r = d$ and h_i are the coordinate functions was solved in [**119**]. See also [**117**, p.57] for the case $r = 2$.

By $\mathcal{B}_c(X)$ and $\mathcal{B}_b(X)$ denote the right hand side of (4.1) with continuous and bounded $g_i : \mathbb{R} \to \mathbb{R}$, $i = 1, \dots, r$, respectively. Our starting point is the well-known superposition theorem of Kolmogorov [**122**]. It states that for the unit cube $\mathbb{I}^d$, $\mathbb{I} = [0, 1]$, $d \geq 2$, there exists $2d + 1$ functions $\{s_q\}_{q=1}^{2d+1} \subset C(\mathbb{I}^d)$ of the form

$$s_q(x_1, \dots, x_d) = \sum_{p=1}^{d} \varphi_{pq}(x_p),\ \varphi_{pq} \in C(\mathbb{I}),\ p = 1, \dots, d,\ q = 1, \dots, 2d + 1 \tag{4.2}$$

such that each function $f \in C(\mathbb{I}^d)$ admits the representation

$$f(x) = \sum_{q=1}^{2d+1} g_q(s_q(x)),\ x = (x_1, \dots, x_d) \in \mathbb{I}^d,\ g_q \in C(\mathbb{R}). \tag{4.3}$$

Note that the functions $g_q(s_q(x))$, involved in the right hand side of (4.3), are generalized ridge functions. In our notation, (4.3) means that $\mathcal{B}_c(s_1, \dots, s_{2d+1}; \mathbb{I}^d) = C(\mathbb{I}^d)$. This surprising and deep result, which solved (negatively) Hilbert's 13-th problem, was improved and generalized in several directions. It was first observed by Lorentz [**142**] that the functions g_q can be replaced by a single continuous function g. Sprecher [**185**] showed that the theorem can be proven with constant multiples of a single function φ and translations. Specifically, φ_{pq} in (4.2) can be chosen as $\lambda^p \varphi(x_p + \varepsilon q)$, where ε and λ are some positive constants. Fridman [**54**] succeeded in showing that the functions φ_{pq} can be constructed to belong to the class $Lip(1)$. Vitushkin and Henkin [**202**] showed that φ_{pq} cannot be taken to be continuously differentiable.

Ostrand [**168**] extended the Kolmogorov theorem to general compact metric spaces. In particular, he proved that for each compact d-dimensional metric space X there exist continuous real functions $\{\alpha_i\}_{i=1}^{2d+1} \subset C(X)$ such that $\mathcal{B}_c(\alpha_1, \dots, \alpha_{2d+1}; X) = C(X)$. Sternfeld [**192**] showed that the number $2d + 1$ cannot be reduced for any d-dimensional space X. Thus the number of terms in the Kolmogorov superposition theorem is the best possible.

Some papers of Sternfeld were devoted to the representation of continuous and bounded functions by linear superpositions. Let $C(X)$ and $B(X)$ denote the space of continuous and bounded functions on some set X respectively (in the first case, X is supposed to be a compact metric space). Let $F = \{h\}$ be a family of functions on X. F is called a uniformly separating family (*u.s.f.*) if there exists a number $0 < \lambda \leq 1$ such that for each pair $\{x_j\}_{j=1}^m$, $\{z_j\}_{j=1}^m$ of disjoint finite sequences in X, there exists some $h \in F$ so that if from the two sequences $\{h(x_j)\}_{j=1}^m$and $\{h(z_j)\}_{j=1}^m$ in $h(X)$ we remove a maximal number of pairs of points $h(x_{j_1})$ and $h(z_{j_2})$ with $h(x_{j_1}) = h(z_{j_2})$, there remains at least λm points in each sequence (or , equivalently, at most $(1 - \lambda)m$ pairs can be removed). Sternfeld [**190**] proved that for a finite family $F = \{h_1, \dots, h_r\}$ of functions on X, being a *u.s.f.* is equivalent to the equality $\mathcal{B}_b(h_1, \dots, h_r; X) = B(X)$, and that in the case where X is a compact metric space and the elements of F are continuous functions on X, the equality $\mathcal{B}_c(h_1,\dots,h_r;X)=C(X)$ implies that F is a *u.s.f.* Thus, in particular, Sternfeld obtained that the formula (4.3) is valid for all bounded functions, where g_q are bounded functions depending on f (see also [**117**, p.21]).

Let X be a compact metric space. The family $F=\{h\}\subset C(X)$ is said to be a measure separating family (*m.s.f.*) if there exists a number $0<\lambda\le 1$ such that for any measure μ in $C(X)^*$, the inequality $\left\|\mu\circ h^{-1}\right\|\ge\lambda\left\|\mu\right\|$ holds for some $h\in F$. Sternfeld [**193**] proved that $\mathcal{B}_c(h_1,\dots,h_r;X)=C(X)$ if and only if the family $\{h_1,\dots,h_r\}$ is a *m.s.f.* In [**190**], it was also shown that if $r=2$, then the properties *u.s.f.* and *m.s.f.* are equivalent. Therefore, the equality $\mathcal{B}_b(h_1,h_2;X)=B(X)$ is equivalent to $\mathcal{B}_c(h_1,h_2;X)=C(X)$. But for $r>2$, these two properties are no longer equivalent. That is, $\mathcal{B}_b(h_1,\dots,h_r;X)=B(X)$ does not always imply $\mathcal{B}_c(h_1,\dots,h_r;X)=C(X)$ (see [**193**]).

Our purpose is to consider the above mentioned problem of representation by linear superpositions without involving any topology (that of continuity or boundedness). We start with characterization of those sets X for which $\mathcal{B}(h_1,\dots,h_r;X)=T(X)$, where $T(X)$ is the space of all functions on X. As in Section 1.2, this will be done in terms of cycles. We claim that nonexistence of cycles in X is equivalent to the equality $\mathcal{B}(X)=T(X)$ for an arbitrary set X. In particular, we show that $\mathcal{B}_c(X)=C(X)$ always implies $\mathcal{B}(X)=T(X)$. This implication will enable us to obtain some new results, namely extensions of the previously known theorems from continuous to discontinuous multivariate functions. For example, we will prove that the formula (4.3) is valid for all discontinuous multivariate functions f defined on the unite cube $\mathbb{I}^d$, where g_q are univariate functions depending on f.

4.1.2. Extension of Kolmogorov's superposition theorem. In this subsection, we show that if some representation by linear superpositions holds for continuous functions, then it holds for all functions. This will lead us to natural extensions of some known superposition theorems (such as Kolmogorov's superposition theorem, Ostrand's superposition theorem, etc) from continuous to discontinuous functions.

In the sequel, by χ_A we will denote the characteristic function of a set $A\subset\mathbb{R}$. That is,

$$\chi_A(y)=\begin{cases}1, & if\ y\in A\\ 0, & if\ y\notin A.\end{cases}$$

The following definition is a generalized version of Definition 1.9 from Section 1.2, where in connection with ridge functions only subsets of $\mathbb{R}^d$ were considered.

DEFINITION 4.1. Given an arbitrary set X and functions $h_i:X\to\mathbb{R}$, $i=1,\dots,r$. A set of points $\{x_1,\dots,x_n\}\subset X$ is called a cycle with respect to the functions $h_1,\dots,h_r$ (or, concisely, a cycle if there is no confusion), if there exists a vector $\lambda=(\lambda_1,\dots,\lambda_n)$ with the nonzero real coordinates λ_i, $i=1,\dots,n$, such that

$$\sum_{j=1}^{n}\lambda_j\chi_{h_i(x_j)}=0,\ i=1,\dots,r. \tag{4.4}$$

A cycle $p=\{x_1,\dots,x_n\}$ is said to be minimal if p does not contain any cycle as its proper subset.

Note that in this definition the vector $\lambda = (\lambda_1, \dots, \lambda_n)$ can be chosen so that it has only integer components. Indeed, let for $i = 1, \dots, r$, the set $\{h_i(x_j),\ j = 1, \dots, n\}$ have k_i different values. Then it is not difficult to see that Eq. (4.4) stands for a system of $\sum_{i=1}^{r} k_i$ homogeneous linear equations in unknowns $\lambda_1, \dots, \lambda_n$. This system can be written in the matrix form $(\lambda_1, \dots, \lambda_n) \times C = 0$, where C is an n by $\sum_{i=1}^{r} k_i$ matrix. The basic property of this matrix is that all of its entries are 0's and 1's and no row or column of C is identically zero. Since Eq. (4.4) has a nontrivial solution $(\lambda_1^{'}, \dots, \lambda_n^{'}) \in \mathbb{R}^n$ and all entries of C are integers, by applying the Gauss elimination method we can see that there always exists a nontrivial solution $(\lambda_1, \dots, \lambda_n)$ with the integer components λ_i, $i = 1, \dots, n$.

For a number of simple examples, see Section 1.2.

Let $T(X)$ denote the set of all functions on X. With each pair $\langle p, \lambda \rangle$, where $p = \{x_1, \dots, x_n\}$ is a cycle in X and $\lambda = (\lambda_1, \dots, \lambda_n)$ is a vector known from Definition 4.1, we associate the functional

$$G_{p,\lambda} : T(X) \to \mathbb{R},\ G_{p,\lambda}(f) = \sum_{j=1}^{n} \lambda_j f(x_j).$$

In the following, such pairs $\langle p, \lambda \rangle$ will be called *cycle-vector pairs* of X. It is clear that the functional $G_{p,\lambda}$ is linear. Besides, $G_{p,\lambda}(g) = 0$ for all functions $g \in \mathcal{B}(h_1, \dots, h_r; X)$. Indeed, assume that (4.4) holds. Given $i \leq r$, let $z = h_i(x_j)$ for some j. Hence, $\sum_{j\ (h_i(x_j)=z)} \lambda_j = 0$ and $\sum_{j\ (h_i(x_j)=z)} \lambda_j g_i(h_i(x_j)) = 0$. A summation yields $G_{p,\lambda}(g_i \circ h_i) = 0$. Since $G_{p,\lambda}$ is linear, we obtain that $G_{p,\lambda}(\sum_{i=1}^{r} g_i \circ h_i) = 0$.

A minimal cycle $p = \{x_1, \dots, x_n\}$ has the following obvious properties:

(a) *The vector λ associated with p by Eq.* (4.4) *is unique up to multiplication by a constant;*

(b) *If in* (4.4), $\sum_{j=1}^{n} |\lambda_j| = 1$, *then all the numbers λ_j, $j = 1, \dots, n$, are rational.*

Thus, a minimal cycle p uniquely (up to a sign) defines the functional

$$G_p(f) = \sum_{j=1}^{n} \lambda_j f(x_j), \quad \sum_{j=1}^{n} |\lambda_j| = 1.$$

PROPOSITION 4.2. 1) *Let X have cycles. A function $f : X \to \mathbb{R}$ belongs to the space $\mathcal{B}(h_1, \dots, h_r; X)$ if and only if $G_p(f) = 0$ for any minimal cycle $p \subset X$ with respect to the functions $h_1, \dots, h_r$.*

2) *Let X has no cycles. Then $\mathcal{B}(h_1, \dots, h_r; X) = T(X)$.*

PROPOSITION 4.3. $\mathcal{B}(h_1, \dots, h_r; X) = T(X)$ *if and only if X has no cycles.*

These propositions are proved by the same way as Theorems 1.13 and 1.14. We use these propositions to obtain our main result (see Theorem 4.5 below).

The condition whether X have cycles or not, depends both on X and the functions $h_1, \dots, h_r$. In the following, we see that if $h_1, \dots, h_r$ are "nice" functions (smooth functions with the simple structure. For example, ridge functions) and $X \subset \mathbb{R}^d$ is a "rich" set (for example, the set with interior points), then X has always cycles. Thus the representability by linear combinations of univariate functions with the fixed "nice" multivariate functions requires at least that X should not possess interior points. The picture is quite different when the functions $h_1, \dots, h_r$ are not "nice". Even in the case when they are continuous, we will see that many sets of $\mathbb{R}^d$ (the unite cube, any compact subset of that, or even the whole space $\mathbb{R}^d$ itself) may have no cycles. If disregard the continuity, there exists even one function h such that every multivariate function is representable as $g \circ h$ over any subset of $\mathbb{R}^d$. First, let us introduce the following definition.

DEFINITION 4.4. Let X be a set and $h_i : X \to \mathbb{R}$, $i = 1, \dots, r$, be arbitrarily fixed functions. A class $A(X)$ of functions on X will be called a "permissible function class" if for any minimal cycle $p \subset X$ with respect to the functions $h_1, \dots, h_r$ (if it exists), there is a function f_0 in $A(X)$ such that $G_p(f_0) \neq 0$.

Clearly, $C(X)$ and $B(X)$ are both permissible function classes (in case of $C(X)$, X is considered to be a normal topological space).

THEOREM 4.5. *Assume $A(X)$ is a permissible function class. If $A(X) \subset \mathcal{B}(h_1, \dots, h_r; X)$, then $\mathcal{B}(h_1, \dots, h_r; X) = T(X)$.*

The proof is simple and based on Propositions 4.2 and 4.3. Assume for a moment that X admits a cycle p. By Proposition 4.2, the functional G_p annihilates all members of the set $B(h_1, \dots, h_r; X)$. By Definition 4.4 of permissible function classes, $A(X)$ contains a function f_0 such that $G_p(f_0) \neq 0$. Therefore, $f_0 \notin B(h_1, \dots, h_r; X)$. We see that the embedding $A(X) \subset B(h_1, \dots, h_r; X)$ is impossible if X has a cycle. Thus X has no cycles. Then by Proposition 4.3, $\mathcal{B}(h_1, \dots, h_r; X) = T(X)$.

In the "if part" of Theorem 4.5, instead of $\mathcal{B}(h_1, \dots, h_r; X)$ and $A(X)$ one can take $\mathcal{B}_c(h_1, \dots, h_r; X)$ and $C(X)$ (or $\mathcal{B}_b(h_1, \dots, h_r; X)$ and $B(X)$) respectively. That is, the following corollaries are valid.

COROLLARY 4.6. *Let X be a set and $h_i : X \to \mathbb{R}$, $i = 1, \dots, r$, be arbitrarily fixed bounded functions. If $\mathcal{B}_b(h_1, \dots, h_r; X) = B(X)$, then $\mathcal{B}(h_1, \dots, h_r; X) = T(X)$.*

COROLLARY 4.7. *Let X be a normal topological space and $h_i : X \to \mathbb{R}$, $i = 1, \dots, r$, be arbitrarily fixed continuous functions. If $\mathcal{B}_c(h_1, \dots, h_r; X) = C(X)$, then $\mathcal{B}(h_1, \dots, h_r; X) = T(X)$.*

The main advantage of Theorem 4.5 is that we need not check directly if the set X has no cycles, which in many cases may turn out to be very tedious task. Using this theorem, we can extend free-of-charge the existing superposition theorems from the classes $B(X)$ or $C(X)$ (or some other permissible function classes) to all functions defined on X. For example, this theorem allows us to extend the Kolmogorov superposition theorem from continuous to all multivariate functions.

THEOREM 4.8. *Let $d \geq 2$, $\mathbb{I} = [-1; 1]$, and φ_{pq}, $p = 1, \dots, d$, $q = 1, \dots, 2d+1$, be the universal continuous functions in (4.2). Then each multivariate function $f : \mathbb{I}^d \to \mathbb{R}$ can be represented in the form*

$$f(x) = \sum_{q=1}^{2d+1} g_q\left(\sum_{p=1}^{d} \varphi_{pq}(x_p)\right), \quad x = (x_1, \dots, x_d) \in \mathbb{I}^d.$$

where g_q are univariate functions depending on f.

It should be remarked that Sternfeld [**190**], in particular, obtained that the formula (4.3) is valid for functions $f \in B(\mathbb{I}^d)$ provided that g_q are bounded functions depending on f (see [**117**, Chapter 1] for more detailed information and interesting discussions).

Let X be a compact metric space and $h_i \in C(X)$, $i = 1, \ldots, r$. The result of Sternfeld about the relation between representations of continuous and bounded functions and Corollary 4.6 give us the implications

$$\mathcal{B}_c(h_1, \ldots, h_r; X) = C(X) \Rightarrow \mathcal{B}_b(h_1, \ldots, h_r; X) = B(X) \Rightarrow \mathcal{B}(h_1, \ldots, h_r; X) = T(X).$$

The first implication is invertible when $r = 2$ (see [**190**]). We want to show that the second is not invertible even in the case $r = 2$. The following interesting example is due to Khavinson [**117**, p.67].

Let $X \subset \mathbb{R}^2$ consist of a broken line whose sides are parallel to the coordinate axis and whose vertices are

$$(0;0), (1;0), (1;1), (1+\frac{1}{2^2};1), (1+\frac{1}{2^2};1+\frac{1}{2^2}), (1+\frac{1}{2^2}+\frac{1}{3^2};1+\frac{1}{2^2}), \ldots$$

We add to this line the limit point of the vertices $(\frac{\pi^2}{6}, \frac{\pi^2}{6})$. Let $r = 2$ and h_1, h_2 be the coordinate functions. Then the set X has no cycles with respect to h_1 and h_2. By Proposition 4.2, every function f on X is of the form $g_1(x_1) + g_2(x_2)$, $(x_1, x_2) \in X$. Now construct a function f_0 on X as follows. On the link joining $(0;0)$ to $(1;0)$ $f_0(x_1, x_2)$ continuously increases from 0 to 1; on the link from $(1;0)$ to $(1;1)$ it continuously decreases from 1 to 0; on the link from $(1;1)$ to $(1+\frac{1}{2^2};1)$ it increases from 0 to $\frac{1}{2}$; on the link from $(1+\frac{1}{2^2};1)$ to $(1+\frac{1}{2^2};1+\frac{1}{2^2})$ it decreases from $\frac{1}{2}$ to 0; on the next link it increases from 0 to $\frac{1}{3}$, etc. At the point $(\frac{\pi^2}{6}, \frac{\pi^2}{6})$ set the value of f_0 equal to 0. Obviously, f_0 is a continuous functions and by the above argument, $f_0(x_1, x_2) = g_1(x_1) + g_2(x_2)$. But g_1 and g_2 cannot be chosen as continuous functions, since they get unbounded as x_1 and x_2 tends to $\frac{\pi^2}{6}$. Thus, $\mathcal{B}(h_1, h_2; X) = T(X)$, but at the same time $\mathcal{B}_c(h_1, h_2; X) \neq C(X)$ (or, equivalently, $\mathcal{B}_b(h_1, h_2; X) \neq B(X)$).

4.1.3. Some other superposition theorems. We have seen in the previous subsection that the unit cube in $\mathbb{R}^d$ has no cycles with respect to some $2d+1$ continuous functions (namely, the Kolmogorov functions s_q (4.2)). From the result of Ostrand [**168**] (see the first subsection of this section) and Corollary 4.7 it follows that compact sets X of finite dimension also lack cycles with respect to a certain family of finitely many continuous functions on X. Namely, the following generalization of Ostrand's theorem is valid.

THEOREM 4.9. *For $p = 1, 2, \ldots, m$ let X_p be a compact metric space of finite dimension d_p and let $n = \sum_{p=1}^m d_p$. There exist continuous functions $\alpha_{pq} : X_p \to [0,1]$, $p = 1, \ldots, m$, $q = 1, \ldots, 2n+1$, such that every real function f defined on $\Pi_{p=1}^m X_p$ is representable in the form*

$$f(x_1, \ldots, x_m) = \sum_{q=1}^{2n+1} g_q(\sum_{p=1}^{m} \alpha_{pq}(x_p)). \tag{4.5}$$

where g_q are real functions depending on f. If f is continuous, then the functions g_q can be chosen continuous.

Note that Ostrand proved "if f is continuous..." part of Theorem 4.9, while we prove the validity of (4.5) for discontinuous f.

One may ask if there exists a finite family of functions $\{h_i : \mathbb{R}^d \to \mathbb{R}\}_{i=1}^n$ such that any subset of $\mathbb{R}^d$ does not admit cycles with respect to this family? The answer is positive. This follows from the result of Demko [**43**]: there exist $2d+1$ continuous functions $\varphi_1, \dots, \varphi_{2d+1}$ defined on $\mathbb{R}^d$ such that every bounded continuous function on $\mathbb{R}^d$ is expressible in the form $\sum_{i=1}^{2d+1} g \circ \varphi_i$ for some $g \in C(\mathbb{R})$. This theorem together with Corollary 4.6 yield that every function on $\mathbb{R}^d$ is expressible in the form $\sum_{i=1}^{2d+1} g_i \circ \varphi_i$ for some $g_i : \mathbb{R} \to \mathbb{R}$, $i = 1, \dots, 2d+1$. We do not yet know if g_i here can be replaced by a single univariate function. We also don't know if the number $2d+1$ can be reduced so that the whole space of $\mathbb{R}^d$ (or any d-dimensional compact subset of that, or at least the unit cube $\mathbb{I}^d$) has no cycles with respect to some continuous functions $\varphi_1, \dots, \varphi_k : \mathbb{R}^d \to \mathbb{R}$, where $k < 2d+1$. One of the basic results of Sternfeld [**192**] says that the dimension of a compact metric space X equals d if and only if there exist functions $\varphi_1, \dots, \varphi_{2d+1} \in C(X)$ such that $\mathcal{B}_c(\varphi_1, \dots, \varphi_{2d+1}; X) = C(X)$ and for any fmily $\{\psi_i\}_{i=1}^k \subset C(X)$, $k < 2d+1$, we have $\mathcal{B}_c(\psi_1, \dots, \psi_k; X) \neq C(X)$. In particular, from this result it follows that the number of terms in the Kolmogorov superposition theorem cannot be reduced. But since the equalities $\mathcal{B}_c(X) = C(X)$ and $\mathcal{B}(X) = T(X)$ are not equivalent, the above question on the nonexistence of cycles in $\mathbb{R}^d$ with respect to less than $2d+1$ continuous functions is far from trivial.

If disregard the continuity, one can construct even one function $\varphi : \mathbb{R}^d \to \mathbb{R}$ such that the whole space $\mathbb{R}^d$ will not possess cycles with respect to φ and therefore, every function $f : \mathbb{R}^d \to \mathbb{R}$ will admit the representation $f = g \circ \varphi$ with some univariate g depending on f. Our argument easily follows from Corollary 4.7 and the result of Sprecher [**183**]: for any natural number d, $d \geq 2$, there exist functions $h_p : \mathbb{I} \to \mathbb{R}$, $p = 1, \dots, d$, such that every function $f \in C(\mathbb{I}^d)$ can be represented in the form

$$f(x_1, \dots, x_d) = g\left(\sum_{p=1}^{d} h_p(x_p)\right), \tag{4.6}$$

where g is a univariate (generally discontinuous) function depending on f.

Note that the function involved in the right hand side of (4.6) is a generalized ridge function. Thus, the result of Sprecher together with our result means that every multivariate function f is representable as a generalized ridge function $g(\cdot)$ and if f is continuous, then g can be chosen continuous as well.

REMARK 4.10. Concerning ordinary ridge functions $g(\mathbf{a} \cdot \mathbf{x})$, representation of every multivariate function by linear combinations of such functions may not be possible over many sets in $\mathbb{R}^d$. For example, this is not possible for sets having interior points. More precisely, assume we are given finitely many nonzero directions $\mathbf{a}^1, \dots, \mathbf{a}^r$ in $\mathbb{R}^d$. Then $\mathcal{R}(\mathbf{a}^1, \dots, \mathbf{a}^r; X) \neq T(X)$ for any set $X \subset \mathbb{R}^d$ with a nonempty interior. Indeed, let $\mathbf{y}$ be a point in the interior of X. Consider vectors $\mathbf{b}^i$, $i = 1, \dots, r$, with sufficiently small coordinates such that $\mathbf{a}^i \cdot \mathbf{b}^i = 0$, $i = 1, \dots, r$. Note that the vectors $\mathbf{b}^i$, $i = 1, \dots, r$, can be chosen pairwise linearly independent. With each vector $\varepsilon = (\varepsilon_1, \dots, \varepsilon_r)$, $\varepsilon_i \in \{0, 1\}$, $i = 1, \dots, r$, we associate the point

$$\mathbf{x}_\varepsilon = \mathbf{y} + \sum_{i=1}^{r} \varepsilon_i \mathbf{b}^i.$$

Since the coordinates of $\mathbf{b}^i$ are sufficiently small, we may assume that all the points $\mathbf{x}_\varepsilon$ are in the interior of X. We correspond each point $\mathbf{x}_\varepsilon$ to the number $(-1)^{|\varepsilon|}$, where $|\varepsilon| = \varepsilon_1 + \cdots + \varepsilon_r$. One may easily verify that the pair $\left\langle \{\mathbf{x}_\varepsilon\}, \{(-1)^{|\varepsilon|}\} \right\rangle$ is a cycle-vector pair of X. Therefore, by Proposition 4.3, $\mathcal{R}\left(\mathbf{a}^1, \ldots, \mathbf{a}^r; X\right) \neq T(X)$.

Note that the above method of construction of the set $\{\mathbf{x}_\varepsilon\}$ is due to Lin and Pinkus [**140**].

REMARK 4.11. A different generalization of ridge functions was considered in Lin and Pinkus [**140**]. This generalization involves multivariate functions of the form $g(A\mathbf{x})$, where $\mathbf{x} \in \mathbb{R}^d$ is the variable, A is a fixed $d \times n$ matrix, $1 \leq n < d$, and g is a real-valued function defined on $\mathbb{R}^n$. For $n = 1$, this reduces to a ridge function.

4.2. Uniqueness theorems

Let Q be a set such that every function on Q can be represented by linear superpositions. This representation is generally not unique. But for some sets it may be unique provided that initial values of the representing functions are prescribed at some point of Q. In this section, we are going to study properties of such sets. All the obtained results are valid, in particular, for linear combinations of generalized ridge functions.

4.2.1. Formulation of the problem. Assume X is an arbitrary set, $h_i : X \to \mathbb{R}$, $i = 1, \ldots, r$, are fixed functions and $\mathcal{B}(X)$ is the set defined in (4.1). Let $T(X)$ denote the set of all real functions on X. Obviously, $\mathcal{B}(X)$ is a linear subspace of $T(X)$. For a set $Q \subset X$, let $T(Q)$ and $\mathcal{B}(Q)$ denote the restrictions of $T(X)$ and $\mathcal{B}(X)$ to Q, respectively. Sets Q with the property $\mathcal{B}(Q) = T(Q)$ will be called *representation sets*. Recall that Proposition 4.3 gives a complete characterization of such sets. For a representation set Q, we will also use the notation $Q \in RS$. Here, RS stands for the set of all representation sets in X.

Let $Q \in RS$. Clearly for a function f defined on Q the representation

$$f(x) = \sum_{i=1}^{r} g_i(h_i(x)), \ x \in Q \tag{4.7}$$

is not unique. We are interested in the uniqueness of such representation under some reasonable restrictions on the functions $g_i \circ h_i$. These restrictions may be various, but in this section, we require that the values of $g_i \circ h_i$ are prescribed at some point $x_0 \in Q$. That is, we require that

$$g_i(h_i(x_0)) = a_i, \ i = 1, \ldots, r-1, \tag{4.8}$$

where a_i are arbitrarily fixed real numbers. Is representation (4.7) subject to initial conditions (4.8) always unique? Obviously, not. We are going to identify those representation sets Q for which representation (4.7) subject to conditions (4.8) is unique for all functions $f : Q \to \mathbb{R}$. In the sequel, such sets Q will be called *unicity sets*.

4.2.2. Complete representation sets. From Proposition 4.3 it is easy to obtain the following set-theoretic properties of representation sets:

(1) $Q \in RS \Longleftrightarrow A \in RS$ for every finite set $A \subset Q$;
(2) The union of any linearly ordered (under inclusion) system of representation sets is also a representation set
(3) For any representation set Q there is a maximal representation set, that is, a set $M \in RS$ such that $Q \subset M$ and for any $P \supset M$, $P \in RS$ we have $P = M$.
(4) If $M \subset X$ is a maximal representation set, then $h_i(M) = h_i(X)$, $i = 1, \ldots, r$.

Properties (1) and (2) are obvious, since any cycle is a finite set. The (3)-rd property follows from (2) and Zorn's lemma. To prove property (4) note that if $x_0 \in X$ and $h_i(x_0) \notin h_i(M)$ for some i, one can construct the representation set $M \cup \{x_0\}$, which is bigger than M. But this is impossible, since M is maximal.

DEFINITION 4.12. A set $Q \subset X$ is called a complete representation set if Q itself is a representation set and there is no other representation set P such that $Q \subset P$ and $h_i(P) = h_i(Q)$, $i = 1, \ldots, r$.

The set of all complete representation sets of X will be denoted by CRS. Obviously, every representation set is contained in a complete representation set. That is, if $A \in RS$, then there exists $B \in CRS$ such that $h_i(B) = h_i(A)$, $i = 1, \ldots, r$. It turns out that for the functions $h_1, \ldots, h_r$, complete representation sets entirely characterize unicity sets. To prove this fact we need some auxiliary lemmas.

LEMMA 4.13. *Let $Q \subset X$ be a representation set and for some point $x_0 \in Q$ the zero function representation*

$$0 = \sum_{i=1}^{r} g_i(h_i(x)), \quad x \in Q,$$

is unique, provided that $g_i(h_i(x_0)) = 0$, $i = 1, \ldots, r-1$. That is, all the functions $g_i \equiv 0$ on the sets $h_i(Q)$, $i = 1, \ldots, r$. Then $Q \in CRS$.

PROOF. Assume that $Q \notin CRS$. Then there exists a point $p \in X$ such that $p \notin Q$, $h_i(p) \in h_i(Q)$, for all $i = 1, \ldots, r$, and $Q' = Q \cup \{p\}$ is also a representation set. Consider a function $f_0 : Q' \to \mathbb{R}$ such that $f_0(q) = 0$, for any $q \in Q$ and $f_0(p) = 1$. Since $Q' \in RS$,

$$f_0(x) = \sum_{i=1}^{r} s_i(h_i(x)), \quad x \in Q'.$$

Then

$$f_0(x) = \sum_{i=1}^{r} g_i(h_i(x)), \quad x \in Q', \tag{4.9}$$

where

$$g_i(h_i(x)) = s_i(h_i(x)) - s_i(h_i(x_0)), \ i = 1, \ldots, r-1$$

and

$$g_r(h_r(x)) = s_r(h_r(x)) + \sum_{i=1}^{r-1} s_i(h_i(x_0)).$$

A restriction of representation (4.9) to the set Q gives the equality

$$\sum_{i=1}^{r} g_i(h_i(x)) = 0, \text{ for all } x \in Q. \tag{4.10}$$

Note that $g_i(h_i(x_0)) = 0$, $i = 1, \ldots, r-1$. It follows from the hypothesis of the lemma that representation (4.10) is unique. Hence, $g_i(h_i(x)) = 0$, for all $x \in Q$ and $i = 1, \ldots, r$. But from (4.9) it follows that

$$\sum_{i=1}^{r} g_i(h_i(p)) = f_0(p) = 1.$$

Since $h_i(p) \in h_i(Q)$ for all $i = 1, \ldots, r$, the above relation contradicts that the functions g_i are identically zero on the sets $h_i(Q)$, $i = 1, \ldots, r$. This means that our assumption is not true and $Q \in CRS$. □

The following lemma is a strengthened version of Lemma 4.13.

LEMMA 4.14. *Let $Q \in RS$ and for some point $x_0 \in Q$, numbers $c_1, c_2, \ldots, c_{r-1} \in \mathbb{R}$ and a function $v \in T(Q)$ the representation*

$$v(x) = \sum_{i=1}^{r} v_i(h_i(x))$$

is unique under the initial conditions $v_i(h_i(x_0)) = c_i$, $i = 1, \ldots, r-1$. Then for any numbers $b_1, b_2 \ldots, b_{r-1} \in \mathbb{R}$ and an arbitrary function $f \in T(Q)$ the representation

$$f(x) = \sum_{i=1}^{r} f_i(h_i(x))$$

is also unique, provided that $f_i(h_i(x_0)) = b_i$, $i = 1, \ldots, r-1$. Besides, $Q \in CRS$.

PROOF. Assume the contrary. Assume that there is a function $f \in T(Q)$ having two different representations subject to the same initial conditions. That is,

$$f(x) = \sum_{i=1}^{r} f_i(h_i(x)) = \sum_{i=1}^{r} f_i^{'}(h_i(x))$$

with $f_i(h_i(x_0)) = f_i^{'}(h_i(x_0)) = b_i$, $i = 1, \ldots, r-1$ and $f_i \neq f_i^{'}$ for some indice $i \in \{1, \ldots, r\}$. In this case, the function $v(x)$ will possess the following two different representations

$$v(x) = \sum_{i=1}^{r} v_i(h_i(x))$$

and

$$v(x) = \sum_{i=1}^{r} \left[v_i(h_i(x)) + f_i(h_i(x)) - f_i^{'}(h_i(x)) \right]$$

both satisfying the initial conditions. The obtained contradiction and above Lemma 4.13 complete the proof. □

In the sequel, we will assume that for any points $t_i \in h_i(X)$, $i = 1, \ldots, r$, the system of equations $h_i(x) = t_i$, $i = 1, \ldots, r$, has at least one solution.

LEMMA 4.15. *Let $Q \in CRS$. Then for any point $x_0 \in Q$ the representation*

$$0 = \sum_{i=1}^{r} g_i(h_i(x)), \; x \in Q, \tag{4.11}$$

subject to the conditions

$$g_i(h_i(x_0)) = 0, \; i = 1, \ldots, r-1, \tag{4.12}$$

is unique. That is, $g_i \equiv 0$ on the sets $h_i(Q)$, $i = 1, \ldots, r$.

PROOF. Assume the contrary. Assume that representation (4.11) subject to (4.12) is not unique, or in other words, not all of g_i are identically zero. Without loss of generality, we may suppose that $g_r(h_r(y)) \neq 0$, for some $y \in Q$. Let $\xi \in X$ be a solution of the system of equations $h_i(x) = h_i(x_0)$, $i = 1, \ldots, r-1$, and $h_r(x) = h_r(y)$. Therefore, $g_i(h_i(\xi)) = 0$, $i = 1, \ldots, r-1$, and $g_r(h_r(\xi)) \neq 0$. Obviously, $\xi \notin Q$. Otherwise, we may have $g_r(h_r(\xi)) = 0$.

We are going to prove that $Q' = Q \cup \{\xi\}$ is a representation set. For this purpose, consider an arbitrary function $f : Q' \to \mathbb{R}$. The restriction of f to the set Q admits a decomposition

$$f(x) = \sum_{i=1}^{r} t_i(h_i(x)), \; x \in Q.$$

One is allowed to fix the values $t_i(h_i(x_0)) = 0$, $i = 1, \ldots, r-1$. Note that then $t_i(h_i(\xi)) = 0$, $i = 1, \ldots, r-1$.

Consider now the functions

$$v_i(h_i(x)) = t_i(h_i(x)) + \frac{f(\xi) - t_r(h_r(\xi))}{g_r(h_r(\xi))} g_i(h_i(x)), \; x \in Q', \; i = 1, \ldots, r.$$

It can be easily verified that

$$f(x) = \sum_{i=1}^{r} v_i(h_i(x)), \; x \in Q'.$$

Since f is arbitrary, we obtain that $Q' \in RS$, where $Q' \supset Q$ and $h_i(Q') = h_i(Q)$, $i = 1, \ldots, r$. But this contradicts the hypothesis of the lemma that $Q \in CRS$. □

The following theorem is valid.

THEOREM 4.16. *$Q \in CRS$ if and only if for any $x_0 \in Q$, any $f \in T(Q)$ and any $a_1, \ldots, a_{r-1} \in \mathbb{R}$ the representation*

$$f(x) = \sum_{i=1}^{r} g_i(h_i(x)), \; x \in Q,$$

subject to the conditions $g_i(h_i(x_0)) = a_i$, $i = 1, \ldots, r-1$, is unique. Equivalently, a set $Q \in CRS$ if and only if it is a unicity set.

Theorem 4.16 is an obvious consequence of Lemmas 4.14 and 4.15.

REMARK 4.17. In Theorem 4.16, all the words "any" can be replaced with the word "some".

REMARK 4.18. For the case $X = X_1 \times \cdots \times X_n$, the possibility and uniqueness of the representation by sums $\sum_{i=1}^{n} u_i(x_i)$, $u_i : X_i \to \mathbb{R}$, $i = 1, \dots, n$, were investigated in [**119, 120**].

EXAMPLE 4.19. Let $r = 2$, $X = \mathbb{R}^2$, $h_1(x_1, x_2) = x_1 + x_2$, $h_2(x_1, x_2) = x_1 - x_2$, Q be the graph of the function $x_2 = \arcsin(\sin x_1)$. The set Q has no cycles with respect to the functions h_1 and h_2. Therefore, by Proposition 4.3, $Q \in RS$. By adding a point $p \notin Q$, we obtain the set $Q \cup \{p\}$, which contains a cycle and hence is not a representation set. Thus, $Q \in CRS$ and hence Q is a unicity set.

EXAMPLE 4.20. Let now $r = 2$, $X = \mathbb{R}^2$, $h_1(x_1, x_2) = x_1$, $h_2(x_1, x_2) = x_2$, and Q be the graph of the function $x_2 = x_1$. Clearly, $Q \in RS$ and $Q \notin CRS$. By the definition of complete representation sets, there is a set $P \supset Q$ such that $P \in RS$ and for any $T \supset P$, T is not a representation set. There are many sets P with this property. One of them can be obtained by adding to Q any straight line l parallel to one of the coordinate axes. Indeed, if $y \notin Q \cup l$, then the set $Q_1 = Q \cup l \cup \{y\}$ contains a four-point cycle (with one vertex as y, two vertices lying on l and one vertex lying on Q). This means that $Q_1 \notin RS$ and hence $Q \cup l \in CRS$.

The following corollary can be easily obtained from Theorem 4.16 and Lemma 4.14.

COROLLARY 4.21. *$Q \in CRS$ if and only if $Q \in RS$ and in the representation*

$$0 = \sum_{i=1}^{r} g_i(h_i(x)), \ x \in Q,$$

all the functions g_i, $i = 1, \dots, r$, are constants.

We have seen that complete representation sets enjoy the unicity property. Let us study some other properties of these sets. The following properties are valid.

(a) If $Q_1, Q_2 \in CRS$, $Q_1 \cap Q_2 \neq \emptyset$ and $Q_1 \cup Q_2 \in RS$, then $Q_1 \cup Q_2 \in CRS$.
(b) Let $\{Q_\alpha\}$, $\alpha \in \Phi$, be a family of complete representation sets such that $\cap_{\alpha \in \Phi} Q_\alpha \neq \emptyset$ and $\cup_{\alpha \in \Phi} Q_\alpha \in RS$. Then $\cup_{\alpha \in \Phi} Q_\alpha \in CRS$.
(c) Let $\{Q_\alpha\}$, $\alpha \in \Phi$, be a totally ordered (under inclusion) family of complete representation sets. Then $\cup_{\alpha \in \Phi} Q_\alpha \in CRS$.

The first and the second properties follow from Corollary 4.21. Note that (b) is a generalization of (a). The third property (c) is a consequence of (b) and property (2) of representation sets.

We know that every representation set A is contained in a complete representation set Q such that $h_i(A) = h_i(Q)$, $i = 1, \dots, r$. What can we say about the set $Q \backslash A$? Clearly, $Q \backslash A \in RS$. But can we chose Q so that $Q \backslash A \in CRS$? The following theorem answers this question.

THEOREM 4.22. *Let $A \in RS$ and $A \notin CRS$. Then there exists a set $B \in CRS$ such that $A \subset B$, $h_i(A) = h_i(B)$, $i = 1, \dots, r$, and $B \backslash A \in CRS$.*

PROOF. Since the representation set A is not complete, there exists a point $p \notin A$ such that $h_i(p) \in h_i(A)$, $i = 1, \ldots, r$, and $A' = A \cup \{p\} \in RS$. By $\mathcal{M}$ denote the collection of sets M such that

1) $A \subset M$ and $M \in RS$;
2) $h_i(M) = h_i(A)$ for all $i = 1, \ldots, r$;
3) $M \backslash A \in CRS$.

Obviously, $\mathcal{M}$ is not empty. It contains the above set A'. Consider the partial order on $\mathcal{M}$ defined by inclusion. Let $\{M_\beta\}$, $\beta \in \Gamma$, be any chain in $\mathcal{M}$. The set $\cup_{\beta \in \Gamma} M_\beta$ is an upper bound for this chain. To see this, let us check that $\cup_{\beta \in \Gamma} M_\beta$ belongs to $\mathcal{M}$. That is, all the above conditions 1)-3) are satisfied. Indeed,

1) $A \subset \cup_{\beta \in \Gamma} M_\beta$ and $\cup_{\beta \in \Gamma} M_\beta \in RS$. This follows from property (2) of representation sets;
2) $h_i(\cup_{\beta \in \Gamma} M_\beta) = \cup_{\beta \in \Gamma} h_i(M_\beta) = \cup_{\beta \in \Gamma} h_i(A) = h_i(A)$, $i = 1, \ldots, r$;
3) $\cup_{\beta \in \Gamma} M_\beta \backslash A \in CRS$. This follows from property (c) of complete representation sets and the facts that $M_\beta \backslash A \in CRS$ for any $\beta \in \Gamma$ and the system $\{M_\beta \backslash A\}$, $\beta \in \Gamma$, is totally ordered under inclusion.

Thus we see that any chain in $\mathcal{M}$ has an upper bound. By Zorn's lemma, there are maximal sets in $\mathcal{M}$. Assume B is one of such sets.

Let us now prove that $B \in CRS$. Assume on the contrary that $B \notin CRS$. Then by Lemma 4.14, for any point $x_0 \in B$ the representation

$$0 = \sum_{i=1}^{r} g_i(h_i(x)), \quad x \in B, \tag{4.13}$$

subject to the conditions $g_i(h_i(x_0)) = 0$, $i = 1, \ldots, r-1$, is not unique. That is, there is a point $y \in B$ such that for some index i, $g_i(h_i(y)) \neq 0$. Without loss of generality we may assume that $g_r(h_r(y)) \neq 0$. Clearly, y cannot belong to $B \backslash A$, since $B \backslash A \in CRS$ and over complete representation sets, the zero function has a trivial representation provided that conditions (4.12) hold. Thus, $y \in A$. Let $\xi \in X$ be a point such that $h_i(\xi) = h_i(x_0)$, $i = 1, \ldots, r-1$, and $h_r(\xi) = h_r(y)$. The point $\xi \notin B$, otherwise from (4.13) we would obtain that $g_r(h_r(y)) = g_r(h_r(\xi)) = 0$. Following the techniques in the proof of Lemma 4.15, it can be shown that $B_1 = B \cup \{\xi\} \in RS$.

Now we prove that $B_1 \backslash A \in CRS$. Consider the representation

$$0 = \sum_{i=1}^{r} g_i'(h_i(x)), \quad x \in B_1 \backslash A, \tag{4.14}$$

subject to the conditions $g_i'(h_i(x_0)) = 0$, $i = 1, \ldots, r-1$, where x_0 is some point in $B \backslash A$. Such representation holds uniquely on $B \backslash A$, since $B \backslash A \in CRS$. That is, all the functions g_i' are identically zero on $h_i(B \backslash A)$, $i = 1, \ldots, r$. On the other hand, since $g_i'(h_i(\xi)) = g_i'(h_i(x_0)) = 0$, for all $i = 1, \ldots, r-1$, we obtain that $g_r'(h_r(\xi)) = 0$. This means that representation (4.14) subject to the conditions $g_i'(h_i(x_0)) = 0$, $i = 1, \ldots, r-1$, is unique on $B_1 \backslash A$. That is, all the functions g_i' in (4.14) are zero functions on $h_i(B_1 \backslash A)$, $i = 1, \ldots, r$. Hence by Lemma 4.13, $B_1 \backslash A \in CRS$. Thus, $B_1 \in \mathcal{M}$. But the set B was chosen as a maximal set in $\mathcal{M}$. We see that our assumption $B \notin CRS$ leads to the contradiction that there is a set $B_1 \in \mathcal{M}$ bigger than the maximal set B. Thus, in fact, $B \in CRS$. $\square$

4.2.3. C-orbits and C-trips. Let A be a representation set. The relation on A defined by setting $x \sim y$ if there is a finite complete representation subset of A containing both x and y, is an equivalence relation. Indeed, it is reflexive and symmetric. It is transitive by property (a) of complete representation sets. The equivalence classes we call *C-orbits*. In the case $r = 2$, C-orbits turn into classical orbits considered by Marshall and O'Farrell [**150, 151**], which have a very nice geometric interpretation in terms of paths (see Section 1.3). A classical orbit consists of all possible traces of an arbitrary point in it traveling alternatively in the level sets of h_1 and h_2. In the general setting, one partial case of C-orbits were introduced by Klopotowski, Nadkarni, Rao [**120**] under the name of *related components*. The case considered in [**120**] requires that $A \subset X = X_1 \times \cdots \times X_n$ and h_i be the canonical projections of X onto X_i, $i = 1, \ldots, r$, respectively.

Finite complete representation sets containing x and y will be called *C-trips* connecting x and y. A C-trip of the smallest cardinality connecting x and y will be called a *minimal C-trip*.

THEOREM 4.23. *Let A be a representation set and x and y be any two points of some C-orbit in A. Then there is only one minimal C-trip connecting them.*

PROOF. Assume that L_1 and L_2 are two minimal C-trips connecting x and y. By the definition, L_1 and L_2 are complete representation sets. Note that $L_1 \cup L_2$ is also complete. Let us prove that the set $L_1 \cap L_2$ is complete. Clearly, $L_1 \cap L_2 \in RS$. Let $x_0 \in L_1 \cap L_2$. In particular, x_0 can be one of the points x and y. Consider the representation

$$0 = \sum_{i=1}^{r} g_i(h_i(x)), \ x \in L_1 \cap L_2, \tag{4.15}$$

subject to $g_i(h_i(x_0)) = 0$, $i = 1, \ldots, r-1$. On the strength of Lemma 4.13, it is enough to prove that this representation is unique. For $i = 1, \ldots, r$, let g_i' be any extension of g_i from the set $h_i(L_1 \cap L_2)$ to the set $h_i(L_1)$. Construct the function

$$f'(x) = \sum_{i=1}^{r} g_i'(h_i(x)), \ x \in L_1. \tag{4.16}$$

Since $f'(x) = 0$ on $L_1 \cap L_2$, the following function is well defined

$$f(x) = \begin{cases} f'(x), & x \in L_1, \\ 0, & x \in L_2. \end{cases}$$

Since $L_1 \cup L_2 \in CRS$, the representation

$$f(x) = \sum_{i=1}^{r} w_i(h_i(x)), \ x \in L_1 \cup L_2. \tag{4.17}$$

subject to

$$w_i(h_i(x_0)) = 0, \ i = 1, \ldots, r-1. \tag{4.18}$$

is unique. Besides, since $L_1 \in CRS$ and $g_i'(h_i(x_0)) = g_i(h_i(x_0)) = 0$, $i = 1, \ldots, r-1$, representation (4.16) is unique. This means that for each function g_i, there is only one extension g'. Note that

$$f(x) = f'(x) = \sum_{i=1}^{r} w_i(h_i(x)), \ x \in L_1.$$

Now from the uniqueness of representation (4.16) we obtain that

$$w_i(h_i(x)) = g_i'(h_i(x)),\ i = 1, \ldots, r,\ x \in L_1. \tag{4.19}$$

A restriction of formula (4.17) to the set L_2 gives

$$0 = \sum_{i=1}^{r} w_i(h_i(x)),\ x \in L_2. \tag{4.20}$$

Since $L_2 \in CRS$, representation (4.20) subject to conditions (4.18) is unique, whence

$$w_i(h_i(x)) = 0,\ i = 1, \ldots, r,\ \ x \in L_2. \tag{4.21}$$

From (4.19) and (4.21) it follows that

$$g_i(h_i(x)) = g_i'(h_i(x)) = 0,\ i = 1, \ldots, r,\ x \in L_1 \cap L_2.$$

Thus, we see that representation (4.15) subject to the conditions $g_i(h_i(x_0)) = 0$, $i = 1, \ldots, r-1$ is unique on the intersection $L_1 \cap L_2$. Therefore by Lemma 4.13, $L_1 \cap L_2 \in CRS$.

Let the cardinalities of L_1 and L_2 be equal to n. Since $x, y \in L_1 \cap L_2$ and $L_1 \cap L_2 \in CRS$, we obtain from the definition of minimal C-trips that the cardinality of $L_1 \cap L_2$ is also n. Hence, $L_1 \cap L_2 = L_1 = L_2$. □

Let Q be a representation set. That is, each function $f : Q \to \mathbb{R}$ enjoys representation (4.7). Can we construct the functions g_i, $i = 1, \ldots, r$, for a given f? There is a procedure for constructing one certain collection of g_i, provided that Q consists of a single C-orbit, that is, any two points of Q can be connected by a C-trip. To describe this procedure, take a point $x_0 \in Q$ and fix it. We are going to find g_i from (4.7) and conditions (4.8). Let y be any point in Q. To find the values of g_i at the points $h_i(y)$, $i = 1, \ldots, r$, connect x_0 and y by a minimal C-trip $S = \{x_1, \ldots, x_n\}$, where $x_1 = x_0$ and $x_n = y$. Since S is a complete representation set, equation (4.7) subject to (4.8) has a unique solution on S. That is, we can find $g_i(h_i(y))$, $i = 1, \ldots, r$, by solving the system of linear equations

$$\sum_{i=1}^{r} g_i(h_i(x_j)) = f(x_j),\ j = 1, \ldots, n.$$

We see that each minimal C-trip containing x_0 generates a system of linear equations, which is uniquely solvable. Since any point in Q can be connected with x_0 by such a trip, we can find $g_i(t)$ at each point $t \in h_i(Q)$, $i = 1, \ldots, r$.

The above procedure can still be effective for some particular representation sets Q consisting of many C-orbits. Let $\{C_\alpha\}$, $\alpha \in \Lambda$, denote the set of all C-orbits of Q. Fix some points $x_\alpha \in C_\alpha$, $\alpha \in \Lambda$, one in each orbit. Let y_α be any points of C_α, $\alpha \in \Lambda$, respectively. We can apply the above procedure of finding the values of g_i at each y_α if $h_i(y_\alpha) \neq h_i(y_\beta)$ for all i and $\alpha \neq \beta$. For $h_i(y_\alpha) = h_i(y_\beta)$, one cannot guarantee that after solving the corresponding systems of linear equations (associated with y_α and y_β), the solutions $g_i(h_i(y_\alpha)$ and $g_i(h_i(y_\beta))$ will be equal. That is, for the case $h_i(y_\alpha) = h_i(y_\beta)$, the constructed functions g_i may not be well defined.

REMARK 4.24. All the results in this section are valid, in particular, for linear combinations of generalized ridge functions.

4.3. An approximation algorithm

The previous sections described existence and unicity conditions for the exact representation by generalized ridge functions. Clearly, such representation is not always possible. This gives rise to the problem of finding efficient approximation methods. In this section we discuss a special algorithmic approximation method called the Diliberto-Straus algorithm. This algorithm was employed by many authors in different approximation problems and under various names. Here we consider the problem of approximation of a continuous function defined on a compact subset of the d-dimensional space by sums of two function compositions, which, in particular, turn into generalized ridge functions. Under reasonable assumptions, we prove the convergence of the Diliberto-Straus algorithm. In the simplest case of approximation by sums of two univariate functions, we see that Diliberto-Straus's original result holds for a large class of convex compact sets.

4.3.1. Problem statement. Let E be a Banach space and U and V be closed subspaces thereof. In addition, let A and B be *proximity maps* (*best approximation operators*) acting from E onto U and V, respectively. In this section, we are interested in algorithmic methods for computing the approximation error of $z \in E$ by elements from $U+V$. Historically, there is a procedure called the *Alternating Algorithm.* This procedure can be described as follows: Starting with $z_1 = z$ compute $z_2 = z_1 - Az_1$, $z_3 = z_2 - Bz_2$, $z_4 = z_3 - Az_3$, and so forth. Clearly, $z - z_n \in U+V$ and the sequence $\{\|z_n\|\}_{n=1}^{\infty}$ is nonincreasing. The question is if and when $\|z_n\|$ converges to the error of approximation by elements from $U+V$?

In 1933, von Neumann [**162**] proved that in Hilbert spaces the alternating algorithm is always effective for any pair of closed subspaces. But in general, for Banach spaces, this method needs favorable conditions to be effective. One of the important results in this direction is due to M. Golomb [**64**] (see also Light and Cheney [**138**, p.57]).

THEOREM 4.25 (Golomb [**64**]). *Assume U and V are subspaces of a Banach space E having central proximity maps A and B, respectively. Assume, in addition, that $U+V$ is closed in E. Then the above sequence $\{z_n\}$ converges in norm to the error of approximation by elements from $U+V$.*

A proximity map A from a Banach space E onto a subspace U is called a *central proximity map* if for all $z \in E$ and $x \in U$,

$$\|z - Az + x\| = \|z - Az - x\| \quad \text{(see [\textbf{138}, Chapter 4])}.$$

It should be noted that in a *uniformly convex* and *uniformly smooth* Banach space it was proved by Deutsch [**44**] that the above algorithm converges for two closed subspaces U and V if the sum $U+V$ is closed. He also noted that in any normed linear space that is not smooth one can always construct two linear subspaces for which the algorithm does not converge.

In the Banach space setting, the alternating algorithm goes under various names. It is quite often called the Diliberto-Straus algorithm. Diliberto and Straus [**47**] were the first to consider this algorithm in the space of continuous functions. They proved that for the uniform approximation of a bivariate function, defined on a rectangle with sides parallel to the coordinate axes, by sums of univariate functions, the sequence produced by the alternating algorithm converges to the desired quantity. Later this algorithm was generalized by Light and Cheney to continuous functions defined on a Cartesian product of two compact spaces (see [**137**, **138**]). Golomb [**64**] observed that the centrality property of proximity maps plays a key role in the analysis of Diliberto and Straus. It should be remarked that the concept *central proximity map* is due to Golomb. We refer the readers to the monographs by Light and Cheney [**138**], and Khavinson [**117**] for interesting discussions around the Diliberto-Straus algorithm.

In this section we discuss the Diliberto-Straus algorithm in the problem of approximating by sums of two function compositions, which may turn into generalized ridge functions. To make the problem more precise, assume X is a compact set in $\mathbb{R}^d$. Assume, in addition, $s : X \to \mathbb{R}$ and $p : X \to \mathbb{R}$ are fixed continuous functions on X. Consider the following sets

$$\begin{aligned} A_1 &= \{\varphi \circ s : \varphi \in C(\mathbb{R})\}, \\ A_2 &= \{\psi \circ p : \psi \in C(\mathbb{R})\}. \end{aligned}$$

That is, we consider all possible compositions $\varphi \circ s$ and $\psi \circ p$ with the proviso that s and p are fixed. Clearly, both A_1 and A_2 are linear subspaces of $C(X)$. Note that, in the particular case in which s and p decompose into sums of univariate functions, A_1 and A_2 consist of generalized ridge functions. Assume any $h \in C(X)$ is given. We are interested in algorithmic methods for computing the error of approximation

$$E(h) = \inf_{g \in A_1 + A_2} \|h - g\|.$$

Does the Diliberto-Straus algorithm converge to $E(h)$? In the next subsections we will discuss this question.

4.3.2. Best approximation operators and convergence of the algorithm. To analyze the Diliberto-Straus algorithm, first of all, we must decribe the best approxmation operators. This task is not difficult. Let us define the following operators

$$F : C(X) \to A_1,\ Fh(a) = \frac{1}{2}\left(\max_{\substack{x \in X \\ s(x)=a}} h(x) + \min_{\substack{x \in X \\ s(x)=a}} h(x) \right), \quad \text{for all } a \in s(X),$$

and

$$G : C(X) \to A_2,\ Gh(b) = \frac{1}{2}\left(\max_{\substack{x \in X \\ p(x)=b}} h(x) + \min_{\substack{x \in X \\ p(x)=b}} h(x) \right), \quad \text{for all } b \in p(X).$$

We see below that, under certain assumption on the max and min functions, these operators are best approximation operators. To this end, let us assume, in the sequel, that the functions s and p obey the following property, which we call "Property C": For any $h \in C(X)$, the functions

$$\begin{aligned} f_1(a) &= \max_{\substack{x \in X \\ s(x)=a}} h(x),\ f_2(a) = \min_{\substack{x \in X \\ s(x)=a}} h(x),\ a \in s(X), \\ g_1(b) &= \max_{\substack{x \in X \\ p(x)=b}} h(x),\ g_2(b) = \min_{\substack{x \in X \\ p(x)=b}} h(x),\ b \in p(X) \end{aligned}$$

are continuous on the appropriate sets $s(X)$ and $p(X)$.

Since the functions s and p both enjoy Property C, for each continuous function h on X, the functions Fh and Gh are continuous on $s(X)$ and $p(X)$, respectively. Via the mappings s and p, the functions Fh and Gh can also be considered as functions defined on X. Since s and p are continuous, the functions Fh and Gh are also continuous on X, and hence belong to the sets A_1 and A_2, respectively.

The following theorem is valid.

THEOREM 4.26. *Assume Property C holds for the functions s and p. Then the operators F and G are central proximity maps onto A_1 and A_2, respectively. In addition, these operators are non-expansive. That is,*

$$\|Fv_1 - Fv_2\| \leq \|v_1 - v_2\| \text{ and } \|Gv_1 - Gv_2\| \leq \|v_1 - v_2\|,$$

for all $v_1, v_2 \in C(X)$.

PROOF. We prove this theorem for the operator F. The proof for G can be carried out in the same way.

Clearly, on the level set $\{x \in X : s(x) = a\}$, the constant $(Fh)(a)$ is a best approximation to h, among all constants. Varying over $a \in s(X)$, we obtain a best approximation $Fh : X \to \mathbb{R}$, which is, due to Property C, belongs to A_1.

Now let us prove that the proximity map F is a central proximity map. In other words, we must prove that for any functions $h \in C(X)$ and $\varphi = \varphi(s(x)) \in A_1$, the following equality holds

$$\|h - Fh - \varphi\| = \|h - Fh + \varphi\|. \tag{4.22}$$

Put $\omega = h - Fh$. There exists a point $x_0 \in X$ such that

$$\|\omega + \varphi\| = |\omega(x_0) + \varphi(s(x_0))|.$$

First assume that $|\omega(x_0) + \varphi(s(x_0))| = \omega(x_0) + \varphi(s(x_0))$. Since $F\omega = 0$,

$$\max_{\substack{x \in X \\ s(x)=a}} \omega(x) = - \min_{\substack{x \in X \\ s(x)=a}} \omega(x), \text{ for all } a \in s(X). \tag{4.23}$$

Let

$$\min_{\substack{x \in X \\ s(x)=s(x_0)}} \omega(x) = \omega(x_1). \tag{4.24}$$

Then from (4.23) and (4.24) it follows that

$$-\omega(x_1) \geq \omega(x_0).$$

Taking the last inequality and the equality $s(x_1) = s(x_0)$ into account we can write that

$$\|\omega - \varphi\| \geq \varphi(s(x_1)) - \omega(x_1) \geq \varphi(s(x_0)) + \omega(x_0) = \|\omega + \varphi\|. \tag{4.25}$$

Changing in (4.25) the function φ to $-\varphi$ gives the reverse inequality $\|\omega + \varphi\| \geq \|\omega - \varphi\|$. Thus (4.22) holds.

Note that if $|\omega(x_0) + \varphi(s(x_0))| = -(\omega(x_0) + \varphi(s(x_0)))$, then by replacing Eq. (4.24) with

$$\max_{\substack{x \in X \\ s(x)=s(x_0)}} \omega(x) = \omega(x_1), \tag{4.26}$$

we derive from (4.23) and (4.26) that $\omega(x_1) \geq -\omega(x_0)$. This inequality can then be used to obtain the estimation

$$\|\omega - \varphi\| \geq -(\varphi(s(x_1) - \omega(x_1)) \geq -(\varphi(s(x_0)) + \omega(x_0)) = \|\omega + \varphi\|,$$

which in turn yields (4.22). The centrality has been proved.

Now we prove that the operator F is non-expansive. First note that it is non-decreasing. That is, if for all $x \in X$, $h_1(x) \leq h_2(x)$, then $Fh_1(s(x)) \leq Fh_2(s(x))$. Besides, $F(h+c) = Fh + c$, for any real number c. Let now v_1 and v_2 be arbitrary functions in $C(X)$. Put $c = \|v_1 - v_2\|$. Then for any $x \in X$, we can write that

$$v_2(x) - c \leq v_1(x) \leq v_2(x) + c$$

and further

$$Fv_2(s(x)) - c \leq Fv_1(s(x)) \leq Fv_2(s(x)) + c.$$

From the last inequality we obtain that

$$\|Fv_1 - Fv_2\| \leq c = \|v_1 - v_2\|.$$

Thus the operator F is non-expansive. □

Consider the iterations

$$h_1(x) = h(x),\ h_{2n} = h_{2n-1} - Fh_{2n-1},\ h_{2n+1} = h_{2n} - Gh_{2n},\ n = 1, 2, \ldots.$$

From Theorem 4.25 and Theorem 4.26 we obtain the convergence of the Diliberto-Straus algorithm.

THEOREM 4.27. *Assume Property C holds for the functions s and p. Assume in addition that $A_1 + A_2$ is closed in $C(X)$. Then $\|h_n\|$ converges to the approximation error $E(h)$.*

Let us now discuss Property C. Note that for a given compact set $X \subset \mathbb{R}^d$ many functions on X may not possess this property. For example, for $X = [0,1] \times [0,2] \cup [1,2] \times [0,1] \subset \mathbb{R}^2$, the functions $s(x,y) = x$ and $p(x,y) = y$ do not possess Property C. To verify this, one can take, for example, $h(x,y) = xy$. In order for Property C to be fulfilled, the functions s and p should satisfy certain conditions. The following proposition provides a sufficient condition of such type.

PROPOSITION 4.28. *The above functions f_1 and f_2 are continuous on $s(X)$ for any continuous h on X if for any two points $x, y \in X$ with $s(x) = s(y)$ and any sequence $\{x_n\}_{n=1}^{\infty}$ tending to x, there exists a sequence $\{y_n\}_{n=1}^{\infty}$ tending to y such that $s(y_n) = s(x_n)$, for all $n = 1, 2, \ldots$*

PROOF. Suppose the contrary. Suppose that the above hypothesis on s holds, but one of the functions f_1 and f_2 is not continuous. Without loss of generality assume that f_1 is not continuous on $s(X)$. Let f_1 be discontinuous at $a_0 \in s(X)$. Then there exists a number $\varepsilon > 0$ and a sequence $\{a_n\}_{n=1}^{\infty} \subset s(X)$ tending to a_0, such that

$$|f_1(a_n) - f_1(a_0)| > \varepsilon, \tag{4.27}$$

for all $n = 1, 2, \ldots$. Consider the level sets

$$L_n = \{x \in X : s(x) = a_n\},\ n = 0, 1, 2 \ldots$$

Since h is continuous, there exist points $y_0 \in L_0$, $x_n \in L_n$, $n = 1, 2, \dots$, such that $h(y_0) = f_1(a_0)$ and $h(x_n) = f_1(a_n)$. Thus the inequality (4.27) can be written as

$$|h(x_n) - h(y_0)| > \varepsilon, \tag{4.28}$$

for all $n = 1, 2, \dots$. Since X is compact, the sequence $\{x_n\}_{n=1}^{\infty}$ has a convergent subsequence. Without loss of generality assume that $\{x_n\}_{n=1}^{\infty}$ itself converges to a point $x_0 \in X$. Then $s(x_n) \to s(x_0)$, as $n \to \infty$. But by the assumption $s(x_n) = a_n \to a_0 = s(y_0)$, as $n \to \infty$. Therefore, $s(x_0) = s(y_0) = a_0$. Note that x_0 and y_0 cannot be the same point, since the equality $x_0 = y_0$ violates inequality (4.28). By the hypothesis of the proposition, we must have a sequence $\{y_n\}_{n=1}^{\infty}$ such that $y_n \to y_0$ and

$$s(x_n) = s(y_n),$$

for $n = 1, 2, \dots$. Since $s(x_n) = a_n$, $n = 1, 2, \dots$, and on each level set L_n, the function h takes its maximum value at x_n, we obtain that

$$h(y_n) \le h(x_n), \ n = 1, 2, \dots$$

Taking the limit in the last inequality, as $n \to \infty$, gives us the new inequality

$$h(y_0) \le h(x_0). \tag{4.29}$$

Recall that on the level set L_0, the function h takes its maximum at y_0. Thus from (4.29) we conclude that $h(x_0) = h(y_0)$. This equality contradicts the choice of the positive ε in (4.28), since $h(x_n) \to h(x_0)$, as $n \to \infty$. The obtained contradiction means that the function f_1 is continuous on $s(X)$. It can be proved in the same way that f_2 is also continuous. □

EXAMPLE 4.29. It is not difficult to see that the inner product function $s(\mathbf{x}) = \mathbf{a} \cdot \mathbf{x}$, where $\mathbf{a}$ is a nonzero vector in $\mathbb{R}^d$, satisfies the hypothesis of Proposition 4.28, provided that $\mathbf{x}$ varies in a convex compact set $Q \subset \mathbb{R}^d$. Indeed, let $\mathbf{x}_0$ and $\mathbf{y}_0$ be any two points in Q such that $s(\mathbf{x}_0) = s(\mathbf{y}_0)$. Take any sequence $\{\mathbf{x}_n\}_{n=1}^{\infty} \subset Q$, $\mathbf{x}_n \to \mathbf{x}_0$, as $n \to \infty$. We must show that there exists a sequence $\{\mathbf{y}_n\}_{n=1}^{\infty} \subset Q$ with the property that $\mathbf{y}_n \to \mathbf{y}_0$, as $n \to \infty$, and $s(\mathbf{x}_n) = s(\mathbf{y}_n)$, for $n = 1, 2, \dots$ To show this, first note that there are points $\mathbf{z}_1, \mathbf{z}_2 \in Q$ such that $s(\mathbf{x}_n) \in [s(\mathbf{z}_1), s(\mathbf{z}_2)]$, for all $n = 1, 2, \dots$ For example, one can always take such points from the set $\{\mathbf{x}_0, \mathbf{x}_1, \mathbf{x}_2, \dots\}$. But there are also other choices. Note that one of the real numbers $s(\mathbf{z}_1)$ or $s(\mathbf{z}_2)$ may equal to $s(\mathbf{x}_0)$ and in this case we will assume that the corresponding $\mathbf{z}_1$ or $\mathbf{z}_2$ coincides with $\mathbf{y}_0$. Consider now the line segments $[\mathbf{z}_1, \mathbf{y}_0]$ and $[\mathbf{y}_0, \mathbf{z}_2]$ (which may be degenerated into the point $\mathbf{y}_0$). Set $L = [\mathbf{z}_1, \mathbf{y}_0] \cup [\mathbf{y}_0, \mathbf{z}_2]$. Since Q is convex, $L \subset Q$. The function s is continuous on L. Hence by the Intermediate Value Theorem, there exists a sequence $\{\mathbf{y}_n\}_{n=1}^{\infty} \subset L$ such that $s(\mathbf{y}_n) = s(\mathbf{x}_n)$, for $n = 1, 2, \dots$ Since $s(\mathbf{x}_n) \to s(\mathbf{x}_0)$ and $s(\mathbf{x}_0) = s(\mathbf{y}_0)$, it follows that $s(\mathbf{y}_n) \to s(\mathbf{y}_0)$, as $n \to \infty$. Now from this fact and the inclusion $\{\mathbf{y}_n\}_{n=1}^{\infty} \subset L$ we easily deduce that $\mathbf{y}_n \to \mathbf{y}_0$, as $n \to \infty$. This example and Proposition 4.28 show that for convex compact sets, inner product functions, in particular, the coordinate functions satisfy Property C.

We conclude from Theorem 4.27 and the above example that for sums of ridge functions the following result is valid.

THEOREM 4.30. *Assume Q is a convex compact set in $\mathbb{R}^d$ and $h \in C(Q)$. Assume for given directions $\mathbf{a}, \mathbf{b} \in \mathbb{R}^d \backslash \{\mathbf{0}\}$ the set*

$$\mathcal{R}(\mathbf{a}, \mathbf{b}) = \{g_1(\mathbf{a} \cdot \mathbf{x}) + g_2(\mathbf{b} \cdot \mathbf{x}) : g_1, g_2 \in C(\mathbb{R})\},$$

where $\mathbf{x}$ varies in Q, is closed in $C(Q)$. Then the sequence $\{h_n\}_{n=1}^{\infty}$ produced by the Diliberto-Straus algorithm converges in norm to the approximation error of h by elements from $\mathcal{R}(\mathbf{a}, \mathbf{b})$.

EXAMPLE 4.31. Note that the function s may satisfy the hypothesis of Proposition 4.28, but for many closed sets $E \subset X$, it may happen that the restriction of s to E no longer satisfies it. For example, let K be the unit square in $\mathbb{R}^2$ and $K_1 = [0, 1] \times [0, \frac{1}{2}] \cup [0, \frac{1}{2}] \times [0, 1]$. Clearly, the coordinate projection $s(x, y) = x$ satisfies the hypothesis if s is considered over K. However, this is not true if s is considered over K_1. Indeed, for the sequence $\{(\frac{1}{2} + \frac{1}{n+1}, \frac{1}{2})\}_{n=1}^{\infty} \subset K_1$, which tends to $(\frac{1}{2}, \frac{1}{2})$, we cannot find a sequence $\{(x_n, y_n)\}_{n=1}^{\infty} \subset K_1$ tending to $(\frac{1}{2}, 1)$ such that $x_n = \frac{1}{2} + \frac{1}{n+1}$, $n = 1, 2, \ldots$

In Theorem 4.27, we use the closedness assumption. For a given compact space X, the closedness of $A_1 + A_2$ in $C(X)$ strictly depends on the internal structure of X. There are several results on closedness of $A_1 + A_2$ even in general case of a sum of two algebras (see, e.g., [**117, 150, 156**]). The most explicit and practical result is due to Medvedev (see [**156**]). He showed that the sum $A_1 + A_2$ is closed in $C(X)$ if and only if the lengths of all irreducible paths in X are uniformly bounded. Recall that a path $\{x_1, \ldots, x_m\}$ is called irreducible if there does not exist another path $\{y_1, \ldots, y_l\}$ with $y_1 = x_1$, $y_l = x_m$ and $l < m$. For example, the set of functions $\varphi(x) + \psi(y)$ defined on a compact set $Q \subset \mathbb{R}^2$ is closed in the space $C(Q)$ if Q has a vertical or horizontal bar (a bar is a closed segment in Q, projection of which into the x or y axis coincides with the projection of Q into the same axis). This is because any two different points of Q can be connected by means of a bolt (that is, a path with respect to the coordinate functions), using at most 2 points of the bar. That is, in this case, the lengths of irreducible bolts are not more than 4. The similar argument can be applied to a compact set Q, which contains a broken line L, with segments parallel to the coordinate axes, such that the projections of Q and L into one of the coordinate axis coincide. At the same time, there are compact sets of simple structure, for which the closedness result does not hold. Take, for example, the triangle ABC with $A = (0, 0)$, $B = (2, 2)$ and $C = (1, 0)$. In this case, there is no natural number bounding the lengths of irreducible bolts $(A, \ldots, X_n) \subset ABC$, $n = 1, 2, \ldots$, provided that X_n tends to B, as $n \to \infty$.

From Theorem 4.27, Proposition 4.28 and the above mentioned closedness result of Medvedev one can obtain the following corollary, which is a generalization of the classical Diliberto and Straus theorem from rectangular sets to certain convex compact sets in $\mathbb{R}^2$.

COROLLARY 4.32. *Assume $Q \subset \mathbb{R}^2$ is a convex compact set with the property that any bolt in Q can be made closed by adding only a fixed number of points of Q. Let $A_1 = \{\varphi(x)\}$ and $A_2 = \{\psi(y)\}$ be the spaces of univariate functions, which are continuous on the projections of Q into the coordinate axes x and y, respectively. Then for a given function $h \in C(Q)$, the sequence $h_1 = h, h_{2n} = h_{2n-1} - Fh_{2n-1}, h_{2n+1} = h_{2n} - Gh_{2n}, n = 1, 2, \ldots$, converges in norm to the error of approximation from $A_1 + A_2$.*

4.3.3. Discussion of the algorithm. Does the Diliberto-Straus algorithm converge without the closedness assumption on the sum $A_1 + A_2$? We do not know a complete answer to this question. It should be remarked that this question was also raised in other works in different contexts (see, e.g., [**174, 175**]). Recall that we obtain Theorem 4.27 by using the general result of Golomb (see Theorem 4.25), in which the "closedness" is a major hypothesis. In this subsection we give a different, and independent of Golomb's result, proof of this theorem, where it is shown how we use the mentioned closedness. It should be remarked that in the simplest case of approximation by sums of univariate functions, the known classical proofs of the Diliberto-Straus algorithm require that one could form a closed bolt from an arbitrary bolt $(x_1, \dots, x_n)$ in a given compact set $Q \subset \mathbb{R}^2$ by adding a single point $y \in Q$, whose first and second coordinates are equal to that of x_1 and x_n, respectively (see, e.g., [**47, 117, 137**]). Note that this point may not lie in Q, unless Q is a Cartesian product of two compact sets. The main idea behind our proof is to make use of weak* cluster points of some sequence of "unclosed bolt functionals" instead of considering only "closed bolt functionals". We hope that this idea can be useful in future attempts to prove the convergence of the Diliberto-Straus algorithm without the closedness assumption.

Consider paths with respect to the functions s and p (see Definition 1.30). With each path $l = \{x_1, \dots, x_n\}$ in X, we associate the following path functional

$$r_l(h) = \frac{1}{n} \sum_{i=1}^{n} (-1)^{n+1} h(x_i).$$

It is an exercise to check that r_l is a linear bounded functional on $C(X)$ with the norm $\|r_l\| \leq 1$ and $\|r_l\| = 1$ if and only if the set of points x_i with odd indices i does not intersect with the set of points with even indices. Besides, if l is closed, then $r_l \in (A_1 + A_2)^\perp$, where $(A_1 + A_2)^\perp$ is the annihilator of the subspace $A_1 + A_2 \subset C(X)$. Note that for l, which is not closed, r_l is generally not an annihilating functional. However, this functional obeys the following important inequality

$$|r_l(f)| \leq \frac{2}{n} \|f\|, \tag{4.30}$$

where $f \in A_i$, $i = 1, 2$. This inequality means that for paths l with sufficiently large number of points, r_l behaves like an annihilating functional.

Now we are able to prove Theorem 4.27 differently, without using Theorem 4.25.

PROOF. First, let us write the above iteration in the following form

$$\begin{aligned} h_1 &= h, \ h_{n+1} = h_n - q_n, \text{ where} \\ q_n &= Fh_n, \text{ if } n \text{ is odd;} \\ q_n &= Gh_n, \text{ if } n \text{ is even.} \end{aligned}$$

Introduce the functions

$$\begin{aligned} u_n &= q_1 + \cdots + q_{2n-1}, \\ v_0 &= 0, \ v_n = q_2 + \cdots + q_{2n}, \ n = 1, 2, \dots \end{aligned}$$

Clearly, $u_n \in A_1$ and $v_n \in A_2$. Besides, $h_{2n} = h - u_n - v_{n-1}$ and $h_{2n+1} = h - u_n - v_n$, for $n = 1, 2, \dots$.

It is easy to see that the following inequalities hold

$$\|h_1\| \geq \|h_2\| \geq \|h_3\| \geq \cdots \geq E(h).$$

Hence, there exists the limit

$$M = \lim_{n\to\infty} \|h_n\| \geq E(h).$$

It is a consequence of the Hahn-Banach extension theorem that

$$E(h) = \sup_{\substack{r\in(A_1+A_2)^\perp \\ \|r\|\leq 1}} |r(h)|,$$

where the sup is attained by some functional. To complete the proof it is enough to show that for any $\varepsilon > 0$, there exists a functional r_0 such that $r_0 \in (A_1 + A_2)^\perp$, $\|r_0\| \leq 1$ and $|r_0(h)| \geq M - \varepsilon$.

Let ε be an arbitrarily small positive number. For each positive integer $k = 1, 2, \ldots$, set $\delta_k = \frac{\varepsilon}{2^{2k}}$. There exists a number n_k such that for all $n \geq n_k$,

$$\|h_n\| \leq M + \delta_k.$$

Without loss of generality we may assume that n_k is even. In the following, for each k, we are going to construct a path $l_k = \{x_1^k, \ldots, x_{2k+1}^k\}$ with the property that $|r_{l_k}(h_{n_k})| \geq M - \varepsilon$. This will lead us to the above mentioned functional r_0.

Since $Fh_{2m} = 0$, for $m = 1, 2, \ldots$, and $n_k + 2k$ is an even number,

$$\max_{\substack{x\in X \\ s(x)=a}} h_{n_k+2k}(x) = -\min_{\substack{x\in X \\ s(x)=a}} h_{n_k+2k}(x), \text{ for all } a \in s(X).$$

Then there exist points x_1 and x_2 such that $h_{n_k+2k}(x_1) = \|h_{n_k+2k}\|$, $h_{n_k+2k}(x_2) = -\|h_{n_k+2k}\|$ and $s(x_1) = s(x_2)$. This can be written in the form

$$h_{n_k+2k-1}(x_1) - Fh_{n_k+2k-1}(s(x_1)) = \|h_{n_k+2k}\|; \tag{4.31}$$

$$h_{n_k+2k-1}(x_2) - Fh_{n_k+2k-1}(s(x_2)) = -\|h_{n_k+2k}\|. \tag{4.32}$$

Therefore,

$$Fh_{n_k+2k-1}(s(x_1)) = Fh_{n_k+2k-1}(s(x_2)) =$$
$$= h_{n_k+2k-1}(x_1) - \|h_{n_k+2k}\| \leq \|h_{n_k+2k-1}\| - M \leq \delta_k$$

and

$$Fh_{n_k+2k-1}(s(x_1)) = Fh_{n_k+2k-1}(s(x_2)) =$$
$$= h_{n_k+2k-1}(x_2) + \|h_{n_k+2k}\| \geq -(\|h_{n_k+2k-1}\| - M) \geq -\delta_k.$$

That is,

$$-\delta_k \leq Fh_{n_k+2k-1}(s(x_1)) = Fh_{n_k+2k-1}(s(x_2)) \leq \delta_k. \tag{4.33}$$

From (4.31)-(4.33) we obtain that

$$h_{n_k+2k-1}(x_1) = Fh_{n_k+2k-1}(s(x_1)) + \|h_{n_k+2k}\| \geq M - \delta_k; \tag{4.34}$$

$$h_{n_k+2k-1}(x_2) = Fh_{n_k+2k-1}(s(x_1)) - \|h_{n_k+2k}\| \leq -M + \delta_k. \tag{4.35}$$

Since $Gh_{2m+1} = 0$, for $m = 1, 2, \ldots$,

$$\max_{\substack{x\in X \\ p(x)=b}} h_{n_k+2k-1}(x) = -\min_{\substack{x\in X \\ p(x)=b}} h_{n_k+2k-1}(x), \text{ for all } b \in p(X).$$

Then from (4.34) and (4.35) it follows that there exist points x_3 and x_3' satisfying the relations

$$h_{n_k+2k-1}(x_3) \geq M - \delta_k, \quad p(x_3) = p(x_2) \tag{4.36}$$

and

$$h_{n_k+2k-1}(x_3') \leq -M + \delta_k, \quad p(x_3') = p(x_1). \tag{4.37}$$

Replacing h_{n_k+2k-1} with $h_{n_k+2k-2} - Gh_{n_k+2k-2}$ in the inequalities (4.34)-(4.37) gives the following estimates

$$h_{n_k+2k-2}(x_1) - Gh_{n_k+2k-2}(p(x_1)) \geq M - \delta_k; \tag{4.38}$$

$$h_{n_k+2k-2}(x_2) - Gh_{n_k+2k-2}(p(x_2)) \leq -M + \delta_k; \tag{4.39}$$

$$h_{n_k+2k-2}(x_3) - Gh_{n_k+2k-2}(p(x_3)) \geq M - \delta_k; \tag{4.40}$$

$$h_{n_k+2k-2}(x_3') - Gh_{n_k+2k-2}(p(x_3')) \leq -M + \delta_k. \tag{4.41}$$

The following inequalities are obvious.

$$\begin{aligned}
h_{n_k+2k-2}(x_1) &\leq \|h_{n_k+2k-2}\| \leq M + \delta_k, \\
h_{n_k+2k-2}(x_3) &\leq \|h_{n_k+2k-2}\| \leq M + \delta_k, \\
h_{n_k+2k-2}(x_2) &\geq -\|h_{n_k+2k-2}\| \geq -M - \delta_k, \\
h_{n_k+2k-2}(x_3') &\geq -\|h_{n_k+2k-2}\| \geq -M - \delta_k.
\end{aligned}$$

Considering these inequalities in (4.38)-(4.41), we can write that

$$-2\delta_k \leq Gh_{n_k+2k-2}(p(x_1)) = Gh_{n_k+2k-2}(p(x_3')) \leq 2\delta_k; \tag{4.42}$$

$$-2\delta_k \leq Gh_{n_k+2k-2}(p(x_2)) = Gh_{n_k+2k-2}(p(x_3)) \leq 2\delta_k. \tag{4.43}$$

Taking into account (4.42) and (4.43) in the estimates (4.38)-(4.41), we obtain that

$$h_{n_k+2k-2}(x_1) \geq M - 3\delta_k,$$

$$h_{n_k+2k-2}(x_2) \leq -M + 3\delta_k,$$

$$h_{n_k+2k-2}(x_3) \geq M - 3\delta_k, \tag{4.44}$$

$$h_{n_k+2k-2}(x_3') \leq -M + 3\delta_k. \tag{4.45}$$

Now since

$$\max_{\substack{x \in X \\ s(x)=a}} h_{n_k+2k-2}(x) = -\min_{\substack{x \in X \\ s(x)=a}} h_{n_k+2k-2}(x), \quad \text{for all } a \in s(X),$$

from (4.44) and (4.45) it follows that there exist points x_4 and x_4' satisfying

$$h_{n_k+2k-2}(x_4) \leq -M + 3\delta_k, \ s(x_4) = s(x_3),$$

and

$$h_{n_k+2k-2}(x_4') \geq M - 3\delta_k, \ s(x_4') = s(x_2).$$

Repeating the above process for the function $h_{n_k+2k-2}(x) = h_{n_k+2k-3}(x) - Fh_{n_k+2k-3}(s(x))$ we obtain that

$$-4\delta_k \leq Fh_{n_k+2k-3}(s(x_i)) \leq 4\delta_k, \ i = 1, 2, 3, 4, \tag{4.46}$$

and

$$\begin{aligned} h_{n_k+2k-3}(x_1) &\geq M - 7\delta_k; \\ h_{n_k+2k-3}(x_2) &\leq -M + 7\delta_k; \\ h_{n_k+2k-3}(x_3) &\geq M - 7\delta_k; \\ h_{n_k+2k-3}(x_4) &\leq -M + 7\delta_k; \\ h_{n_k+2k-3}(x_4') &\geq M - 7\delta_k. \end{aligned}$$

In the same way as above, we can find a point x_5 such that $p(x_5) = p(x_4)$ and

$$h_{n_k+2k-3}(x_5) \geq M - 7\delta_k.$$

Continuing this process until we reach the function h_{n_k}, we obtain the points $x_1, x_2, \ldots, x_{2k+2}$ satisfying $s(x_1) = s(x_2)$, $p(x_2) = p(x_3), \ldots, s(x_{2k+1}) = s(x_{2k+2})$. Clearly, these points, in the given order, form a path, which we denote by l_k. Note that in the considered process, we also deal with the points x_3', x_4', and so on, but these points play only an auxiliary role: they are needed in obtaining the inequalities (4.42), (4.43), (4.46), etc. At the points of l_k, the values of h_{n_k} obey the estimates

$$h_{n_k}(x_i) \geq M - (2^{2k} - 1)\delta_k \geq M - \varepsilon, \text{ for } i = 1, 3, \ldots, 2k+1$$

and

$$h_{n_k}(x_j) \leq -M + (2^{2k} - 1)\delta_k \leq -M + \varepsilon, \text{ for } j = 2, 4, \ldots, 2k+2.$$

Using these inequalities, we can estimate the absolute value of $r_{l_k}(h_{n_k})$ as follows

$$|r_{l_k}(h_{n_k})| \geq \frac{(k+1)(M-\varepsilon) - (k+1)(-M+\varepsilon)}{2k+2} = M - \varepsilon. \tag{4.47}$$

We also can write the following obvious inequality

$$\begin{aligned} |r_{l_k}(h)| &= \left|r_{l_k}(h_{n_k} + u_{n_k/2} + v_{n_k/2-1})\right| \\ &\geq |r_{l_k}(h_{n_k})| - \left|r_{l_k}(u_{n_k/2})\right| - \left|r_{l_k}(v_{n_k/2-1})\right|. \end{aligned} \tag{4.48}$$

Considering (4.47) and (4.30) in (4.48) we obtain that

$$|r_{l_k}(h)| \geq M - \varepsilon - \frac{2}{2k+2}\left(\left\|u_{n_k/2}\right\| + \left\|v_{n_k/2-1}\right\|\right). \tag{4.49}$$

Now we use the closedness assumption on $A_1 + A_2$. Note that a subspace $B = B_1 + B_2$ of a Banach space is closed if and only if there exists a $K < \infty$ such that each element b in B has a representation $b = b_1 + b_2$, where $b_1 \in B_1$, $b_2 \in B_2$ and $\max(\|b_1\|, \|b_2\|) \leq K\|b\|$ (see, e.g., [**150**]). Thus, since $A_1 + A_2$ is closed, there exists a constant K such that

$$\|u_n\| + \|v_{n-1}\| \leq K\|u_n + v_{n-1}\|, \; n = 1, 2, \ldots \tag{4.50}$$

Note that the sequence $\{\|u_n + v_{n-1}\|\}_{n=1}^{\infty}$ is uniformly bounded. Indeed,

$$\|u_n + v_{n-1}\| = \|h - h_{2n}\| \leq 2\|h\|, \text{ for all } n = 1, 2, \ldots \tag{4.51}$$

From (4.49), (4.50) and (4.51) it follows that there exists a constant $C > 0$ such that

$$|r_{l_k}(h)| \geq M - \varepsilon - \frac{C}{2k+2}. \tag{4.52}$$

Thus, for each positive integer k we have constructed paths l_k and path functionals r_{l_k}, for which the inequality (4.52) holds. Note that $\|r_{l_k}\| \leq 1$, for all k. By the

Banach-Alaoglu theorem, the sequence $\{r_{l_k}\}_{k=1}^{\infty}$ has a weak* cluster point. Denote this point by r_0. Then from (4.52) it follows that

$$|r_0(h)| \geq M - \varepsilon.$$

The above inequality completes the proof. □

4.3.4. Some remarks.

REMARK 4.33. Let us explain why we consider the sum of only two function compositions. The problem is that for a sum of more than two compositions, the appropriate generalization of the Diliberto-Straus algorithm does not hold. In fact, it does not hold even in the simplest case of approximation by sums of univariate functions. To be more precise, let E be the unit cube in $\mathbb{R}^d$ and A_i be a best approximation operator from the space of continuous functions $C(E)$ to the subspace of univariate functions $G_i = \{g_i(x_i) : g_i \in C[0,1]\}$, $i = 1, \ldots, d$.

That is, for each function $f \in C(X)$, the function $A_i f$ is a best approximation to f from G_i. Set

$$Tf = (I - A_d)(I - A_{d-1}) \cdots (I - A_1)f,$$

where I is the identity operator. It is clear that

$$Tf = f - g_1 - g_2 - \cdots - g_d,$$

where g_k is a best approximation from G_k to the function $f - g_1 - g_2 - \cdots - g_{k-1}$, $k = 1, \ldots, d$. Consider powers of the operator T: T^2, T^3 and so on. What is the limit of $\|T^n f\|$ as $n \to \infty$? One may expect that the sequence $\{\|T^n f\|\}_{n=1}^{\infty}$ converges to $E(f)$ (the error of approximation from $G_1 + \cdots + G_d$), as in the case $d = 2$. This conjecture was first proposed in 1951 by Diliberto and Straus [**47**]. But later it was shown by Aumann [**10**], and independently by Medvedev [**155**], that the algorithm may not converge to $E(f)$ in the case $d > 2$.

REMARK 4.34. There is a difference between the convergence $\|h_n\| \to E(h)$, as $n \to \infty$, and the convergence of the sequence $\{h - h_n\}_{n=1}^{\infty}$ to a best approximation to h from $A_1 + A_2$. The latter is much stronger. Theorem 4.27 does not guarantee that the sequence $\{h - h_n\}$ converges in the latter strong sense. We do not know if it converges to a best approximation to h from $A_1 + A_2$. The question, for which compact spaces X and subalgebras A_1 and A_2, the sequence of functions produced by the Diliberto-Straus algorithm converges to a best approximation is fair, but too difficult to solve in this generality. It should be noted that Aumann [**10**] proved that for a rectangle $S = [a, b] \times [c, d]$ in $\mathbb{R}^2$ and for any function $h \in C(S)$, the sequence $\{h - h_n\}$ converges uniformly to a best approximation to h from the subspace

$$D = \{\varphi(x) + \psi(y) : \ \varphi \in C[a, b], \ \psi \in C[c, d]\}.$$

Aumann's proof is based on the equicontinuity of some families of functions, namely the families $\{u_n\}$ and $\{v_n\}$ (see the proof of Theorem 4.27). To show the equicontinuity, Aumann substantially uses the non-expansivness of the averaging operators

$$(M_x f)(y) = \frac{\max_x f + \min_x f}{2} \text{ and } (M_y f)(x) = \frac{\max_y f + \min_y f}{2}.$$

In case of the rectangle S, for any function $f \in C(S)$ and two fixed points x_1, $x_2 \in [a, b]$ we can always write

$$|M_y f_1 - M_y f_2| \leq \|f_1 - f_2\|, \tag{4.53}$$

where $f_1 = f(x_1, y)$, $f_2 = f(x_2, y)$. But inequality (4.53) is not generally valid for other compact sets $Q \subset \mathbb{R}^2$. This is because for some fixed points x_1 and x_2, the projections of the sets $\{(x_1, y) \in Q\}$ and $\{(x_2, y) \in Q\}$ into the y axis do not coincide; hence $M_y f_1$ and $M_y f_2$ are the averaging operators over two different sets.

REMARK 4.35. It should be noted that in Halperin [**71**] it was proved that the algorithm converges as desired for any finite number of closed subspaces in the Hilbert space setting (without demanding closure of the sum of the subspaces). In fact, in a *uniformly convex* and *uniformly smooth* Banach space it was proved in Pinkus [**174**] that if the sum is closed then the algorithm converges (in the strong sense of the previous remark) for any finite number of closed subspaces.

CHAPTER 5

Applications to neural networks

Neural networks have increasingly been used in many areas of applied sciences. Most of the applications employ neural networks to approximate complex nonlinear functional dependencies on a high dimensional data set. The theoretical justification for such applications is that any continuous function can be approximated within an arbitrary precision by carefully selecting parameters in the network. The most commonly used model of neural networks is the *multilayer feedforward perceptron* (MLP) model. This model consists of a finite number of successive layers. The first and the last layers are called the input and the output layers, respectively. The intermediate layers are called hidden layers. MLP models are usually classified not by their number of layers, but by their number of hidden layers. In this chapter, we study approximation properties of the single and two hidden layer feedforward perceptron models. Our analysis is based on ridge functions and the Kolmogorov superposition theorem.

The material of this chapter may be found in [**68, 88, 89, 92**].

5.1. Universal approximation property of neural networks

In this section, we discuss the *universal approximation property* of neural networks. This property shows the ability of neural networks to approximate continuous multivariate functions on any compact set in $\mathbb{R}^d$ with arbitrary precision.

5.1.1. History of the problem. Approximation capabilities of neural networks have been investigated in a great deal of works over the last 30 years (see, e.g., [**6, 7, 17, 26, 29, 31, 38, 39, 41, 45, 67–69, 73, 102, 103, 107, 108, 135, 148, 157, 172, 178, 194**]). In this subsection we briefly review various results on the universal approximation property or *density property* of single hidden layer feedforward neural networks.

A single hidden layer feedforward perceptron model with r units in the hidden layer and input $\mathbf{x} = (x_1, \ldots, x_d)$ evaluates a function of the form

$$\sum_{i=1}^{r} c_i \sigma(\mathbf{w}^i \cdot \mathbf{x} - \theta_i), \tag{5.1}$$

where the *weights* $\mathbf{w}^i$ are vectors in $\mathbb{R}^d$, the *thresholds* θ_i and the coefficients c_i are real numbers and the *activation function* σ is a univariate function, which is considered to be continuous here. Note that in Eq. (5.1) each function $\sigma(\mathbf{w}^i \cdot \mathbf{x} - \theta_i)$ is a ridge function with the direction $\mathbf{w}^i$. The following are common examples of activation functions:

$$\sigma(t)=\frac{1}{1+e^{-t}} \qquad \text{(the squashing function),}$$

$$\sigma(t)=\begin{cases}0, & t\le -1,\\ \dfrac{t+1}{2}, & -1\le t\le 1,\\ 1, & t\ge 1\end{cases} \qquad \text{(the piecewise linear function),}$$

$$\sigma(t)=\frac{1}{\pi}\arctan t+\frac{1}{2} \qquad \text{(the arctan sigmoid function),}$$

$$\sigma(t)=\frac{1}{\sqrt{2\pi}}\int\limits_{-\infty}^{t} e^{-x^2/2}dx \qquad \text{(the Gaussian function).}$$

In many applications, it is convenient to take the activation function σ as a *sigmoidal function* which is defined as

$$\lim_{t\to-\infty}\sigma(t)=0 \quad \text{and} \quad \lim_{t\to+\infty}\sigma(t)=1.$$

The literature on neural networks abounds with the use of such functions and their superpositions. Note that all the above activation functions are sigmoidal.

The possibility of approximating a continuous function on a compact subset of the d-dimensional space by single hidden layer neural networks has been well studied in a number of papers. Different methods were used. Carroll and Dickinson [**28**] used the inverse Radon transformation to prove the universal approximation property of single hidden layer neural networks. Gallant and White [**59**] constructed a specific continuous, nondecreasing sigmoidal function, called a cosine squasher, from which it was possible to obtain any Fourier series. Thus their activation function had the density property. Cybenko [**41**] and Funahashi, [**57**] independently from each other, established that feedforward neural networks with a continuous sigmoidal activation function can approximate any continuous function within any degree of accuracy on compact subsets of $\mathbb{R}^d$. Cybenko's proof uses the functional analysis method, combining the Hahn-Banach theorem and the Riesz representation theorem, whiles Funahashi's proof applies the result of Irie and Miyake [**100**] on the integral representation of $f\in L_1(\mathbb{R}^n)$ functions, using a kernel which can be expressed as a difference of two sigmoidal functions. Hornik, Stinchcombe and White [**74**] applied the Stone-Weierstrass theorem, using trigonometric functions.

Kůrková [**134**] proved that staircase like functions of any sigmoidal type can approximate continuous functions on any compact subset of the real line within an arbitrary accuracy. This is effectively used in Kůrková's subsequent results, which show that a continuous multivariate function can be approximated arbitrarily well by two hidden layer neural networks with a sigmoidal activation function (see [**133**, **134**]).

Chen, Chen and Liu [**30**] extended the result of Cybenko by proving that any continuous function on a compact subset of $\mathbb{R}^d$ can be approximated by a single hidden layer feedforward network with a bounded (not necessarily continuous) sigmoidal activation function. Almost the same result was independently obtained by Jones [**104**].

Costarelli and Spigler [**38**] reconsidered Cybenko's approximation theorem and for a given function $f\in C[a,b]$ constructed certain sums of the form (5.1), which approximate f within any degree of accuracy. In their result, similar to [**30**], σ is bounded and sigmoidal. Therefore, when $\sigma \in C(\mathbb{R})$, the result can be viewed as a density result in $C[a,b]$ for the set of all functions of the form (5.1).

Chui and Li [**32**] proved that a single hidden layer network with a continuous sigmoidal activation function having integer weights and thresholds can approximate an arbitrary continuous function on a compact subset of $\mathbb{R}$. Ito [**101**] established a density result for continuous functions on a compact subset of $\mathbb{R}$ by neural networks with a sigmoidal function having only unit weights. Density properties of a single hidden layer network with a restricted set of weights were studied also in other papers (for a detailed discussion see [**172**]).

In many subsequent papers, which dealt with the density problem, nonsigmoidal activation functions were allowed. Among them are the papers by Stinchcombe and White [**194**], Cotter [**40**], Hornik [**73**], Mhaskar and Micchelli [**158**], and other researchers. The most general result in this direction belongs to Leshno, Lin, Pinkus and Schocken [**135**]. They proved that the necessary and sufficient condition for any continuous activation function to have the density property is that it not be a polynomial. This result shows the efficacy of the single hidden layer perceptron model within all possible choices of the activation function σ, provided that σ is continuous. In the next subsection, we discuss this result in detail. We see that ridge functions play an essential role in its proof.

5.1.2. The universal approximation theorem. The following theorem establishes the universal approximation property of feedforward neural networks.

THEOREM 5.1 (Leshno, Lin, Pinkus and Schocken [**135**]). *Assume $\sigma \in C(\mathbb{R})$. Then the set*

$$\mathcal{M}_d(\sigma) = span\ \{\sigma(\mathbf{w}\cdot\mathbf{x}-\theta) : \ \theta \in \mathbb{R},\ \mathbf{w} \in \mathbb{R}^d\}$$

is dense in $C(\mathbb{R}^d)$, in the topology of uniform convergence on compact sets, if and only if σ is not a polynomial.

PROOF. It is almost clear that if σ is a polynomial, then density does not hold. Indeed, if σ is a polynomial of degree k, then $\sigma(\mathbf{w}\cdot\mathbf{x}-\theta)$ is a multivariate polynomial of total degree at most k for every $\mathbf{w}$ and θ, and thus $\mathcal{M}(\sigma)$ is exactly the space of algebraic polynomials of total degree at most k, which cannot span $C(\mathbb{R}^d)$. The essence of this theorem is the converse result.

Assume σ is not a polynomial. The proof can be carried out in three steps.

Step 1. If $\sigma \in C^\infty(\mathbb{R})$, then the set $\mathcal{M}_1(\sigma)$ is dense in $C(\mathbb{R})$ in the topology of uniform convergence on compact sets.

Note that $[\sigma((w+h)x-\theta)-\sigma(wx-\theta)]/h \in \mathcal{M}_1(\sigma)$ for every $w,\theta \in \mathbb{R}$ and $h \neq 0$. Since $\sigma \in C^\infty(\mathbb{R})$, it follows that $\frac{d}{dw}\sigma(wx-\theta) \in \overline{\mathcal{M}_1(\sigma)}$. By the same argument $\frac{d^k}{dw^k}\sigma(wx-\theta) \in \overline{\mathcal{M}_1(\sigma)}$ for all $k \in \mathbb{N}$ and all $w,\theta \in \mathbb{R}$. Obviously,

$$\frac{d^k}{dw^k}\sigma(wx-\theta) = x^k\sigma^{(k)}(wx-\theta)$$

and since σ is not a polynomial, there exists $-\theta_k \in \mathbb{R}$ such that $\sigma^{(k)}(-\theta_k) \neq 0$. Therefore,

$$x^k\sigma^{(k)}(-\theta_k) = x^k\sigma^{(k)}(wx-\theta)|_{w=0,\theta=\theta_k} \in \overline{\mathcal{M}_1(\sigma)}$$

We see that $\overline{\mathcal{M}_1(\sigma)}$ contains all monomials and thus all algebraic polynomials. By the Weierstrass theorem this implies that $\overline{\mathcal{M}_1(\sigma)}$ is dense in $C(K)$ for every compact set $K \subset \mathbb{R}$.

Step 2. If $\sigma \in C(\mathbb{R})$, then the set $\mathcal{M}_1(\sigma)$ is dense in $C(\mathbb{R})$ in the topology of uniform convergence on compact sets.

We prove this assertion by contradiction. Assume, on the contrary, that $\mathcal{M}_1(\sigma)$ is not dense in $C(\mathbb{R})$ in the above mentioned topology. Consider the convolution of σ with a function $\phi \in C_0^\infty(\mathbb{R})$, that is, an infinitely differentiable function with compact support

$$(\sigma * \varphi)(t) = \int_{-\infty}^{+\infty} \sigma(t-y)\phi(y)dy.$$

Since σ and φ are continuous functions and φ has compact support the above integral converges for all t. Taking Riemann sums, it is easy to understand that $(\sigma * \varphi)(t)$ is contained in $\overline{\mathcal{M}_1(\sigma)}$. By the same way, we obtain that $(\sigma * \varphi)(wt-\theta)$ is contained in $\overline{\mathcal{M}_1(\sigma)}$. Thus, $\overline{\mathcal{M}_1(\sigma * \varphi)} \subset \overline{\mathcal{M}_1(\sigma)}$. Since $\sigma * \varphi \in C^\infty(\mathbb{R})$, we have from Step 1 that $t^k(\sigma * \varphi)^{(k)}(-\theta) \in \overline{\mathcal{M}_1(\sigma * \varphi)}$ for all $k \in \mathbb{N}$ and all $\theta \in \mathbb{R}$. Since $\mathcal{M}_1(\sigma)$ is not dense in $C(\mathbb{R})$, the monomial t^k is not in $\overline{\mathcal{M}_1(\sigma * \varphi)}$ for some k. Therefore, t^k is not in $\overline{\mathcal{M}_1(\sigma * \varphi)}$ for each $\varphi \in C_0^\infty(\mathbb{R})$. This means that $(\sigma * \varphi)^{(k)}(-\theta) = 0$ for all $\theta \in \mathbb{R}$. Hence for any $\varphi \in C_0^\infty(\mathbb{R})$ the convolution $\sigma * \varphi$ is a polynomial of degree at most $k-1$.

It is well known that there exists a sequence $\{\varphi_n\}_{n=1}^\infty \subset C_0^\infty(\mathbb{R})$ for which $\sigma * \varphi_n$ converges uniformly to σ on any compact set in $\mathbb{R}$. Since all $\sigma * \varphi_n$ are polynomials of degree at most $k-1$, the limit must be also a polynomial of degree at most $k-1$. Thus we obtain that σ is a polynomial. This contradicts our assumption.

Step 3. The linear span of continuous d-variate ridge functions is dense in $C(\mathbb{R}^d)$ in the topology of uniform convergence in compact sets.

This is the final step showing the importance of ridge functions in approximation theory of neural networks. The space

$$\mathcal{R} = span\left\{g(\mathbf{a}\cdot\mathbf{x}) : \mathbf{a} \in \mathbb{R}^d, g \in C(\mathbb{R})\right\}$$

is indeed dense in $C(\mathbb{R}^d)$. This fact follows in various ways, for example, it easily follows from the result of Vostrecov and Kreines, and Lin and Pinkus (see Theorem 1.4). Using this fact, we can finish the proof of the theorem as follows. Since ridge functions are dense, for any compact set $K \subset \mathbb{R}^d$, any $\varepsilon > 0$ and any $f \in C(\mathbb{R}^d)$, there exist functions $g_i \in C(\mathbb{R})$ and directions $\mathbf{a}^i \in \mathbb{R}^d$, $i = 1, \ldots, k$, such that

$$\max_{\mathbf{x}\in K}\left|f(\mathbf{x}) - \sum_{i=1}^k g_i(\mathbf{a}^i\cdot\mathbf{x})\right| < \varepsilon/2.$$

Let S stand for a compact set containing all compact sets $\left\{\mathbf{a}^i\cdot\mathbf{x} : \mathbf{x}\in K\right\}$, $i = 1, .., k$. Now since for any nonpolynomial $\sigma \in C(\mathbb{R})$ the set $\mathcal{M}_1(\sigma)$ is dense in $C(\mathbb{R})$, we obtain that

$$\left|g_i(t) - \sum_{j=1}^{m_i} c_{ij}\sigma(w^{ij}t - \theta_{ij})\right| < \varepsilon/2k,$$

for all $t \in S$, $i = 1, \dots, k$ and for some $m_i \in \mathbb{N}$, $c_{ij}, w^{ij}, \theta_{ij} \in \mathbb{R}$. Therefore,

$$\max_{\mathbf{x} \in K} \left| f(\mathbf{x}) - \sum_{i=1}^{k} \sum_{j=1}^{m_i} c_{ij} \sigma(w^{ij} \mathbf{a}^i \cdot \mathbf{x} - \theta_{ij}) \right| < \varepsilon.$$

This means that $\mathcal{M}_d(\sigma)$ is dense in $C(\mathbb{R}^d)$. □

REMARK 5.2. It is possible to make various restrictions on the activation function σ and the sets of weights and/or thresholds so that the universal approximation property (that is, Theorem 5.1) still holds. For details see **[172]**. For example, $\mathcal{M}_d(\sigma)$ is dense in $C(\mathbb{R}^d)$, if the weights $\mathbf{w}$ vary in $\{\lambda \mathbf{a} : \lambda \in \mathbb{R}, \mathbf{a} \in A\}$, where A is any open set on the sphere

$$S^{d-1} = \left\{ \mathbf{x} : \|\mathbf{x}\|_2 = \left(\sum_{i=1}^{d} x_i^2 \right)^{\frac{1}{2}} = 1 \right\}.$$

Indeed, since no nontrivial homogeneous polynomial vanishes on A, by Theorem 1.4, the linear span of ridge functions with directions $\mathbf{a} \in A$ is dense in $C(\mathbb{R}^d)$. Now replacing in Step 3 $\mathbf{a} \in \mathbb{R}^d$ with $\mathbf{a} \in A$ we obtain the desired assertion.

The above remark contains a simple sharpening of the universal approximation property of single hidden layer neural networks, which uses the density property of ridge functions. Some different kind of restrictions on the set of weights will be considered in the following section.

5.2. Single hidden layer neural networks with restricted weights

In this section, we consider single hidden layer neural networks with a set of weights consisting of a finite number of directions or straight lines. For certain activation functions, we characterize compact sets X in the d-dimensional space such that the corresponding neural network can approximate any continuous function on X.

5.2.1. Problem statement. In a number of papers it has been shown that a single hidden layer perceptron with a suitably restricted set of weights can also have the density property. For example, White and Stinchcombe **[194]** proved that a single layer network with a polygonal, polynomial spline or analytic activation function and a bounded set of weights has the density property. Ito **[103]** investigated this property of networks using monotone sigmoidal functions, with only weights located on the unit sphere. We see that weights required for the density property are not necessary to be of an arbitrarily large magnitude. But what if they are too restricted. How can one learn approximation properties of networks with an arbitrarily restricted set of weights? This problem is too difficult to be solved completely in this general formulation. But there are some cases that deserve a special attention. The most interesting case is, of course, neural networks with weights varying on a finite set of directions or lines. More precisely, let W be a set of weights consisting of a finite number of vectors (or straight lines) in $\mathbb{R}^d$. It is clear that if w varies only in W, the set $\mathcal{M}(\sigma)$ can not be dense in $C(\mathbb{R}^d)$ in the topology of uniform convergence on compacta (compact sets). In this case, one may want to determine boundaries of efficacy of the model. Over which compact sets $X \subset \mathbb{R}^d$ does the model preserve its general propensity to approximate arbitrarily well every continuous multivariate function? In the following subsections, we will consider this problem and give both sufficient and necessary conditions for well approximation (approximation with arbitrary precision) by networks with weights from a finite set of directions or lines. For a set W of weights consisting of two vectors, we show that there is a geometrically explicit solution to the problem. Then we discuss some aspects of the exact representation by neural networks with weights varying on finitely many straight lines. We will also obtain a formula for the approximation error of a multivariate continuous function by single hidden layer neural networks with weights from the union of two fixed straight lines.

5.2.2. Density results. In this subsection we give a sufficient and also a necessary conditions for approximation by neural networks with finitely many weights and with weights varying on a finite set of straight lines (through the origin).

Let X be a compact subset of $\mathbb{R}^d$. Consider the following set functions

$$\tau_i(Z) = \{\mathbf{x} \in Z : \ |p_i^{-1}(p_i(\mathbf{x})) \bigcap Z| \geq 2\}, \quad Z \subset X, \ i = 1, \dots, k,$$

where $p_i(\mathbf{x}) = \mathbf{a}^i \cdot \mathbf{x}$, $|Y|$ denotes the cardinality of a considered set Y. Define $\tau(Z)$ to be $\bigcap_{i=1}^k \tau_i(Z)$ and define $\tau^2(Z) = \tau(\tau(Z))$, $\tau^3(Z) = \tau(\tau^2(Z))$ and so on inductively. These functions first appeared in the work [**190**] by Sternfeld, where he investigated problems of representation by linear superpositions. Clearly, $\tau(Z) \supseteq \tau^2(Z) \supseteq \tau^3(Z) \supseteq \dots$ It is possible that for some n, $\tau^n(Z) = \emptyset$. In this case, one can see that Z does not contain a cycle. In general, if some set $Z \subset X$ forms a cycle, then $\tau^n(Z) = Z$. But the reverse is not true. Indeed, assume $Z = X = \{(0,0,\frac{1}{2}), (0,0,1), (0,1,0), (1,0,1), (1,1,0), (\frac{1}{2},\frac{1}{2},0), (\frac{1}{2},\frac{1}{2},\frac{1}{2})\}$ and $\mathbf{a}^i$, $i = 1,2,3$, are the coordinate directions in $\mathbb{R}^3$. It is not difficult to verify that X does not possess cycles with respect to these directions and at the same time $\tau(X) = X$ (and so $\tau^n(X) = X$ for every n).

Consider the linear combinations of ridge functions with fixed directions $\mathbf{a}^1, \dots, \mathbf{a}^k$

$$\mathcal{R}\left(\mathbf{a}^1, \dots, \mathbf{a}^k\right) = \left\{ \sum_{i=1}^k g_i\left(\mathbf{a}^i \cdot \mathbf{x}\right) : g_i \in C(\mathbb{R}), \ i = 1, \dots, k \right\}. \tag{5.2}$$

Let K be a family of functions defined on $\mathbb{R}^d$ and X be a subset of $\mathbb{R}^d$. By K_X we will denote the restriction of this family to X. Thus $\mathcal{R}_X\left(\mathbf{a}^1, \dots, \mathbf{a}^k\right)$ stands for the set of sums of ridge functions in (5.2) defined on X.

The following theorem is a particular case of the known general result of Sproston and Strauss [**187**] established for the sum of subalgebras of $C(X)$.

THEOREM 5.3. *Let X be a compact subset of $\mathbb{R}^d$. If $\cap_{n=1,2,\dots} \tau^n(X) = \emptyset$, then the set $\mathcal{R}_X\left(\mathbf{a}^1, \dots, \mathbf{a}^k\right)$ is dense in $C(X)$.*

In our analysis, we need the following lemma.

LEMMA 5.4. *If $\mathcal{R}_X\left(\mathbf{a}^1, \dots, \mathbf{a}^k\right)$ is dense in $C(X)$, then the set X does not contain a cycle with respect to the directions $\mathbf{a}^1, \dots, \mathbf{a}^k$.*

PROOF. Suppose the contrary. Suppose that the set X contains cycles. Each cycle $l = (x_1, \dots, x_n)$ and the associated vector $\lambda = (\lambda_1, \dots, \lambda_n)$ generate the functional

$$G_{l,\lambda}(f) = \sum_{j=1}^{n} \lambda_j f(x_j), \quad f \in C(X).$$

Clearly, $G_{l,\lambda}$ is linear and continuous with the norm $\sum_{j=1}^{n} |\lambda_j|$.It is not difficult to verify that $G_{l,\lambda}(g) = 0$ for all functions $g \in \mathcal{R}\left(\mathbf{a}^1, \dots, \mathbf{a}^k\right)$. Let f_0 be a continuous function such that $f_0(x_j) = 1$ if $\lambda_j > 0$ and $f_0(x_j) = -1$ if $\lambda_j < 0$, $j = 1, \dots, n$. For this function, $G_{l,\lambda}(f_0) \neq 0$. Thus, we have constructed a nonzero linear functional which belongs to the annihilator of the manifold $\mathcal{R}_X\left(\mathbf{a}^1, \dots, \mathbf{a}^k\right)$. This means that $\mathcal{R}_X\left(\mathbf{a}^1, \dots, \mathbf{a}^k\right)$ is not dense in $C(X)$. The obtained contradiction proves the lemma. □

Now we are ready to step forward from ridge function approximation to neural networks. Let $\sigma \in C(\mathbb{R})$ be a continuous activation function. For a subset $W \subset \mathbb{R}^d$, let $\mathcal{M}(\sigma; W, \mathbb{R})$ stand for the set of neural networks with weights from W. That is,

$$\mathcal{M}(\sigma; W, \mathbb{R}) = span\{\sigma(\mathbf{w} \cdot \mathbf{x} - \theta) : \ \mathbf{w} \in W, \ \theta \in \mathbb{R}\}.$$

THEOREM 5.5. *Let $\sigma \in C(\mathbb{R}) \cap L_p(\mathbb{R})$, where $1 \le p < \infty$, or σ be a continuous, bounded, nonconstant function, which has a limit at infinity (or minus infinity). Let $W = \{\mathbf{a}^1, \dots, \mathbf{a}^k\} \subset \mathbb{R}^d$ be the given set of weights and X be a compact subset of $\mathbb{R}^d$. The following assertions are valid:*

1. *If $\cap_{n=1,2,\dots} \tau^n(X) = \emptyset$, then the set $\mathcal{M}_X(\sigma; W, \mathbb{R})$ is dense in the space of all continuous functions on X.*
2. *If $\mathcal{M}_X(\sigma; W, \mathbb{R})$ is dense in $C(X)$, then the set X does not contain cycles.*

PROOF. *Part (1).* Let X be a compact subset of $\mathbb{R}^d$ for which $\cap_{n=1,2,\dots} \tau^n(X) = \emptyset$. By Theorem 5.3, the set $\mathcal{R}_X\left(\mathbf{a}^1, \dots, \mathbf{a}^k\right)$ is dense in $C(X)$. This means that for any positive real number ε there exist continuous univariate functions g_i, $i = 1, \dots, k$ such that

$$\left| f(\mathbf{x}) - \sum_{i=1}^{k} g_i\left(\mathbf{a}^i \cdot \mathbf{x}\right) \right| < \frac{\varepsilon}{k+1} \tag{5.3}$$

for all $\mathbf{x} \in X$. Since X is compact, the sets $Y_i = \{\mathbf{a}^i \cdot \mathbf{x} : \ \mathbf{x} \in X\}$, $i = 1, 2, \dots, k$ are also compact. In 1947, Schwartz [**180**] proved that continuous and p-th degree Lebesgue integrable univariate functions or continuous, bounded, nonconstant functions having a limit at infinity (or minus infinity) are not mean-periodic. Note that a function $f \in C(\mathbb{R}^d)$ is called mean periodic if the set *span* $\{f(\mathbf{x} - \mathbf{b}) : \ \mathbf{b} \in \mathbb{R}^d\}$ is not dense in $C(\mathbb{R}^d)$ in the topology of uniform convergence on compacta (see [**180**]). Thus, Schwartz proved that the set

$$span \ \{\sigma(y - \theta) : \ \theta \in \mathbb{R}\}$$

is dense in $C(\mathbb{R})$ in the topology of uniform convergence. We learned about this result from Pinkus [**172**, p.162]. This density result means that for the given ε

there exist numbers $c_{ij}, \theta_{ij} \in \mathbb{R}$, $i = 1, 2, \ldots, k$, $j = 1, \ldots, m_i$ such that

$$\left| g_i(y) - \sum_{j=1}^{m_i} c_{ij}\sigma(y - \theta_{ij}) \right| < \frac{\varepsilon}{k+1} \tag{5.4}$$

for all $y \in Y_i$, $i = 1, 2, \ldots, k$. From (5.3) and (5.4) we obtain that

$$\left\| f(\mathbf{x}) - \sum_{i=1}^{k}\sum_{j=1}^{m_i} c_{ij}\sigma(\mathbf{a}^i{\cdot}\mathbf{x} - \theta_{ij}) \right\|_{C(X)} < \varepsilon. \tag{5.5}$$

Hence $\overline{\mathcal{M}_X(\sigma; W, \mathbb{R})} = C(X)$.

Part (2). Let X be a compact subset of $\mathbb{R}^d$ and the set $\mathcal{M}_X(\sigma; W, \mathbb{R})$ be dense in $C(X)$. Then for an arbitrary positive real number ε, inequality (5.5) holds with some coefficients c_{ij}, θ_{ij}, $i = 1, 2$, $j = 1, \ldots, m_i$. Since for each $i = 1, 2, \ldots, k$, the function $\sum_{j=1}^{m_i} c_{ij}\sigma(\mathbf{a}^i{\cdot}\mathbf{x} - \theta_{ij})$ is a function of the form $g_i(\mathbf{a}^i{\cdot}\mathbf{x})$, the subspace $\mathcal{R}_X\left(\mathbf{a}^1, \ldots, \mathbf{a}^k\right)$ is dense in $C(X)$. Then by Lemma 5.4, the set X contains no cycles. □

The above theorem still holds if the set of weights $W = \{\mathbf{a}^1, \ldots, \mathbf{a}^k\}$ is replaced by the set $W_1 = \{t_1\mathbf{a}^1, \ldots, t_k\mathbf{a}^k : t_1, \ldots, t_k \in \mathbb{R}\}$. In fact, for W_1, the above restrictions on the activation function σ may be weakened.

THEOREM 5.6. *Assume $\sigma \in C(\mathbb{R})$ is not a polynomial. Let $W_1 = \{t_1\mathbf{a}^1, \ldots, t_k\mathbf{a}^k : t_1, \ldots, t_k \in \mathbb{R}\}$ be the given set of weights and X be a compact subset of $\mathbb{R}^d$. The following assertions are valid:*

1. *If $\bigcap_{n=1,2,\ldots} \tau^n(X) = \emptyset$, then the set $\mathcal{M}_X(\sigma; W_1, \mathbb{R})$ is dense in the space of all continuous functions on X.*
2. *If $\mathcal{M}_X(\sigma; W_1, \mathbb{R})$ is dense in $C(X)$, then the set X does not contain cycles.*

The proof of this theorem is similar to that of Theorem 5.5 and based on the above result of Leshno, Lin, Pinkus and Schocken (see Theorem 5.1).

The above example with the set

$$\{(0,0,\tfrac{1}{2}), (0,0,1), (0,1,0), (1,0,1), (1,1,0), (\tfrac{1}{2},\tfrac{1}{2},0), (\tfrac{1}{2},\tfrac{1}{2},\tfrac{1}{2})\}$$

shows that the sufficient condition in part (1) of Theorem 5.5 is not necessary. The necessary condition in part (2), in general, is not sufficient. But this is not easily seen. The following is the nontrivial example, which shows that nonexistence of cycles is not sufficient for the density $\overline{\mathcal{M}_X(\sigma; W, \mathbb{R})} = C(X)$.

EXAMPLE 5.7. For the sake of simplicity, we restrict ourselves to $\mathbb{R}^2$. Let $\mathbf{a}^1 = (1; 1)$, $\mathbf{a}^2 = (1; -1)$ and the set of weights $W = \{\mathbf{a}^1, \mathbf{a}^2\}$. The set X can be constructed as follows. Let X_1 be the union of the four line segments $[(-3; 0), (-1; 0)]$, $[(-1; 2), (1; 2)]$, $[(1; 0), (3; 0)]$ and $[(-1; -2), (1; -2)]$. Rotate one segment in X_1 90° about its center and remove the middle one-third from each line segment. The obtained set denote by X_2. By the same way, one can construct X_3, X_4, and so on. It is clear that the set X_i has 2^{i+1} line segments. Let X be a limit of the sets X_i, $i = 1, 2, \ldots$. Note that there are no cycles.

By S_i, $i = \overline{1, 4}$, denote the closed discs with the unit radius and centered at the points $(-2; 0)$, $(0; 2)$, $(2; 0)$ and $(0; -2)$ respectively. Consider a continuous function

f_0 such that $f_0(\mathbf{x}) = 1$ for $\mathbf{x} \in (S_1 \cup S_3) \cap X$, $f_0(\mathbf{x}) = -1$ for $\mathbf{x} \in (S_2 \cup S_4) \cap X$, and $-1 < f_0(\mathbf{x}) < 1$ elsewhere on $\mathbb{R}^2$. Let $p = (\mathbf{y}^1, \mathbf{y}^2, \dots)$ be any infinite path in X. Note that the points $\mathbf{y}^i$, $i = 1, 2, \dots$, are alternatively in the sets $(S_1 \cup S_3) \cap X$ and $(S_2 \cup S_4) \cap X$. Obviously,

$$E(f_0, X) \overset{def}{=} \inf_{g \in \mathcal{R}_X(\mathbf{a}^1, \mathbf{a}^2)} \|f_0 - g\|_{C(X)} \le \|f_0\|_{C(X)} = 1. \tag{5.6}$$

For each positive integer $k = 1, 2, \dots$, set $p_k = (\mathbf{y}^1, \dots, \mathbf{y}^k)$ and consider the path functionals

$$G_{p_k}(f) = \frac{1}{k} \sum_{i=1}^{k} (-1)^{i-1} f(\mathbf{y}^i).$$

G_{p_k} is a continuous linear functional obeying the following obvious properties:

(1) $\|G_{p_k}\| = G_{p_k}(f_0) = 1$;

(2) $G_{p_k}(g_1 + g_2) \le \frac{2}{k}(\|g_1\| + \|g_2\|)$ for ridge functions $g_1 = g_1\left(\mathbf{a}^1 \cdot \mathbf{x}\right)$ and $g_2 = g_2\left(\mathbf{a}^2 \cdot \mathbf{x}\right)$.

By property (1), the sequence $\{G_{p_k}\}_{k=1}^{\infty}$ has a weak* cluster point. Let us denote this point by G. By property (2), $G \in \mathcal{R}_X\left(\mathbf{a}^1, \mathbf{a}^2\right)^{\perp}$. Therefore,

$$1 = G(f_0) = G(f_0 - g) \le \|f_0 - g\|_{C(X)} \quad \text{for any } g \in \mathcal{R}_X\left(\mathbf{a}^1, \mathbf{a}^2\right).$$

Taking inf over g in the right-hand side of the last inequality, we obtain that $1 \le E(f_0, X)$. Now it follows from (5.6) that $E(f_0, X) = 1$. Recall that $\mathcal{M}_X(\sigma; W, \mathbb{R}) \subset \mathcal{R}_X\left(\mathbf{a}^1, \mathbf{a}^2\right)$. Thus

$$\inf_{h \in \mathcal{M}_X(\sigma; W, \mathbb{R})} \|f - h\|_{C(X)} \ge 1.$$

The last inequality finally shows that $\overline{\mathcal{M}_X(\sigma; W, \mathbb{R})} \neq C(X)$.

For neural networks with weights consisting of only two vectors (or directions) the problem of density becomes more clear. In this case, under some minor restrictions on X, the necessary condition in part (2) of Theorem 5.5 (nonexistence of cycles) is also sufficient for the density of $\mathcal{M}_X(\sigma; W, \mathbb{R})$ in $C(X)$. These restrictions are imposed on the following equivalent classes of X induced by paths. The relation $\mathbf{x} \sim \mathbf{y}$ when $\mathbf{x}$ and $\mathbf{y}$ belong to some path in a given compact set $X \subset \mathbb{R}^d$ defines an equivalence relation. Recall that the equivalence classes are called orbits (see Section 1.3).

THEOREM 5.8. *Let $\sigma \in C(\mathbb{R}) \cap L_p(\mathbb{R})$, where $1 \le p < \infty$, or σ be a continuous, bounded, nonconstant function, which has a limit at infinity (or minus infinity). Let $W = \{\mathbf{a}^1, \mathbf{a}^2\} \subset \mathbb{R}^d$ be the given set of weights and X be a compact subset of $\mathbb{R}^d$ with all its orbits closed. Then $\mathcal{M}_X(\sigma; W, \mathbb{R})$ is dense in the space of all continuous functions on X if and only if X contains no closed paths with respect to the directions $\mathbf{a}^1$ and $\mathbf{a}^2$.*

PROOF. *Sufficiency.* Let X be a compact subset of $\mathbb{R}^d$ with all its orbits closed. Besides, let X contain no closed paths. By Theorem 1.27 (see Section 1.3), the set $\mathcal{R}_X\left(\mathbf{a}^1, \mathbf{a}^2\right)$ is dense in $C(X)$. This means that for any positive real number ε there exist continuous univariate functions g_1 and g_2 such that

$$\left|f(\mathbf{x}) - g_1\left(\mathbf{a}^1 \cdot \mathbf{x}\right) - g_2\left(\mathbf{a}^2 \cdot \mathbf{x}\right)\right| < \frac{\varepsilon}{3} \tag{5.7}$$

for all $\mathbf{x} \in X$. Since X is compact, the sets $Y_i = \{\mathbf{a}^i \cdot \mathbf{x} : \mathbf{x} \in X\}$, $i = 1, 2$, are also compacts. As mentioned above, Schwartz [**180**] proved that continuous and p-th degree Lebesgue integrable univariate functions or continuous, bounded, nonconstant functions having a limit at infinity (or minus infinity) are not mean-periodic. Thus, the set

$$span \{\sigma(y - \theta) : \theta \in \mathbb{R}\}$$

is dense in $C(\mathbb{R})$ in the topology of uniform convergence. This density result means that for the given ε there exist numbers $c_{ij}, \theta_{ij} \in \mathbb{R}$, $i = 1, 2$, $j = 1, \ldots, m_i$ such that

$$\left| g_i(y) - \sum_{j=1}^{m_i} c_{ij}\sigma(y - \theta_{ij}) \right| < \frac{\varepsilon}{3} \tag{5.8}$$

for all $y \in Y_i$, $i = 1, 2$. From (5.7) and (5.8) we obtain that

$$\left\| f(\mathbf{x}) - \sum_{i=1}^{2} \sum_{j=1}^{m_i} c_{ij}\sigma(\mathbf{a}^i \cdot \mathbf{x} - \theta_{ij}) \right\|_{C(X)} < \varepsilon. \tag{5.9}$$

Hence $\overline{\mathcal{M}_X(\sigma; W, \mathbb{R})} = C(X)$.

Necessity. Let X be a compact subset of $\mathbb{R}^n$ with all its orbits closed and the set $\mathcal{M}_X(\sigma; W, \mathbb{R})$ be dense in $C(X)$. Then for an arbitrary positive real number ε, inequality (5.9) holds with some coefficients c_{ij}, θ_{ij}, $i = 1, 2$, $j = 1, \ldots, m_i$. Since for $i = 1, 2$, $\sum_{j=1}^{m_i} c_{ij}\sigma(\mathbf{a}^i \cdot \mathbf{x} - \theta_{ij})$ is a function of the form $g_i(\mathbf{a}^i \cdot \mathbf{x})$, the subspace $\mathcal{R}_X(\mathbf{a}^1, \mathbf{a}^2)$ is dense in $C(X)$. Then by Theorem 1.27, the set X contains no closed paths. □

REMARK 5.9. It can be shown that the necessity of the theorem is valid without any restriction on orbits of X. Indeed if X contains a closed path, then it contains a closed path $p = (\mathbf{x}^1, \ldots, \mathbf{x}^{2m})$ with different points. The functional $G_p = \sum_{i=1}^{2m} (-1)^{i-1} f(\mathbf{x}^i)$ belongs to the annihilator of the subspace $\mathcal{R}_X(\mathbf{a}^1, \mathbf{a}^2)$. There exist nontrivial continuous functions f_0 on X such that $G_p(f_0) \neq 0$. Take, for example, any continuous function f_0 taking values $+1$ at $\{\mathbf{x}^1, \mathbf{x}^3, \ldots, \mathbf{x}^{2m-1}\}$, -1 at $\{\mathbf{x}^2, \mathbf{x}^4, \ldots, \mathbf{x}^{2m}\}$ and $-1 < f_0(\mathbf{x}) < 1$ elsewhere. This shows that the subspace $\mathcal{R}_X(\mathbf{a}^1, \mathbf{a}^2)$ is not dense in $C(X)$. But in this case, the set $\mathcal{M}_X(\sigma; W, \mathbb{R})$ cannot be dense in $C(X)$. The obtained contradiction means that our assumption is not true and X contains no closed paths.

Theorem 5.8 remains valid if the set of weights $W = \{\mathbf{a}^1, \mathbf{a}^2\}$ is replaced by the set $W_1 = \{t_1\mathbf{a}^1, t_2\mathbf{a}^2 : t_1, t_2 \in \mathbb{R}\}$. In fact, for the set W_1, the required conditions on σ may be weakened. As in Theorem 5.6, the activation function σ can be taken only non-polynomial.

THEOREM 5.10. *Assume $\sigma \in C(\mathbb{R})$ is not a polynomial. Let $\mathbf{a}^1$ and $\mathbf{a}^2$ be fixed vectors and $W_1 = \{t_1\mathbf{a}^1, t_2\mathbf{a}^2 : t_1, t_2 \in \mathbb{R}\}$ be the set of weights. Let X be a compact subset of $\mathbb{R}^d$ with all its orbits closed. Then $\mathcal{M}_X(\sigma; W_1, \mathbb{R})$ is dense in the space of all continuous functions on X if and only if X contains no closed paths with respect to the directions $\mathbf{a}^1$ and $\mathbf{a}^2$.*

The proof is analogous to that of Theorem 5.8 and based on the above result of Leshno, Lin, Pinkus and Schocken [**135**].

EXAMPLES:

(a) Let $\mathbf{a}^1$ and $\mathbf{a}^2$ be two noncollinear vectors in $\mathbb{R}^2$. Let $B = B_1 \dots B_k$ be a broken line with the sides B_iB_{i+1}, $i = 1, \dots, k-1$, alternatively perpendicular to $\mathbf{a}^1$ and $\mathbf{a}^2$. Besides, let B does not contain vertices of any parallelogram with sides perpendicular to these vectors. Then the set $\mathcal{M}_B(\sigma; \{\mathbf{a}^1, \mathbf{a}^2\}, \mathbb{R})$ is dense in $C(B)$.

(b) Let $\mathbf{a}^1$ and $\mathbf{a}^2$ be two noncollinear vectors in $\mathbb{R}^2$. If X is the union of two parallel line segments, not perpendicular to any of the vectors $\mathbf{a}^1$ and $\mathbf{a}^2$, then the set $\mathcal{M}_X(\sigma; \{\mathbf{a}^1, \mathbf{a}^2\}, \mathbb{R})$ is dense in $C(X)$.

(c) Let now $\mathbf{a}^1$ and $\mathbf{a}^2$ be two collinear vectors in $\mathbb{R}^2$. Note that in this case any path consisting of two points is automatically closed. Thus the set $\mathcal{M}_X(\sigma; \{\mathbf{a}^1, \mathbf{a}^2\}, \mathbb{R})$ is dense in $C(X)$ if and only if X contains no path different from a singleton. A simple example is a line segment not perpendicular to the given direction.

(d) Let X be a compact set with an interior point. Then Theorem 5.8 fails, since any such set contains vertices of some parallelogram with sides perpendicular to the given directions $\mathbf{a}^1$ and $\mathbf{a}^2$, that is a closed path.

5.2.3. A necessary condition for the representation by neural networks. In this subsection we give a necessary condition for the representation of functions by neural networks with weights from a finitely many straight lines. Before formulating our result, we introduce new objects, namely *semicycles* with respect to directions $\mathbf{a}^1, \dots, \mathbf{a}^k \in \mathbb{R}^d \setminus \{\mathbf{0}\}$.

DEFINITION 5.11. A set of points $l = (\mathbf{x}^1, \dots, \mathbf{x}^n) \subset \mathbb{R}^d$ is called a semicycle with respect to the directions $\mathbf{a}^1, \dots, \mathbf{a}^k$ if there exists a vector $\lambda = (\lambda_1, \dots, \lambda_n) \in \mathbf{Z}^n \setminus \{\mathbf{0}\}$ such that for any $i = 1, \dots, k$, we have

$$\sum_{j=1}^{n} \lambda_j \delta_{\mathbf{a}^i \cdot \mathbf{x}^j} = \sum_{s=1}^{r_i} \lambda_{i_s} \delta_{\mathbf{a}^i \cdot \mathbf{x}^{i_s}}, \quad \text{where } r_i \leq k. \tag{5.10}$$

Here δ_a is the characteristic function of the single point set $\{a\}$.

Note that for $i = 1, \dots, k$, the set $\{\lambda_{i_s},\ s = 1, \dots, r_i\}$ is a subset of the set $\{\lambda_j,\ j = 1, \dots, n\}$. Thus, Eq. (5.10) means that for each i, we actually have at most k terms in the sum $\sum_{j=1}^{n} \lambda_j \delta_{\mathbf{a}^i \cdot \mathbf{x}^j}$.

Recall that if in (5.10) for any $i = 1, \dots, k$, we have

$$\sum_{j=1}^{n} \lambda_j \delta_{\mathbf{a}^i \cdot \mathbf{x}^j} = 0,$$

then the set $l = (\mathbf{x}^1, \dots, \mathbf{x}^n)$ is a cycle with respect to the directions $\mathbf{a}^1, \dots, \mathbf{a}^k$ (see Section 1.2). Thus a cycle is a special case of a semicycle.

Let us give a simple example of a semicycle. Assume $k = 2$ and $\mathbf{a}^1 \cdot \mathbf{x}^1 = \mathbf{a}^1 \cdot \mathbf{x}^2$, $\mathbf{a}^2 \cdot \mathbf{x}^2 = \mathbf{a}^2 \cdot \mathbf{x}^3$, $\mathbf{a}^1 \cdot \mathbf{x}^3 = \mathbf{a}^1 \cdot \mathbf{x}^4$,..., $\mathbf{a}^2 \cdot \mathbf{x}^{n-1} = \mathbf{a}^2 \cdot \mathbf{x}^n$. Then it is not difficult to see that for a vector $\lambda = (\lambda_1, \dots, \lambda_n)$ with the components $\lambda_j = (-1)^j$, the following equalities hold:

$$\sum_{j=1}^{n} \lambda_j \delta_{\mathbf{a}^1 \cdot \mathbf{x}^j} = \lambda_n \delta_{\mathbf{a}^1 \cdot \mathbf{x}^n},$$
$$\sum_{j=1}^{n} \lambda_j \delta_{\mathbf{a}^2 \cdot \mathbf{x}^j} = \lambda_1 \delta_{\mathbf{a}^2 \cdot \mathbf{x}^1}.$$

Thus, by Definition 5.11, the set $l = \{\mathbf{x}^1, \dots, \mathbf{x}^n\}$ is a semicycle with respect to the directions $\mathbf{a}^1$ and $\mathbf{a}^2$. Note that this set, in the given order of its points, forms a path with respect to the directions $\mathbf{a}^1$ and $\mathbf{a}^2$ (see Section 1.3). It is not difficult to see that any path with respect to $\mathbf{a}^1$ and $\mathbf{a}^2$ is a semicycle with respect to these directions. But semicycles may also involve some union of paths.

Note that one can construct many semicycles by adding not more than k arbitrary points to a cycle with respect to the directions $\mathbf{a}^1, \mathbf{a}^2, \dots, \mathbf{a}^k$.

A cycle (or semicycle) l is called a *q-cycle* (*q-semicycle*) if the vector λ associated with l can be chosen so that $|\lambda_i| \le q$, $i = 1, \dots, n$, and q is the minimal number with this property. Note that here we consider only associated vectors with nonzero integer components.

The semicycle considered above is a 1-semicycle. If in that example, $\mathbf{a}^2 \cdot \mathbf{x}^{n-1} = \mathbf{a}^2 \cdot \mathbf{x}^1$, then the set $\{x_1, x_2, \dots, x_{n-1}\}$ is a 1-cycle. Let us give a simple example of a 2-cycle with respect to the directions $\mathbf{a}^1 = (1, 0)$ and $\mathbf{a}^2 = (0, 1)$. Consider the union

$$\{0, 1\}^2 \cup \{0, 2\}^2 = \{(0, 0), (1, 1), (2, 2), (0, 1), (1, 0), (0, 2), (2, 0)\}.$$

It is not difficult to see that this set is a 2-cycle with the associated vector $(2, 1, 1, -1, -1, -1, -1)$. Similarly, one can construct a q-cycle or q-semicycle for any positive integer q.

THEOREM 5.12. *Assume $W = \{t_1\mathbf{a}^1, \dots, t_k\mathbf{a}^k : t_1, \dots, t_k \in \mathbb{R}\}$ is the given set of weights. If $\mathcal{M}_X(\sigma; W, \mathbb{R}) = C(X)$, then X contains no cycles and the lengths (number of points) of all q-semicycles in X are bounded by some positive integer.*

PROOF. Let $\mathcal{M}_X(\sigma; W, \mathbb{R}) = C(X)$. Then $\mathcal{R}_1 + \mathcal{R}_2 + \cdots + \mathcal{R}_k = C(X)$, where

$$\mathcal{R}_i = \{g_i(\mathbf{a}^i \cdot \mathbf{x}) : g_i \in C(\mathbb{R})\},\ i = 1, 2, \dots, k.$$

Consider the linear space

$$\mathcal{U} = \prod_{i=1}^{k} \mathcal{R}_i = \{(g_1, \dots, g_k) : \ g_i \in \mathcal{R}_i,\ i = 1, \dots, k\}$$

endowed with the norm

$$\|(g_1, \dots, g_k)\| = \|g_1\| + \cdots + \|g_k\|.$$

By $\mathcal{U}^*$ denote the dual space of $\mathcal{U}$. Each functional $F \in \mathcal{U}^*$ can be written as

$$F = F_1 + \cdots + F_k,$$

where the functionals $F_i \in \mathcal{R}_i^*$ and

$$F_i(g_i) = F[(0, \dots, g_i, \dots, 0)], \quad i = 1, \dots, k.$$

We see that the functional F determines the collection $(F_1,\dots,F_k)$. Conversely, every collection $(F_1,\dots,F_k)$ of continuous linear functionals $F_i\in\mathcal{R}_i^*$, $i=1,\dots,k$, determines the functional $F_1+\cdots+F_k$ on $\mathcal{U}$. Considering this, in what follows, elements of $\mathcal{U}^*$ will be denoted by $(F_1,\dots,F_k)$.

It is not difficult to verify that

$$\|(F_1,\dots,F_k)\|=\max\{\|F_1\|,\dots,\|F_k\|\}. \tag{5.11}$$

Let $l=(\mathbf{x}^1,\dots,\mathbf{x}^n)$ be any q-semicycle (with respect to the directions $\mathbf{a}^1,\dots,\mathbf{a}^k$) in X and $\lambda=(\lambda_1,\dots,\lambda_n)$ be a vector associated with it. Consider the following functional

$$G_{l,\lambda}(f)=\sum_{j=1}^n\lambda_j f(\mathbf{x}^j),\quad f\in C(X).$$

Since l satisfies (5.10), for each function $g_i\in\mathcal{R}_i$, $i=1,\dots,k$, we have

$$G_{l,\lambda}(g_i)=\sum_{j=1}^n\lambda_j g_i(\mathbf{a}^i\cdot\mathbf{x}^j)=\sum_{s=1}^{r_i}\lambda_{i_s}g_i(\mathbf{a}^i\cdot\mathbf{x}^{i_s}), \tag{5.12}$$

where $r_i\le k$. That is, for each set $\mathcal{R}_i$, $G_{l,\lambda}$ can be reduced to a functional defined with the help of not more than k points of the semicycle l.

Consider the operator

$$A:\mathcal{U}\to C(X),\quad A[(g_1,\dots,g_k)]=g_1+\cdots+g_k.$$

Clearly, A is a linear continuous operator with the norm $\|A\|=1$. Besides, since $\mathcal{R}_1+\mathcal{R}_2+\cdots+\mathcal{R}_k=C(X)$, A is a surjection. Consider also the conjugate operator

$$A^*:C(X)^*\to\mathcal{U}^*,\ A^*[H]=(F_1,\dots,F_k),$$

where $F_i(g_i)=H(g_i)$, for any $g_i\in\mathcal{R}_i$, $i=1,\dots,k$. Set $A^*[G_{l,\lambda}]=(G_1,\dots,G_k)$. From (5.12) it follows that

$$|G_i(g_i)|=|G_{l,\lambda}(g_i)|\le\|g_i\|\sum_{s=1}^{r_i}|\lambda_{i_s}|\le kq\|g_i\|,\quad i=1,\dots,k,$$

Therefore,

$$\|G_i\|\le kq,\quad i=1,\dots,k.$$

From (5.11) we obtain that

$$\|A^*[G_{l,\lambda}]\|=\|(G_1,\dots,G_k)\|\le kq. \tag{5.13}$$

Since A is a surjection, there exists a positive real number δ such that

$$\|A^*[H]\|>\delta\|H\|$$

for any functional $H\in C(X)^*$(see [**177**, p.100]). Taking into account that $\|G_{l,\lambda}\|=\sum_{j=1}^n|\lambda_j|$, for the functional $G_{l,\lambda}$ we have

$$\|A^*[G_{l,\lambda}]\|>\delta\sum_{j=1}^n|\lambda_j|. \tag{5.14}$$

It follows from (5.13) and (5.14) that

$$\delta<\frac{kq}{\sum_{j=1}^n|\lambda_j|}.$$

The last inequality shows that n (the length of the arbitrarily chosen q-semicycle l) cannot be as great as possible, otherwise $\delta = 0$. This simply means that there must be some positive integer bounding the lengths of all q-semicycles in X.

It remains to show that there are no cycles in X. Indeed, if $l = (\mathbf{x}^1, \dots, \mathbf{x}^n)$ is a cycle in X and $\lambda = (\lambda_1, \dots, \lambda_n)$ is a vector associated with it, then the above functional $G_{l,\lambda}$ annihilates all functions from $\mathcal{R}_1 + \mathcal{R}_2 + \cdots + \mathcal{R}_k$. On the other hand, $G_{l,\lambda}(f) = \sum_{j=1}^{n} |\lambda_j| \neq 0$ for a continuous function f on X satisfying the conditions $f(\mathbf{x}^j) = 1$ if $\lambda_j > 0$ and $f(\mathbf{x}^j) = -1$ if $\lambda_j < 0$, $j = 1, \dots, n$. This implies that $\mathcal{R}_1 + \mathcal{R}_2 + \cdots + \mathcal{R}_k \neq C(X)$. Since $\mathcal{M}_X(\sigma; W, \mathbb{R}) \subseteq \mathcal{R}_1 + \mathcal{R}_2 + \cdots + \mathcal{R}_k$, we obtain that $\mathcal{M}_X(\sigma; W, \mathbb{R}) \neq C(X)$ on the contrary to our assumption. □

REMARK 5.13. Assume $\mathcal{M}_X(\sigma; W, \mathbb{R})$ is dense in $C(X)$. Is it necessarily closed? Theorem 5.12 may describe cases when it is not. For example, let $\mathbf{a}^1 = (1; -1)$, $\mathbf{a}^2 = (1; 1)$, $W = \{\mathbf{a}^1, \mathbf{a}^2\}$ and σ be any continuous, bounded and nonconstant function, which has a limit at infinity. Consider the set

$$\begin{aligned} X &= \{(2; \frac{2}{3}), (\frac{2}{3}; \frac{2}{3}), (0; 0), (1; 1), (1 + \frac{1}{2}; 1 - \frac{1}{2}), (1 + \frac{1}{2} + \frac{1}{4}; 1 - \frac{1}{2} + \frac{1}{4}), \\ &\quad (1 + \frac{1}{2} + \frac{1}{4} + \frac{1}{8}; 1 - \frac{1}{2} + \frac{1}{4} - \frac{1}{8}), \dots\}. \end{aligned}$$

It is clear that X is a compact set with all its orbits closed. (In fact, there is only one orbit, which coincides with X). Hence, by Theorem 5.8, $\overline{\mathcal{M}_X(\sigma; W, \mathbb{R})} = C(X)$. But by Theorem 5.12, $\mathcal{M}_X(\sigma; W, \mathbb{R}) \neq C(X)$. Therefore, the set $\mathcal{M}_X(\sigma; W, \mathbb{R})$ is not closed in $C(X)$.

5.2.4. Approximation error. In this subsection, we obtain an exact formula for the approximation error of a multivariate continuous function by single hidden layer neural networks with weights varying on two fixed straight lines (through the origin).

Let $\sigma \in C(\mathbb{R})$, $l_1 = \{t\mathbf{a} : t \in \mathbb{R}\}$, $l_2 = \{t\mathbf{b} : t \in \mathbb{R}\}$, where $\mathbf{a}$ and $\mathbf{b}$ be two nonzero vectors in $\mathbb{R}^d$. Consider the set

$$\mathcal{N}(\sigma) = \left\{ \sum_{i=1}^{r} c_i \sigma(\mathbf{w}^i \cdot \mathbf{x} - \theta_i) : r \in \mathbb{N},\ c_i, \theta_i \in \mathbb{R},\ \mathbf{w}^i \in l_1 \cup l_2 \right\}.$$

This is the set of single hidden layer neural networks with weights from these two straight lines l_1 and l_2. Let Q be a compact subset of $\mathbb{R}^d$ and $f \in C(Q)$. Consider the approximation of f by neural networks from $\mathcal{N}(\sigma)$. The approximation error is defined as

$$E(f, \mathcal{N}(\sigma)) \stackrel{def}{=} dist(f, \mathcal{N}(\sigma)) = \inf_{g \in \mathcal{N}(\sigma)} \|f - g\|.$$

In the following, we assume that a compact set $Q \subset \mathbb{R}^d$ contains a closed path. Such sets exist in profusion. For example, any set with interior contains closed paths. The case when Q does not contain a closed path was considered in the previous subsections. Note that for most such Q, the set $\mathcal{N}(\sigma)$ is dense in $C(Q)$ (see Theorem 5.10); hence $E(f, \mathcal{N}(\sigma)) = 0$ for any continuous function f on Q.

We associate a closed path $p = (\mathbf{p}_1, \mathbf{p}_2, \dots, \mathbf{p}_{2n})$ in Q with the functional

$$G_p(f) = \frac{1}{2n} \sum_{k=1}^{2n} (-1)^{k+1} f(\mathbf{p}_k).$$

The following theorem establishes a lower bound error estimate in approximation with elements of $\mathcal{N}(\sigma)$.

THEOREM 5.14. *Let σ be any continuous activation function. Then*

$$\sup_{p \subset Q} |G_p(f)| \leq E(f, \mathcal{N}(\sigma)), \tag{5.15}$$

where the sup is taken over all closed paths. Moreover, inequality (5.15) *is sharp, i.e. there exist functions for which* (5.15) *turns into equality.*

PROOF. Consider an element of $\mathcal{N}(\sigma)$. This is a sum of the functions $f_i(\mathbf{x}) = c_i\sigma(\mathbf{w}^i{\cdot}\mathbf{x} - \theta_i)$, $i = 1, \ldots, r$. Note that for each $c_i, \theta_i \in \mathbb{R}$, $f_i(\mathbf{x})$ is a ridge function $g(\mathbf{w}^i{\cdot}\mathbf{x})$ with the direction $\mathbf{w}^i$. Thus, we see that any network $\sum_{i=1}^{r} c_i\sigma(\mathbf{w}^i{\cdot}\mathbf{x} - \theta_i)$ in $\mathcal{N}(\sigma)$ can be written as a sum $g(\mathbf{a} \cdot \mathbf{x}) + h(\mathbf{b} \cdot \mathbf{x})$. Introduce the set of ridge functions

$$\mathcal{R}(\mathbf{a}, \mathbf{b}) = \{g(\mathbf{a} \cdot \mathbf{x}) + h(\mathbf{b} \cdot \mathbf{x}) : g, h \in C(\mathbb{R})\}.$$

Since $\mathcal{N}(\sigma) \subset \mathcal{R}(\mathbf{a}, \mathbf{b})$, we obtain that $E(f, \mathcal{R}(\mathbf{a}, \mathbf{b})) \leq E(f, \mathcal{N}(\sigma))$. Now (5.15) follows from Lemma 1.19.

Note that the sharpness of (5.15) does not follow directly from the sharpness of inequality (1.13) in Lemma 1.19. But it can be proven by the same method. To see this, let $p = (\mathbf{p}_1, \ldots, \mathbf{p}_{2n})$ be any closed path in Q with distinct points. By Urysohn's lemma, there exists a continuous function $s = s(\mathbf{x})$ on Q such that $s(\mathbf{p}_i) = 1$, $i = 1, 3, \ldots, 2n - 1$, $s(\mathbf{p}_i) = -1$, $i = 2, 4, \ldots, 2n$ and $-1 < s(\mathbf{x}) < 1$ elsewhere. For this function we have

$$G_p(s) = \|s\| = 1.$$

On the other hand, since $0 \in \mathcal{N}(\sigma)$,

$$E(s, \mathcal{N}(\sigma)) \leq \|s\|.$$

Therefore,

$$E(s, \mathcal{N}(\sigma)) \leq G_p(s).$$

Considering the last inequality, we obtain from (5.15) that

$$\sup_{p \subset Q} |G_p(s)| = E(s, \mathcal{N}(\sigma)).$$

This shows the sharpness of (5.15). Theorem 5.14 has been proved. □

In Theorem 5.14 one can take an arbitrary continuous univariate function σ as an activation function. This raises a question whether there is σ for which (5.15) turns into equality for all $f \in C(Q)$. In [**97**] we employed a specific activation function φ, which answered this question positively. The φ was algorithmically constructed in Guliyev and Ismailov [**67**], and used further in [**68,69**]. This function was able to give an exact equality in (5.15). Below we show that, in fact, a much wider class of functions, namely the class of nonpolynomial activation functions possesses such a property.

In Theorem 5.15 below, we assume the *proximinality* of the space $\mathcal{R}(\mathbf{a}, \mathbf{b})$. Recall that the proximinality of $\mathcal{R}(\mathbf{a}, \mathbf{b})$ in $C(Q)$ was investigated in Section 1.5.

THEOREM 5.15. *Assume $Q \subset \mathbb{R}^d$ is a convex compact set, $f \in C(Q)$ and the space $\mathcal{R}(\mathbf{a}, \mathbf{b})$ is proximinal in $C(Q)$. Then for any nonpolynomial activation function σ the error of approximation from the class of single hidden layer neural networks $\mathcal{N}(\sigma)$ can be computed by the formula*

$$E(f, \mathcal{N}(\sigma)) = \sup_{p \subset Q} |G_p(f)|,$$

where the sup is taken over all closed paths.

PROOF. Since $\mathcal{R}(\mathbf{a},\mathbf{b})$ is proximinal in $C(Q)$, for the function f there exists an extremal element $v_0(\mathbf{x}) = g_0(\mathbf{a} \cdot \mathbf{x}) + h_0(\mathbf{b} \cdot \mathbf{x})$ in $\mathcal{R}(\mathbf{a},\mathbf{b})$. Let us concentrate on extremal paths for the function $f_1 = f - v_0$. Theorem 1.23 says that regarding such paths there may be only two possible cases.

Case 1. There exists a closed path $p_0 = (\mathbf{p}_1, \dots, \mathbf{p}_{2n})$ extremal for the function f_1.

In this case, we can write that

$$|G_{p_0}(f)| = |G_{p_0}(f - v_0)| = \|f - v_0\|. \tag{5.16}$$

It follows from Theorem 5.1 that for any $\varepsilon > 0$ there exist natural numbers m_1, m_2 and real numbers $c_{ij}, w_{ij}, \theta_{ij}$, $i = 1, 2$, $j = 1, \dots, m_i$, for which

$$\left| g_0(t) - \sum_{j=1}^{m_1} c_{1j}\sigma(w_{1j}t - \theta_{1j}) \right| < \frac{\varepsilon}{2} \tag{5.17}$$

and

$$\left| h_0(t) - \sum_{j=1}^{m_2} c_{2j}\sigma(w_{2j}t - \theta_{2j}) \right| < \frac{\varepsilon}{2} \tag{5.18}$$

for all $t \in [a, b]$. Here $[a, b]$ is a sufficiently large interval which contains both the sets $\{\mathbf{a} \cdot \mathbf{x} : \mathbf{x} \in Q\}$ and $\{\mathbf{b} \cdot \mathbf{x} : \mathbf{x} \in Q\}$.

Taking $t_1 = \mathbf{a} \cdot \mathbf{x}$ in (5.17) and $t_2 = \mathbf{b} \cdot \mathbf{x}$ in (5.18), we obtain from (5.17) and (5.18) that

$$\left| g_0(\mathbf{a} \cdot \mathbf{x}) + h_0(\mathbf{b} \cdot \mathbf{x}) - \sum_{i=1}^{m} c_i \sigma\left(\mathbf{w}^i \cdot \mathbf{x} - \theta_i\right) \right| < \varepsilon, \tag{5.19}$$

for all $\mathbf{x} \in Q$ and some $c_i, \theta_i \in \mathbb{R}$ and $\mathbf{w}^i \in l_1 \cup l_2$. For the simplicity of notation set

$$\Lambda(\mathbf{x}) = \sum_{i=1}^{m} c_i \sigma\left(\mathbf{w}^i \cdot \mathbf{x} - \theta_i\right)$$

Clearly,

$$\|f - \Lambda\| \le \|f - g_0 - h_0\| + \|g_0 + h_0 - \Lambda\|. \tag{5.20}$$

It follows from (5.20) that

$$E(f, \mathcal{N}(\sigma)) \le \|f - g_0 - h_0\| + \|g_0 + h_0 - \Lambda\|. \tag{5.21}$$

The last inequality together with (5.16) and (5.19) yield

$$E(f, \mathcal{N}(\sigma)) \le |G_{p_0}(f)| + \varepsilon.$$

Now since ε is arbitrarily small, we obtain that

$$E\left(f,\mathcal{N}(\sigma)\right)\leq\left|G_{p_0}(f)\right|.$$

From this and Theorem 5.14 it follows that

$$E\left(f,\mathcal{N}(\sigma)\right)=\sup_{p\subset Q}\left|G_p(f)\right|,$$

where the sup is taken over all closed paths.

Case 2. There exists an infinite path $p=(\mathbf{p}_1,\mathbf{p}_2,\dots,\mathbf{p}_n,\dots)$ extremal for f_1. We assume that all the points $\mathbf{p}_i$ are distinct; otherwise we can form a closed extremal path and return to Case 1. Since $\mathcal{R}\left(\mathbf{a},\mathbf{b}\right)$ is proximinal in $C(Q)$, the lengths of irreducible paths are uniformly bounded (see Section 1.5). Recall that a path $(\mathbf{p}_1,\dots,\mathbf{p}_n)$ is irreducible if there is not a path connecting $\mathbf{p}_1$ and $\mathbf{p}_n$ with the length less than n. Let a positive integer n_0 bounds the lengths of all irreducible paths in Q. Then for $n>n_0$, any extremal path $(\mathbf{p}_1,\mathbf{p}_2,\dots,\mathbf{p}_n)$ (with respect to f_1) must be made closed by adding not more than n_0 points. Thus for each finite extremal path $p_n=(\mathbf{p}_1,\mathbf{p}_2,\dots,\mathbf{p}_n)$, $n>n_0$, there exists a closed path $p_n^{m_n}=(\mathbf{p}_1,\mathbf{p}_2,\dots,\mathbf{p}_n,\mathbf{q}_{n+1},\dots,\mathbf{q}_{n+m_n})$, where $m_n\leq n_0$. The functional $G_{p_n^{m_n}}$ obeys the inequalities

$$(5.22)\qquad \left|G_{p_n^{m_n}}(f)\right|=\left|G_{p_n^{m_n}}(f-v_0)\right|\leq\frac{n\left\|f-v_0\right\|+m_n\left\|f-v_0\right\|}{n+m_n}=\left\|f-v_0\right\|$$

and

$$(5.23)\qquad \left|G_{p_n^{m_n}}(f)\right|\geq\frac{n\left\|f-v_0\right\|-m_n\left\|f-v_0\right\|}{n+m_n}=\frac{n-m_n}{n+m_n}\left\|f-v_0\right\|.$$

We obtain from (5.22) and (5.23) that

$$(5.24)\qquad \sup_{p_n^{m_n}}\left|G_{p_n^{m_n}}(f)\right|=\left\|f-v_0\right\|=\left\|f-g_0-h_0\right\|.$$

As in Case 1, we can apply the universal approximation theorem to the functions g_0 and h_0. Using the above function $\Lambda(\mathbf{x})=\sum_{i=1}^{m}c_i\sigma\left(\mathbf{w}^i\cdot\mathbf{x}-\theta_i\right)$ here, we obtain from (5.19), (5.21) and (5.24) that

$$(5.25)\qquad E\left(f,\mathcal{N}(\sigma)\right)\leq\sup_{p_n^{m_n}}\left|G_{p_n^{m_n}}(f)\right|+\varepsilon.$$

Since ε is sufficiently small, the inequality (5.25) together with (5.15) yield that

$$E\left(f,\mathcal{N}(\sigma)\right)=\sup_{p\subset Q}\left|G_p(f)\right|,$$

where the sup is taken over all closed paths. The theorem has been proved. □

REMARK 5.16. Since $\mathcal{N}(\sigma)\subset\mathcal{R}(\mathbf{a},\mathbf{b})$, we have $E\left(f,\mathcal{N}(\sigma)\right)\geq E\left(f,\mathcal{R}(\mathbf{a},\mathbf{b})\right)$. Theorem 5.15 yields that, in fact, $E\left(f,\mathcal{N}(\sigma)\right)=E\left(f,\mathcal{R}(\mathbf{a},\mathbf{b})\right)$.

Remark 5.16 implies that the existing methods for computation of the approximation error of ridge functions with two fixed directions can be directly applied to the class of neural networks considered here. For example, Theorem 1.24 from Section 1.3 is also valid for the class of neural networks with weights varying on two fixed straight lines in the plane. That is, the following theorem holds.

THEOREM 5.17. *Consider the set*

$$\Omega = \{\mathbf{x} \in \mathbb{R}^2 : c_1 \le \mathbf{a} \cdot \mathbf{x} \le d_1, \quad c_2 \le \mathbf{b} \cdot \mathbf{x} \le d_2\},$$

where $\mathbf{a} = (a_1, a_2)$ *and* $\mathbf{b} = (b_1, b_2)$ *are linearly independent vectors in* $\mathbb{R}^2$, $c_1 < d_1$ *and* $c_2 < d_2$. *Assume* $f \in C^2(\Omega)$ *and for any* $\mathbf{x} = (x_1, x_2) \in \Omega$,

$$\left[a_1 \frac{\partial}{\partial x_1} - a_2 \frac{\partial}{\partial x_2}\right]\left[b_1 \frac{\partial}{\partial x_1} - b_2 \frac{\partial}{\partial x_2}\right] f(x_1, x_2) \le 0.$$

Then the error of approximation from the class of single hidden layer neural networks $\mathcal{N}(\sigma)$ *with a continuous nonpolynomial activation function* σ *can be computed by the formula*

$$E(f, \mathcal{N}(\sigma)) = \frac{1}{4}\left[g(c_1, c_2) + g(d_1, d_2) - g(c_1, d_2) - g(d_1, c_2)\right],$$

where

$$g(y_1, y_2) = f\left(\frac{y_1 b_2 - y_2 a_2}{a_1 b_2 - a_2 b_1}, \frac{y_2 a_1 - y_1 b_1}{a_1 b_2 - a_2 b_1}\right).$$

EXAMPLE 5.18. Let (a_1, a_2) and (b_1, b_2) be the coordinate vectors $(1, 0)$ and $(0, 1)$, correspondingly. As a set Ω we take the unit square $[0, 1]^2$. Let σ be any continuous nonpolynomial function on $[0, 1]$ and $f_0(x_1, x_2) = (x_1 - \frac{1}{2})(x_2 - \frac{1}{2})$. The function f_0 satisfies all the conditions of Theorem 5.17. Note that the approximating set of networks $\mathcal{N}(\sigma)$ has members of the form

$$\sum_{i=1}^{n_1} c_i \sigma(s_i x_1 - \theta_i) + \sum_{j=1}^{n_2} d_j \sigma(t_j x_2 - \lambda_j),$$

where $c_i, d_j, \theta_i, \lambda_j, s_i, t_j$ are real numbers and n_1, n_2 are positive integers. Applying Theorem 5.17, we obtain that the error of approximation $E(f_0, \mathcal{N}(\sigma)) = \frac{1}{4}$.

REMARK 5.19. The problem of generalization of Theorem 5.15 to single hidden layer neural networks with weights from the union of arbitrarily many straight lines needs a careful approach. We do not yet know how to obtain exact formulas for the approximation error in that case. Note that Theorem 5.15 is based on the characteristic property of an extremal sum of ridge functions with two fixed directions. No such property for a sum of ridge functions with more than two directions is known. The existing difficulties are delineated in Section 1.3. However, by the above technique, one can prove that the sharp inequality (5.15) holds also in the general case, with the convention that the sup is taken over all minimal cycles with respect to a given set of directions. For minimal cycles and for the corresponding functionals G_p see Section 1.2.

5.3. Two hidden layer neural networks

A single hidden layer perceptron is able to approximate a given data with any degree of accuracy. But in applications it is necessary to define how many neurons one should take in a hidden layer. The more the number of neurons, the more the probability of the network to give precise results. Unfortunately, practicality decreases with the increase of the number of neurons in the hidden layer. In other words, single hidden layer perceptrons are not always effective if the number of neurons in the hidden layer is prescribed. In this section, we show that this phenomenon is no longer true for perceptrons with two hidden layers. We prove that a two hidden layer neural network with d inputs, d neurons in the first hidden layer, $2d + 2$ neurons in the second hidden layer and with a specifically constructed sigmoidal and infinitely differentiable activation function can approximate any continuous multivariate function with arbitrary accuracy.

5.3.1. Relation of the Kolmogorov superposition theorem to two hidden layer neural networks. Note that if in (5.1) r is fixed, then for any activation function σ the set

$$\mathcal{M}_r(\sigma) = \left\{ \sum_{i=1}^{r} c_i \sigma(\mathbf{w}^i \cdot \mathbf{x} - \theta_i) : \ c_i, \theta_i \in \mathbb{R}, \mathbf{w} \in \mathbb{R}^d \right\}$$

is no longer dense in the space $C(\mathbb{R}^d)$ (in the topology of uniform convergence on compact sets). The set $\mathcal{M}_r(\sigma)$ will not be dense even if we variate over all univariate continuous functions σ (see Subsection 5.3.3). In the following, we will see that this negative property of single hidden layer neural networks does not carry over to networks with more than one hidden layer.

A two hidden layer network is defined by iteration of the single hidden layer neural network model. The output of two hidden layer perceptron with r units in the first layer, s units in the second layer and the input $x = (x_1, \ldots, x_d)$ is

$$\sum_{i=1}^{s} d_i \sigma \left(\sum_{j=1}^{r} c_{ij} \sigma(\mathbf{w}^{ij} \cdot \mathbf{x} - \theta_{ij}) - \gamma_i \right).$$

Here $d_i, c_{ij}, \theta_{ij}, \gamma_i$ are real numbers, $\mathbf{w}^{ij}$ are vectors of $\mathbb{R}^d$ and σ is a fixed univariate function.

In approximation by neural networks, there are two main problems. The first is the density problem of determining the conditions under which an arbitrary target function can be approximated arbitrarily well by neural networks. The second problem, called the *complexity problem*, is to determine how many neurons in hidden layers are necessary to give a prescribed degree of approximation. This problem is almost the same as the problem of degree of approximation (see, e.g., [**16**]). In this section, we will prove that there exists a two hidden layer neural network model with d units in the first layer and $2d + 2$ units in the second layer such that it has the ability to approximate any d-variable continuous function with arbitrary accuracy. Thus our result will solve both the density and the complexity problems for the considered model. As an activation function for this model we take a specific sigmoidal function. The idea behind the proof of this result is very much connected to the Kolmogorov superposition theorem (see Section 4.1). This theorem has been much discussed in neural network literature (see, e.g., [**72, 106, 133, 134, 159, 172, 179, 181, 186**]). In our opinion, the most remarkable application of the Kolmogorov superposition theorem to neural networks was given by Maiorov and Pinkus [**147**]. They showed that there exists a sigmoidal, strictly increasing, analytic activation function, for which a fixed number of units in both hidden layers are sufficient to approximate arbitrarily well any continuous multivariate function. Namely, the authors of [**147**] proved the following theorem.

THEOREM 5.20 (Maiorov and Pinkus [**147**]). *There exists an activation function σ which is analytic, strictly increasing and sigmoidal and has the following property: For any $f \in C[0,1]^d$ and $\varepsilon > 0$, there exist constants d_i, c_{ij}, θ_{ij}, γ_i, and vectors $\mathbf{w}^{ij} \in \mathbb{R}^d$ for which*

$$\left| f(\mathbf{x}) - \sum_{i=1}^{6d+3} d_i \sigma \left(\sum_{j=1}^{3d} c_{ij} \sigma(\mathbf{w}^{ij} \cdot \mathbf{x} - \theta_{ij}) - \gamma_i \right) \right| < \varepsilon \tag{5.26}$$

for all $\mathbf{x} = (x_1, \dots, x_d) \in [0,1]^d$.

This theorem is based on the following version of the Kolmogorov superposition theorem given by Lorentz [**142**] and Sprecher [**185**].

THEOREM 5.21 (Kolmogorov's superposition theorem). *For the unit cube $\mathbb{I}^d$, $\mathbb{I} = [0,1]$, $d \geq 2$, there exists constants $\lambda_q > 0$, $q = 1, \dots, d$, $\sum_{q=1}^d \lambda_q = 1$, and nondecreasing continuous functions $\phi_p : [0,1] \to [0,1]$, $p = 1, \dots, 2d+1$, such that every continuous function $f : \mathbb{I}^d \to \mathbb{R}$ admits the representation*

$$f(x_1, \dots x_d) = \sum_{p=1}^{2d+1} g \left(\sum_{q=1}^{d} \lambda_q \phi_p(x_q) \right) \tag{5.27}$$

for some $g \in C[0,1]$ depending on f.

In the next subsection, using the general ideas developed in [**147**], we show that the bounds of units in hidden layers in (5.26) may be chosen even equal to the bounds in the Kolmogorov superposition theorem. More precisely, these bounds can be taken as $2d+2$ and d instead of $6d+3$ and $3d$. To attain this purpose, we change the "analyticity" of σ to "infinite differentiability". In addition, near infinity we assume that σ is "λ-strictly increasing" instead of being "strictly increasing".

5.3.2. The main result.

We begin this subsection with a definition of a λ-*monotone function*. Let λ be any nonnegative number. A real function f defined on (a,b) is called λ-*increasing* (λ-*decreasing*) if there exists an increasing (decreasing) function $u : (a,b) \to \mathbb{R}$ such that $|f(x) - u(x)| \leq \lambda$, for all $x \in (a,b)$. If u is strictly increasing (or strictly decreasing), then the above function f is called a λ-*strictly increasing* (or λ-*strictly decreasing*) function. Clearly, 0-monotonicity coincides with the usual concept of monotonicity and a λ_1-monotone function is λ_2-monotone if $\lambda_1 \leq \lambda_2$. It is also clear from the definition that a λ-monotone function behaves like a usual monotone function as λ gets very small.

Our purpose is to prove the following theorem.

THEOREM 5.22. *For any positive numbers α and λ, there exists a $C^\infty(\mathbb{R})$, sigmoidal activation function $\sigma : \mathbb{R} \to \mathbb{R}$ which is strictly increasing on $(-\infty, \alpha)$, λ-strictly increasing on $[\alpha, +\infty)$, and satisfies the following property: For any $f \in C[0,1]^d$ and $\varepsilon > 0$, there exist constants d_p, c_{pq}, θ_{pq}, γ_p, and vectors $\mathbf{w}^{pq} \in \mathbb{R}^d$ for which*

$$\left| f(\mathbf{x}) - \sum_{p=1}^{2d+2} d_p \sigma \left(\sum_{q=1}^{d} c_{pq} \sigma(\mathbf{w}^{pq} \cdot \mathbf{x} - \theta_{pq}) - \gamma_p \right) \right| < \varepsilon \tag{5.28}$$

for all $\mathbf{x} = (x_1, \dots, x_d) \in [0,1]^d$.

PROOF. Let α be any positive number. Divide the interval $[\alpha, +\infty)$ into the segments $[\alpha, 2\alpha]$, $[2\alpha, 3\alpha], \dots$. Let $h(t)$ be any strictly increasing, infinitely differentiable function on $[\alpha, +\infty)$ with the properties

(1) $0 < h(t) < 1$ for all $t \in [\alpha, +\infty)$;
(2) $1 - h(\alpha) \le \lambda$;
(3) $h(t) \to 1$, as $t \to +\infty$.

The existence of a strictly increasing smooth function satisfying these properties is easy to verify. Note that from conditions (1)-(3) it follows that any function $f(t)$ satisfying the inequality $h(t) < f(t) < 1$ for all $t \in [\alpha, +\infty)$, is λ-strictly increasing and $f(t) \to 1$, as $t \to +\infty$.

We are going to construct σ obeying the required properties in stages. Let $\{u_n(t)\}_{n=1}^{\infty}$ be the sequence of all polynomials with rational coefficients defined on $[0, 1]$. First, we define σ on the closed intervals $[(2m-1)\alpha, 2m\alpha]$, $m = 1, 2, \dots$, as the function

$$\sigma(t) = a_m + b_m u_m(\frac{t}{\alpha} - 2m + 1),\ t \in [(2m-1)\alpha, 2m\alpha], \tag{5.29}$$

or equivalently,

$$\sigma(\alpha t + (2m-1)\alpha) = a_m + b_m u_m(t),\ t \in [0, 1], \tag{5.30}$$

where a_m and $b_m \neq 0$ are appropriately chosen constants. These constants are determined from the condition

$$h(t) < \sigma(t) < 1, \tag{5.31}$$

for all $t \in [(2m-1)\alpha, 2m\alpha]$. There is a simple procedure for determining a suitable pair of a_m and b_m. Indeed, let

$$M = \max h(t),\ A_1 = \min u_m(\frac{t}{\alpha} - 2m + 1),\ A_2 = \max u_m(\frac{t}{\alpha} - 2m + 1),$$

where in all the above max and min, the variable t runs over the closed interval $[(2m-1)\alpha, 2m\alpha]$. Note that $M < 1$. If $A_1 = A_2$ (that is, if the function u_m is constant on $[0, 1]$), then we can set $\sigma(t) = (1 + M)/2$ and easily find a suitable pair of a_m and b_m from (5.29). Let now $A_1 \neq A_2$ and $y = a + bx$, $b \neq 0$, be a linear function mapping the segment $[A_1, A_2]$ into $(M, 1)$. Then it is enough to take $a_m = a$ and $b_m = b$.

At the second stage we define σ on the intervals $[2m\alpha, (2m+1)\alpha]$, $m = 1, 2, \dots$, so that it is in $C^\infty(\mathbb{R})$ and satisfies the inequality (5.31). Finally, in all of $(-\infty, \alpha)$ we define σ while maintaining the C^∞ strict monotonicity property, and also in such a way that $\lim_{t \to -\infty} \sigma(t) = 0$. We obtain from the properties of h and the condition (5.31) that $\sigma(t)$ is a λ-strictly increasing function on the interval $[\alpha, +\infty)$ and $\sigma(t) \to 1$, as $t \to +\infty$.

From the above construction of σ, that is, from (5.30) it follows that for each $m = 1, 2, \dots$, there exists numbers A_m, B_m and r_m such that

$$u_m(t) = A_m \sigma(\alpha t - r_m) - B_m, \tag{5.32}$$

where $A_m \neq 0$.

Let f be any continuous function on the unit cube $[0, 1]^d$. By the Kolmogorov superposition theorem the expansion (5.27) is valid for f. For the exterior continuous univariate function $g(t)$ in (5.27) and for any $\varepsilon > 0$ there exists a polynomial $u_m(t)$ of the above form such that

$$|g(t) - u_m(t)| < \frac{\varepsilon}{2(2d+1)},$$

for all $t \in [0,1]$. This together with (5.32) means that

$$|g(t) - [a\sigma(\alpha t - r) - b]| < \frac{\varepsilon}{2(2d+1)}, \tag{5.33}$$

for some $a, b, r \in \mathbb{R}$ and all $t \in [0,1]$.

Substituting (5.33) in (5.27) we obtain that

$$\left| f(x_1, \dots, x_d) - \sum_{p=1}^{2d+1} \left(a\sigma \left(\alpha \cdot \sum_{q=1}^{d} \lambda_q \phi_p(x_q) - r \right) - b \right) \right| < \frac{\varepsilon}{2} \tag{5.34}$$

for all $(x_1, \dots, x_d) \in [0,1]^d$.

For each $p \in \{1, 2, \dots, 2d+1\}$ and $\delta > 0$ there exist constants a_p, b_p and r_p such that

$$|\phi_p(x_q) - [a_p\sigma(\alpha x_q - r_p) - b_p]| < \delta, \tag{5.35}$$

for all $x_q \in [0,1]$. Since $\lambda_q > 0$, $q = 1, \dots, d$, $\sum_{q=1}^{d} \lambda_q = 1$, it follows from (5.35) that

$$\left| \sum_{q=1}^{d} \lambda_q \phi_p(x_q) - \left[\sum_{q=1}^{d} \lambda_q a_p \sigma(\alpha x_q - r_p) - b_p \right] \right| < \delta, \tag{5.36}$$

for all $(x_1, \dots, x_d) \in [0,1]^d$.

Now since the function $a\sigma(\alpha t - r)$ is uniformly continuous on every closed interval, we can choose δ sufficiently small and obtain from (5.36) that

$$\left| \sum_{p=1}^{2d+1} a\sigma \left(\alpha \sum_{q=1}^{d} \lambda_q \phi_p(x_q) - r \right) - \sum_{p=1}^{2d+1} a\sigma \left(\alpha \left[\sum_{q=1}^{d} \lambda_q a_p \sigma(\alpha x_q - r_p) - b_p \right] - r \right) \right| < \frac{\varepsilon}{2}.$$

This inequality may be rewritten as

$$\left| \sum_{p=1}^{2d+1} a\sigma \left(\alpha \sum_{q=1}^{d} \lambda_q \phi_p(x_q) - r \right) - \sum_{p=1}^{2d+1} d_p\sigma \left(\sum_{q=1}^{d} c_{pq}\sigma(\mathbf{w}^{pq} \cdot \mathbf{x} - \theta_{pq}) - \gamma_p \right) \right| < \frac{\varepsilon}{2}. \tag{5.37}$$

From (5.34) and (5.37) it follows that

$$\left| f(\mathbf{x}) - \left[\sum_{p=1}^{2d+1} d_p\sigma \left(\sum_{q=1}^{d} c_{pq}\sigma(\mathbf{w}^{pq} \cdot \mathbf{x} - \theta_{pq}) - \gamma_p \right) - s \right] \right| < \varepsilon, \tag{5.38}$$

where $s = (2d+1)b$. Since the constant s can be written in the form

$$s = d\sigma \left(\sum_{q=1}^{d} c_q\sigma(\mathbf{w}^q \cdot \mathbf{x} - \theta_q) - \gamma \right),$$

from (5.38) we finally obtain the validity of (5.28). □

REMARK 5.23. It is easily seen in the proof of Theorem 5.22 that all the weights $\mathbf{w}^{ij}$ are fixed (see (5.37)). Namely, $\mathbf{w}^{ij} = \alpha \mathbf{e}^j$, for all $i = 1, \ldots, 2d+2$, $j = 1, \ldots, d$, where $\mathbf{e}^j$ is the j-th coordinate vector of the space $\mathbb{R}^d$.

The next theorem follows from Theorem 5.22 easily, since the Kolmogorov superposition theorem is valid for all compact sets of $\mathbb{R}^d$.

THEOREM 5.24. *Let Q be a compact set in $\mathbb{R}^d$. For any numbers $\alpha \in \mathbb{R}$ and $\lambda > 0$, there exists a $C^\infty(\mathbb{R})$, sigmoidal activation function $\sigma : \mathbb{R} \to \mathbb{R}$ which is strictly increasing on $(-\infty, \alpha)$, λ-strictly increasing on $[\alpha, +\infty)$, and satisfies the following property: For any $f \in C(Q)$ and $\varepsilon > 0$ there exist real numbers d_i, c_{ij}, θ_{ij}, γ_i, and vectors $\mathbf{w}^{ij} \in \mathbb{R}^d$ for which*

$$\left| f(\mathbf{x}) - \sum_{i=1}^{2d+2} d_i \sigma \left(\sum_{j=1}^{d} c_{ij} \sigma(\mathbf{w}^{ij} \cdot \mathbf{x} - \theta_{ij}) - \gamma_i \right) \right| < \varepsilon$$

for all $\mathbf{x} = (x_1, \ldots, x_d) \in Q$.

REMARK 5.25. In some literature, a single hidden layer perceptron is defined as the function

$$\sum_{i=1}^{r} c_i \sigma(\mathbf{w}^i \cdot \mathbf{x} - \theta_i) - c_0.$$

A two hidden layer network then takes the form

$$\sum_{i=1}^{s} d_i \sigma \left(\sum_{j=1}^{r} c_{ij} \sigma(\mathbf{w}^{ij} \cdot \mathbf{x} - \theta_{ij}) - \gamma_i \right) - d_0. \tag{5.39}$$

The proof of Theorem 5.22 shows that for networks of type (5.39) the theorem is valid if we take $2d+1$ neurons in the second hidden layer (instead of $2d+2$ neurons as above). That is, there exist networks of type (5.39) having the universal approximation property and for which the number of units in the hidden layers is equal to the number of summands in the Kolmogorov superposition theorem.

REMARK 5.26. It is known that the $2d+1$ in the Kolmogorov superposition theorem is minimal (see Sternfeld [**192**]). Thus it is doubtful if the number of neurons in Theorems 5.22 and 5.24 can be reduced.

REMARK 5.27. Inequality (5.33) shows that single hidden layer neural networks of the form (5.39) with the activation function σ and with only one neuron in the hidden layer can approximate any continuous function on the interval $[0, 1]$ with arbitrary precision. Since the number b in (5.33) can always be written as $b = a_1 \sigma(0 \cdot t - r_1)$ for some a_1 and r_1, we see that two neurons in the hidden layer are sufficient for traditional single hidden layer neural networks with the activation function σ to approximate continuous functions on $[0, 1]$. Applying the linear transformation $x = a + (b-a)t$ it can be proven that the same argument holds for any interval $[a, b]$.

5.3.3. Why two hidden layers are better than one. In neural network literature, feedforward neural networks with two or more hidden layers take the name of *deep neural networks* in contrast with *shallow neural networks* that comprise only one hidden layer. Deep neural networks are being extensively used in a wide variety of applications, such as image classification, speech recognition, natural language processing, machine translation, game intelligence, robotics, and autonomous systems. For an in-depth overview, we refer the reader to the book by I. Goodfellow, Y. Bengio, and A. Courville [**65**]. In a number of papers it was shown that for certain approximation tasks deep networks can perform fundamentally better than shallow networks. For example, for the popular activation function $\sigma(x) = \max\{0, x\}$, which is called ReLU, it was shown in [**169, 204**] that deep networks approximate sufficiently smooth functions with rates higher than those achieved by shallow networks.

In this subsection, using ridge functions, we demonstrate theoretical advantages of two hidden layer neural networks over single hidden layer networks. For comparison with single hidden layer networks, we bring into consideration two properties of two hidden layer networks. First, two hidden layer networks with bounded number of neurons are able to approximate any continuous multivariate function with arbitrary precision (see Theorems 5.22 and 5.24). Note that results of this type change previous thoughts about the universal approximation property of neural networks. It was formerly believed and particularly emphasized in many works that for achieving high degree of accuracy in approximating multivariate functions, large networks with sufficiently many hidden neurons are needed (see, e.g., [**65**, Chapter 6.4.1]). However, Theorem 5.24 reveals that there exist neural networks with very few hidden neurons, which can approximate arbitrarily well all continuous multivariate functions. Moreover, the number of hidden neurons does not depend on the approximation accuracy and is precisely known in advance. For example, for d-variate functions this number is $3d + 2$. Second, this property of networks holds even if we fix all the weights (see Remark 5.23). Since fixed weights reduce the computational expense and training time, this phenomenon is of particular interest. It is natural to ask whether these properties can be carried over to single hidden layer neural networks. We will see below that the answer to this question is negative.

We already know that single hidden layer networks with only two neurons in the hidden layer can approximate any continuous univariate function on any compact subset of the real line (see Remark 5.27). That is, for any compact set $Q \subset \mathbb{R}$, any function $f \in C(Q)$ and any $\varepsilon > 0$ there exist numbers c_i, w_i, θ_i, $i = 1, 2$, such that

$$|f(x) - c_1\sigma(w_1 x - \theta_1) - c_2\sigma(w_2 x - \theta_2)| < \varepsilon$$

for all $x \in Q$. The existence of such a σ has been proven above. Its practical construction will be implemented in the next section. Nevertheless, single hidden layer neural networks with bounded number of hidden neurons cannot approximate multivariate functions. To see this, consider the set of single hidden layer neural networks with at most k neurons in the hidden layer. That is, consider the set

$$\mathcal{M}_k(\sigma) = \left\{ \sum_{i=1}^{k} c_i \sigma(\mathbf{w}^i \cdot \mathbf{x} - \theta_i) : \mathbf{w}^i \in \mathbb{R}^d, c_i, \theta_i \in \mathbb{R} \right\}.$$

Recall that each summand $c_i\sigma(\mathbf{w}^i \cdot \mathbf{x} - \theta_i)$ is a ridge function of the form $g(\mathbf{w}^i \cdot \mathbf{x})$. Thus, $\mathcal{M}_k(\sigma)$ is a subset of the more general set

$$\mathcal{R}_k = \left\{ \sum_{i=1}^{k} g_i(\mathbf{w}^i \cdot \mathbf{x}) : \mathbf{w}^i \in \mathbb{R}^d,\ g_i \in C(\mathbb{R}) \right\}.$$

Note that in $\mathcal{R}_k$ we vary over both the vectors $\mathbf{w}^i$ and the functions g_i, whilst k is fixed. In [**140**], Lin and Pinkus proved that for any $k \in \mathbb{N}$, there exists a function $f \in C(\mathbb{R}^d)$ and a compact set $Q \subset \mathbb{R}^d$ such that

$$\inf_{g \in \mathcal{R}_k} \|f - g\| > 0.$$

It follows from this result that for each $k \in \mathbb{N}$ the set $\mathcal{R}_k$ is not dense in $C(\mathbb{R}^d)$ in the topology of uniform convergence on compacta. Since $\mathcal{M}_k(\sigma) \subset \mathcal{R}_k$, we obtain that the set $\mathcal{M}_k(\sigma)$ cannot be dense either.

Let us now analyze if we can fix weights in shallow networks while preserving their universal approximation property. We have seen that this is possible for the simplest type of deep networks, namely for two hidden layer networks. In the univariate setting, this problem is solved positively. Inequality (5.33) together with Remark 5.27 say that in the case $d = 1$ there exist activation functions σ for which the universal approximation property holds even if we fix the weights. But this is no longer true in cases when the space dimension $d > 1$. That is, shallow networks of the form

$$h(\mathbf{x}) = \sum_{i=1}^{k} c_i\sigma(\mathbf{w} \cdot \mathbf{x} - \theta_i),$$

where the weight $\mathbf{w} \in \mathbb{R}^d$ is fixed for all neurons of the hidden layer, but which may be different for different networks h, cannot approximate all continuous multivariate functions $f(x_1, \dots, x_d)$ if $d > 1$. This is almost clear if $\mathbf{w}$ is fixed for all h. Indeed, in this special case, the linear functional

$$F(f) = f(\mathbf{x}_1) - f(\mathbf{x}_2),$$

where $\mathbf{x}_1$ and $\mathbf{x}_2$ are selected in such a way that $\mathbf{w} \cdot \mathbf{x}_1 = \mathbf{w} \cdot \mathbf{x}_2$, annihilates all networks h. Since the functional F is nontrivial, the set of all networks h, which we denote in the sequel by $\mathcal{H}$, is not dense in $C(Q)$ for compact sets Q containing the points $\mathbf{x}_1$ and $\mathbf{x}_2$; hence approximation to all continuous functions cannot be possible on such compact sets Q. This means that $\mathcal{H}$ is not dense in $C(\mathbb{R}^d)$ in the topology of uniform convergence on compacta. Note that in $\mathcal{H}$ we do not fix k.

The general case, where $\mathbf{w}$ is different for different networks h, is rather complicated to approach straightforwardly, but can be easily investigated by applying the above negative result of approximation by ridge functions from $\mathcal{R}_k$. It is not difficult to see that in this case, $\mathcal{H} \subset \mathcal{R}_1$. Thus $\mathcal{H}$ cannot be dense, since $\mathcal{R}_1$ is not dense.

Thus we see that there are continuous multivariate functions, which cannot be approximated arbitrarily well by shallow networks with fixed weights. The above analysis leads us to the following negative result on the approximation by shallow networks with limited weights.

THEOREM 5.28. *For any continuous function $\sigma\colon \mathbb{R} \to \mathbb{R}$, there is a multivariate continuous function, which cannot be approximated arbitrarily well by neural* *networks of the form*

$$\sum_{i=1}^{k} c_i \sigma(\mathbf{w}^i \cdot \mathbf{x} - \theta_i), \tag{5.40}$$

where we vary over all $k \in \mathbb{N}$, $c_i, \theta_i \in \mathbb{R}$, $\mathbf{w}^i \in \mathbb{R}^d$, *but the number of pairwise linearly independent vectors (weights)* $\mathbf{w}^i$ *in each network* (5.40) *is uniformly bounded by some positive integer* N *(which is the same for all networks).*

The proof follows from the fact that for linearly dependent $\mathbf{w}^1$ and $\mathbf{w}^2$, the sum $c_1\sigma(\mathbf{w}^1 \cdot \mathbf{x} - \theta_1) + c_2\sigma(\mathbf{w}^2 \cdot \mathbf{x} - \theta_2)$ is a single ridge function; hence the whole sum (5.40) is a member of the set $\mathcal{R}_N$.

REMARK 5.29. Theorem 5.28 shows a particular limitation of neural networks with one hidden layer. We refer the reader to [**33, 139**] for interesting results and discussions around other limitations of such networks.

5.4. Construction of a universal sigmoidal function

In the preceding section, we considered two theorems (Theorem 5.20 of Maiorov and Pinkus, and Theorem 5.22) on the approximation capabilities of the MLP model of neural networks with a prescribed number of hidden neurons. Note that both results are more theoretical than practical, as they indicate only the existence of the corresponding activation functions.

In this section, we construct algorithmically a smooth, sigmoidal, almost monotone activation function σ providing approximation to an arbitrary continuous function within any degree of accuracy. This algorithm is implemented in a computer program, which computes the value of σ at any reasonable point of the real axis.

5.4.1. A construction algorithm. In this subsection, we construct algorithmically a sigmoidal function σ which we use in our results in Subsection 5.4.3.

To start with the construction of σ, assume that we are given a closed interval $[a, b]$ and a sufficiently small real number λ. We construct σ algorithmically, based on two numbers, namely λ and $d := b - a$. The following steps describe the algorithm.

Step 1. Introduce the function

$$h(x) := 1 - \frac{\min\{1/2, \lambda\}}{1 + \log(x - d + 1)}.$$

Note that this function is strictly increasing on the real line and satisfies the following properties:

(1) $0 < h(x) < 1$ for all $x \in [d, +\infty)$;
(2) $1 - h(d) \le \lambda$;
(3) $h(x) \to 1$, as $x \to +\infty$.

We want to construct σ satisfying the inequalities

$$h(x) < \sigma(x) < 1 \tag{5.41}$$

for $x \in [d, +\infty)$. Then our σ will tend to 1 as x tends to $+\infty$ and obey the inequality

$$|\sigma(x) - h(x)| \le \lambda,$$

i.e., it will be a λ-increasing function.

Step 2. Before proceeding to the construction of σ, we need to enumerate the monic polynomials with rational coefficients. Let q_n be the Calkin–Wilf sequence (see [**23**]). Then we can enumerate all the rational numbers by setting

$$r_0 := 0, \quad r_{2n} := q_n, \quad r_{2n-1} := -q_n, \ n = 1, 2, \ldots.$$

Note that each monic polynomial with rational coefficients can uniquely be written as $r_{k_0} + r_{k_1}x + \ldots + r_{k_{l-1}}x^{l-1} + x^l$, and each positive rational number determines a unique finite continued fraction

$$[m_0; m_1, \ldots, m_l] := m_0 + \cfrac{1}{m_1 + \cfrac{1}{m_2 + \cfrac{1}{\ddots + \cfrac{1}{m_l}}}}$$

with $m_0 \geq 0$, $m_1, \ldots, m_{l-1} \geq 1$ and $m_l \geq 2$. We now construct a bijection between the set of all monic polynomials with rational coefficients and the set of all positive rational numbers as follows. To the only zeroth-degree monic polynomial 1 we associate the rational number 1, to each first-degree monic polynomial of the form $r_{k_0} + x$ we associate the rational number $k_0 + 2$, to each second-degree monic polynomial of the form $r_{k_0} + r_{k_1}x + x^2$ we associate the rational number $[k_0; k_1 + 2] = k_0 + 1/(k_1 + 2)$, and to each monic polynomial

$$r_{k_0} + r_{k_1}x + \ldots + r_{k_{l-2}}x^{l-2} + r_{k_{l-1}}x^{l-1} + x^l$$

of degree $l \geq 3$ we associate the rational number $[k_0; k_1 + 1, \ldots, k_{l-2} + 1, k_{l-1} + 2]$. In other words, we define $u_1(x) := 1$,

$$u_n(x) := r_{q_n - 2} + x$$

if $q_n \in \mathbb{Z}$,

$$u_n(x) := r_{m_0} + r_{m_1 - 2}x + x^2$$

if $q_n = [m_0; m_1]$, and

$$u_n(x) := r_{m_0} + r_{m_1 - 1}x + \ldots + r_{m_{l-2} - 1}x^{l-2} + r_{m_{l-1} - 2}x^{l-1} + x^l$$

if $q_n = [m_0; m_1, \ldots, m_{l-2}, m_{l-1}]$ with $l \geq 3$. For example, the first few elements of this sequence are

$$1, \quad x^2, \quad x, \quad x^2 - x, \quad x^2 - 1, \quad x^3, \quad x - 1, \quad x^2 + x, \quad \ldots.$$

Step 3. We start with constructing σ on the intervals $[(2n-1)d, 2nd]$, $n = 1, 2, \ldots$. For each monic polynomial $u_n(x) = \alpha_0 + \alpha_1 x + \ldots + \alpha_{l-1}x^{l-1} + x^l$, set

$$B_1 := \alpha_0 + \frac{\alpha_1 - |\alpha_1|}{2} + \ldots + \frac{\alpha_{l-1} - |\alpha_{l-1}|}{2}$$

and

$$B_2 := \alpha_0 + \frac{\alpha_1 + |\alpha_1|}{2} + \ldots + \frac{\alpha_{l-1} + |\alpha_{l-1}|}{2} + 1.$$

Note that the numbers B_1 and B_2 depend on n. To avoid complication of symbols, we do not indicate this in the notation.

Introduce the sequence

$$M_n := h((2n+1)d), \qquad n = 1, 2, \ldots.$$

Clearly, this sequence is strictly increasing and converges to 1.

Now we define σ as the function

$$\sigma(x) := a_n + b_n u_n\left(\frac{x}{d} - 2n + 1\right), \quad x \in [(2n-1)d, 2nd], \tag{5.42}$$

where

$$a_1 := \frac{1}{2}, \qquad b_1 := \frac{h(3d)}{2}, \tag{5.43}$$

and

$$a_n := \frac{(1+2M_n)B_2 - (2+M_n)B_1}{3(B_2 - B_1)}, \quad b_n := \frac{1 - M_n}{3(B_2 - B_1)}, \quad n = 2, 3, \dots. \tag{5.44}$$

It is not difficult to notice that for $n > 2$ the numbers a_n, b_n are the coefficients of the linear function $y = a_n + b_n x$ mapping the closed interval $[B_1, B_2]$ onto the closed interval $[(1+2M_n)/3, (2+M_n)/3]$. Besides, for $n = 1$, i.e. on the interval $[d, 2d]$,

$$\sigma(x) = \frac{1 + M_1}{2}.$$

Therefore, we obtain that

$$h(x) < M_n < \frac{1 + 2M_n}{3} \le \sigma(x) \le \frac{2 + M_n}{3} < 1, \tag{5.45}$$

for all $x \in [(2n-1)d, 2nd]$, $n = 1, 2, \dots$.

Step 4. In this step, we construct σ on the intervals $[2nd, (2n+1)d]$, $n = 1, 2, \dots$. For this purpose we use the *smooth transition function*

$$\beta_{a,b}(x) := \frac{\widehat{\beta}(b - x)}{\widehat{\beta}(b - x) + \widehat{\beta}(x - a)},$$

where

$$\widehat{\beta}(x) := \begin{cases} e^{-1/x}, & x > 0, \\ 0, & x \le 0. \end{cases}$$

Obviously, $\beta_{a,b}(x) = 1$ for $x \le a$, $\beta_{a,b}(x) = 0$ for $x \ge b$, and $0 < \beta_{a,b}(x) < 1$ for $a < x < b$.

Set

$$K_n := \frac{\sigma(2nd) + \sigma((2n+1)d)}{2}, \qquad n = 1, 2, \dots.$$

Note that the numbers $\sigma(2nd)$ and $\sigma((2n+1)d)$ have already been defined in the previous step. Since both the numbers $\sigma(2nd)$ and $\sigma((2n+1)d)$ lie in the interval $(M_n, 1)$, it follows that $K_n \in (M_n, 1)$.

First we extend σ smoothly to the interval $[2nd, 2nd + d/2]$. Take $\varepsilon := (1 - M_n)/6$ and choose $\delta \le d/2$ such that

$$\left| a_n + b_n u_n\left(\frac{x}{d} - 2n + 1\right) - (a_n + b_n u_n(1)) \right| \le \varepsilon, \quad x \in [2nd, 2nd + \delta]. \tag{5.46}$$

One can choose this δ as

$$\delta := \min\left\{\frac{\varepsilon d}{b_n C}, \frac{d}{2}\right\},$$

where $C > 0$ is a number satisfying $|u_n'(x)| \le C$ for $x \in (1, 1.5)$. For example, for $n = 1$, δ can be chosen as $d/2$. Now define σ on the first half of the interval

$[2nd, (2n+1)d]$ as the function

$$\sigma(x) := K_n - \beta_{2nd,2nd+\delta}(x)$$
$$\times \left(K_n - a_n - b_n u_n \left(\frac{x}{d} - 2n + 1\right)\right), x \in \left[2nd, 2nd + \frac{d}{2}\right]. \tag{5.47}$$

Let us prove that $\sigma(x)$ satisfies the condition (5.41). Indeed, if $2nd + \delta \leq x \leq 2nd + d/2$, then there is nothing to prove, since $\sigma(x) = K_n \in (M_n, 1)$. If $2nd \leq x < 2nd + \delta$, then $0 < \beta_{2nd,2nd+\delta}(x) \leq 1$ and hence from (5.47) it follows that for each $x \in [2nd, 2nd + \delta)$, $\sigma(x)$ is between the numbers K_n and $A_n(x) := a_n + b_n u_n \left(\frac{x}{d} - 2n + 1\right)$. On the other hand, from (5.46) we obtain that

$$a_n + b_n u_n(1) - \varepsilon \leq A_n(x) \leq a_n + b_n u_n(1) + \varepsilon,$$

which together with (5.42) and (5.45) yields $A_n(x) \in \left[\frac{1+2M_n}{3} - \varepsilon, \frac{2+M_n}{3} + \varepsilon\right]$ for $x \in [2nd, 2nd + \delta)$. Since $\varepsilon = (1 - M_n)/6$, the inclusion $A_n(x) \in (M_n, 1)$ is valid. Now since both K_n and $A_n(x)$ belong to $(M_n, 1)$, we finally conclude that

$$h(x) < M_n < \sigma(x) < 1, \quad \text{for } x \in \left[2nd, 2nd + \frac{d}{2}\right].$$

We define σ on the second half of the interval in a similar way:

$$\sigma(x) := K_n - (1 - \beta_{(2n+1)d - \overline{\delta},(2n+1)d}(x))$$
$$\times \left(K_n - a_{n+1} - b_{n+1} u_{n+1} \left(\frac{x}{d} - 2n - 1\right)\right), \quad x \in \left[2nd + \frac{d}{2}, (2n+1)d\right],$$

where

$$\overline{\delta} := \min\left\{\frac{\overline{\varepsilon} d}{b_{n+1}\overline{C}}, \frac{d}{2}\right\}, \qquad \overline{\varepsilon} := \frac{1 - M_{n+1}}{6}, \qquad \overline{C} \geq \sup_{[-0.5,0]} |u'_{n+1}(x)|.$$

One can easily verify, as above, that the constructed $\sigma(x)$ satisfies the condition (5.41) on $[2nd + d/2, 2nd + d]$ and

$$\sigma\left(2nd + \frac{d}{2}\right) = K_n, \qquad \sigma^{(i)}\left(2nd + \frac{d}{2}\right) = 0, \quad i = 1, 2, \ldots.$$

Steps 3 and 4 construct σ on the interval $[d, +\infty)$.

Step 5. On the remaining interval $(-\infty, d)$, we define σ as

$$\sigma(x) := \left(1 - \widehat{\beta}(d - x)\right) \frac{1 + M_1}{2}, \quad x \in (-\infty, d).$$

It is not difficult to verify that σ is a strictly increasing, smooth function on $(-\infty, d)$. Note also that $\sigma(x) \to \sigma(d) = (1 + M_1)/2$, as x tends to d from the left and $\sigma^{(i)}(d) = 0$ for $i = 1, 2, \ldots$. This final step completes the construction of σ on the whole real line.

5.4.2. Properties of the constructed sigmoidal function. It should be noted that the above algorithm allows one to compute the constructed σ at any point of the real axis instantly. The code of this algorithm is available at `http://sites.google.com/site/njguliyev/papers/monic-sigmoidal`. As a practical example, we give here the graph of σ (see Figure 1) and a numerical table (see Table 1) containing several computed values of this function on the interval $[0, 20]$. Figure 2 shows how the graph of λ-increasing function σ changes on the interval $[0, 100]$ as the parameter λ decreases.

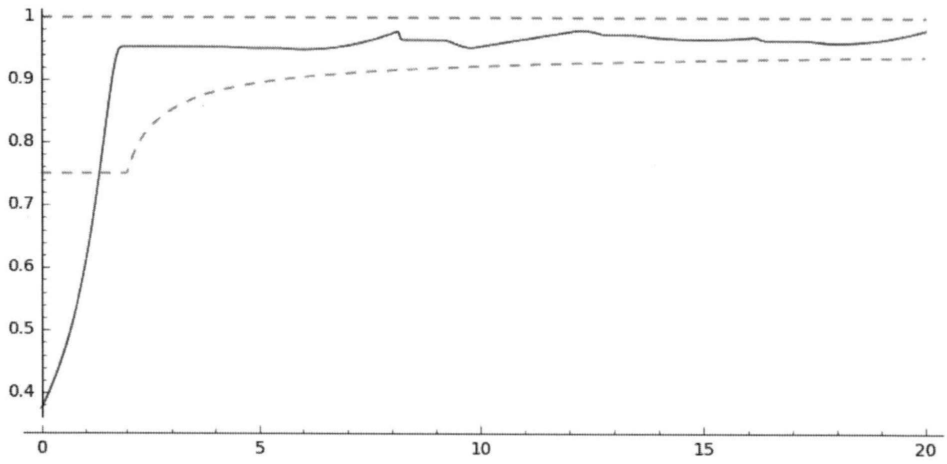

FIGURE 1. The graph of σ on $[0,20]$ ($d=2$, $\lambda=1/4$)

TABLE 1. Some computed values of σ ($d=2$, $\lambda=1/4$)

t	σ	t	σ	t	σ	t	σ	t	σ
0.0	0.37462	4.0	0.95210	8.0	0.97394	12.0	0.97662	16.0	0.96739
0.4	0.44248	4.4	0.95146	8.4	0.96359	12.4	0.97848	16.4	0.96309
0.8	0.53832	4.8	0.95003	8.8	0.96359	12.8	0.97233	16.8	0.96309
1.2	0.67932	5.2	0.95003	9.2	0.96314	13.2	0.97204	17.2	0.96307
1.6	0.87394	5.6	0.94924	9.6	0.95312	13.6	0.97061	17.6	0.96067
2.0	0.95210	6.0	0.94787	10.0	0.95325	14.0	0.96739	18.0	0.95879
2.4	0.95210	6.4	0.94891	10.4	0.95792	14.4	0.96565	18.4	0.95962
2.8	0.95210	6.8	0.95204	10.8	0.96260	14.8	0.96478	18.8	0.96209
3.2	0.95210	7.2	0.95725	11.2	0.96727	15.2	0.96478	19.2	0.96621
3.6	0.95210	7.6	0.96455	11.6	0.97195	15.6	0.96565	19.6	0.97198

The above σ obeys the following properties:

(1) σ is sigmoidal;
(2) $\sigma \in C^\infty(\mathbb{R})$;
(3) σ is strictly increasing on $(-\infty, d)$ and λ-strictly increasing on $[d, +\infty)$;
(4) σ is easily computable in practice.

All these properties are easily seen from the above exposition. But the essential property of our sigmoidal function is its ability to approximate an arbitrary continuous function using only a fixed number of translations and scalings of σ. More precisely, only two translations and scalings are sufficient. We formulate this important property as a theorem in the next section.

5.4.3. Theoretical results. The following theorems are valid.

THEOREM 5.30. *Assume that f is a continuous function on a finite segment $[a,b]$ of $\mathbb{R}$ and σ is the sigmoidal function constructed in Subsection* 5.4.1 *Then for*

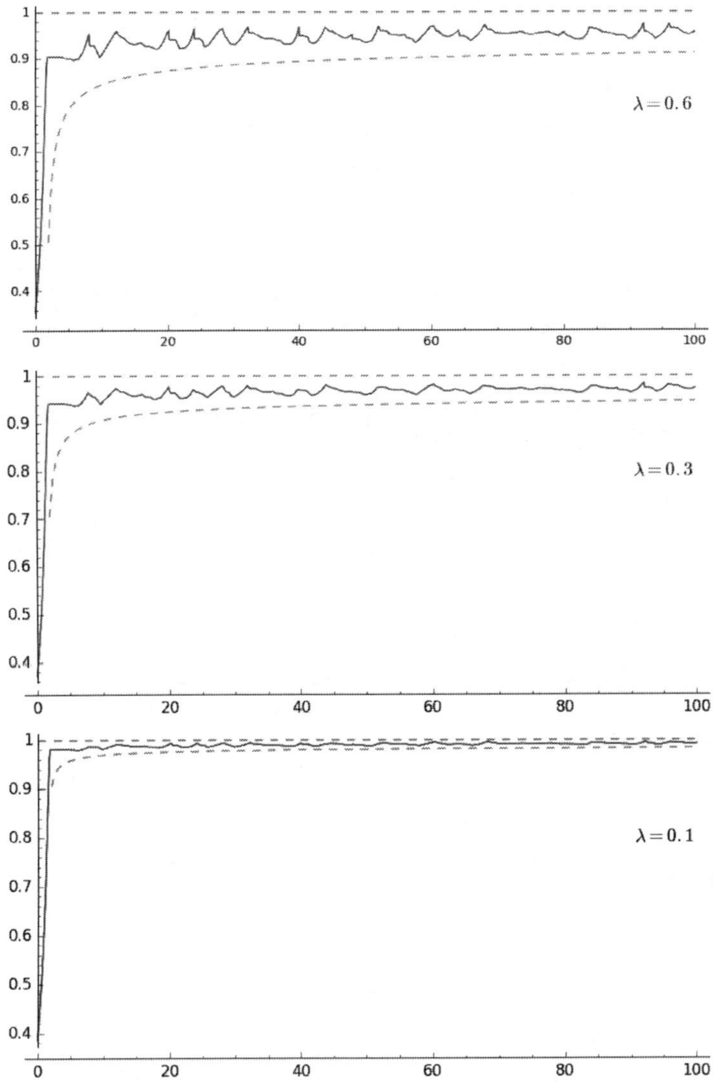

FIGURE 2. The graph of σ on $[0, 100]$ $(d = 2)$

any sufficiently small $\varepsilon > 0$ *there exist constants* c_1, c_2, θ_1 *and* θ_2 *such that*

$$|f(x) - c_1\sigma(x - \theta_1) - c_2\sigma(x - \theta_2)| < \varepsilon$$

for all $x \in [a, b]$.

PROOF. Set $d := b - a$ and divide the interval $[d, +\infty)$ into the segments $[d, 2d]$, $[2d, 3d]$, ... It follows from (5.42) that

$$\sigma(dx + (2n - 1)d) = a_n + b_n u_n(x), \quad x \in [0, 1] \tag{5.48}$$

for $n = 1, 2, \ldots$. Here a_n and b_n are computed by (5.43) and (5.44) for $n = 1$ and $n > 1$, respectively.

From (5.48) it follows that for each $n = 1, 2, \ldots,$

$$u_n(x) = \frac{1}{b_n}\sigma(dx + (2n-1)d) - \frac{a_n}{b_n}. \tag{5.49}$$

Let now g be any continuous function on the unit interval $[0, 1]$. By the density of polynomials with rational coefficients in the space of continuous functions on any compact subset of $\mathbb{R}$, for any $\varepsilon > 0$ there exists a polynomial $p(x)$ of the above form such that

$$|g(x) - p(x)| < \varepsilon$$

for all $x \in [0, 1]$. Denote by p_0 the leading coefficient of p. If $p_0 \neq 0$ (i.e., $p \not\equiv 0$) then we define u_n as $u_n(x) := p(x)/p_0$, otherwise we just set $u_n(x) := 1$. In both cases

$$|g(x) - p_0 u_n(x)| < \varepsilon, \qquad x \in [0, 1].$$

This together with (5.49) means that

$$|g(x) - c_1\sigma(dx - s_1) - c_0| < \varepsilon \tag{5.50}$$

for some c_0, c_1, $s_1 \in \mathbb{R}$ and all $x \in [0, 1]$. Namely, $c_1 = p_0/b_n$, $s_1 = d - 2nd$ and $c_0 = p_0 a_n/b_n$. On the other hand, we can write $c_0 = c_2\sigma(dx - s_2)$, where $c_2 := 2c_0/(1 + h(3d))$ and $s_2 := -d$. Hence,

$$|g(x) - c_1\sigma(dx - s_1) - c_2\sigma(dx - s_2)| < \varepsilon. \tag{5.51}$$

Note that (5.51) is valid for the unit interval $[0, 1]$. Using linear transformation it is not difficult to go from $[0, 1]$ to the interval $[a, b]$. Indeed, let $f \in C[a, b]$, σ be constructed as above, and ε be an arbitrarily small positive number. The transformed function $g(x) = f(a + (b - a)x)$ is well defined on $[0, 1]$ and we can apply the inequality (5.51). Now using the inverse transformation $x = (t-a)/(b-a)$, we can write

$$|f(t) - c_1\sigma(t - \theta_1) - c_2\sigma(t - \theta_2)| < \varepsilon$$

for all $t \in [a, b]$, where $\theta_1 = a + s_1$ and $\theta_2 = a + s_2$. The last inequality completes the proof. □

Since any compact subset of the real line is contained in a segment $[a, b]$, the following generalization of Theorem 5.30 holds.

THEOREM 5.31. *Let Q be a compact subset of the real line and d be its diameter. Let λ be any positive number. Then one can algorithmically construct a computable sigmoidal activation function $\sigma\colon \mathbb{R} \to \mathbb{R}$, which is infinitely differentiable, strictly increasing on $(-\infty, d)$, λ-strictly increasing on $[d, +\infty)$, and satisfies the following property: For any $f \in C(Q)$ and $\varepsilon > 0$ there exist numbers c_1, c_2, θ_1 and θ_2 such that*

$$|f(x) - c_1\sigma(x - \theta_1) - c_2\sigma(x - \theta_2)| < \varepsilon$$

for all $x \in Q$.

REMARK 5.32. Theorems 5.30 and 5.31 show that single hidden layer neural networks with the constructed sigmoidal activation function σ and only two neurons in the hidden layer can approximate any continuous univariate function. Moreover, in this case, one can fix the weights equal to 1. For the approximation of continuous multivariate functions two hidden layer neural networks with $3d+2$ hidden neurons can be taken. Namely, Theorem 5.22 (and hence Theorem 5.24) is valid with the constructed in Subsection 5.4.1 activation function σ. Indeed, the proof of this theorem shows that any activation function with the property (5.33) suffices. But the activation function constructed in Subsection 5.4.1 satisfies this property (see (5.50)).

5.4.4. Numerical results. We prove in Theorem 5.30 that any continuous function on $[a, b]$ can be approximated arbitrarily well by single hidden layer neural networks with the fixed weight 1 and with only two neurons in the hidden layer. An activation function σ for such a network is constructed in Subsection 5.4.1. We have seen from the proof that our approach is totally constructive. One can evaluate the value of σ at any point of the real axis and draw its graph instantly, using the programming interface at the URL shown at the beginning of Subsection 5.4.2. In the current subsection, we demonstrate our result in various examples. For different error bounds we find the parameters c_1, c_2, θ_1 and θ_2 in Theorem 5.30. All computations were done in SageMath [**188**]. For computations, we use the following algorithm, which works well for analytic functions. Assume f is a function, whose Taylor series around the point $(a+b)/2$ converges uniformly to f on $[a, b]$, and $\varepsilon > 0$.

(1) Consider the function $g(t) := f(a + (b-a)t)$, which is well-defined on $[0, 1]$;
(2) Find k such that the k-th Taylor polynomial

$$T_k(x) := \sum_{i=0}^{k} \frac{g^{(i)}(1/2)}{i!}\left(x - \frac{1}{2}\right)^i$$

satisfies the inequality $|T_k(x) - g(x)| \le \varepsilon/2$ for all $x \in [0, 1]$;
(3) Find a polynomial p with rational coefficients such that

$$|p(x) - T_k(x)| \le \frac{\varepsilon}{2}, \qquad x \in [0, 1],$$

and denote by p_0 the leading coefficient of this polynomial;
(4) If $p_0 \neq 0$, then find n such that $u_n(x) = p(x)/p_0$. Otherwise, set $n := 1$;
(5) For $n = 1$ and $n > 1$ evaluate a_n and b_n by (5.43) and (5.44), respectively;
(6) Calculate the parameters of the network as

$$c_1 := \frac{p_0}{b_n}, \qquad c_2 := \frac{2p_0 a_n}{b_n(1 + h(3d))}, \qquad \theta_1 := b - 2n(b-a), \qquad \theta_2 := 2a - b;$$

(7) Construct the network $\mathcal{N} = c_1\sigma(x - \theta_1) + c_2\sigma(x - \theta_2)$. Then $\mathcal{N}$ gives an ε-approximation to f.

In the sequel, we give four practical examples. To be able to make comparisons between these examples, all the considered functions are given on the same interval $[-1, 1]$. First we select the polynomial function $f(x) = x^3 + x^2 - 5x + 3$ as a target function. We investigate the sigmoidal neural network approximation to $f(x)$. This function was considered in [**70**] as well. Note that the authors of [**70**] chose the sigmoidal function as

$$\sigma(x) = \begin{cases} 1, & \text{if } x \ge 0, \\ 0, & \text{if } x < 0, \end{cases}$$

TABLE 2. The Heaviside function as a sigmoidal function

N	Number of neurons ($2N^2$)	Maximum error
2	8	0.666016
4	32	0.165262
8	128	0.041331
16	512	0.010333

TABLE 3. Several ε-approximators of the function $1 + x + x^2/2 + x^3/6 + x^4/24 + x^5/120 + x^6/720$

Number of neurons	Parameters of the network				Maximum error
	c_1	c_2	θ_1	θ_2	
2	2.0619×10^2	2.1131×10^2	-1979	-3	0.95
2	5.9326×10^2	6.1734×10^2	-1.4260×10^8	-3	0.60
2	1.4853×10^3	1.5546×10^3	-4.0140×10^{22}	-3	0.35
2	5.1231×10^2	5.3283×10^2	-3.2505×10^7	-3	0.10
2	4.2386×10^3	4.4466×10^3	-2.0403×10^{65}	-3	0.04
2	2.8744×10^4	3.0184×10^4	-1.7353×10^{442}	-3	0.01

and obtained the numerical results (see Table 2) for single hidden layer neural networks with 8, 32, 128, 532 neurons in the hidden layer (see also [**27**] for an additional constructive result concerning the error of approximation in this example).

As it is seen from the table, the number of neurons in the hidden layer increases as the error bound decreases in value. This phenomenon is no longer true for our sigmoidal function. Using Theorem 5.30, we can construct explicitly a single hidden layer neural network model with only two neurons in the hidden layer, which approximates the above polynomial with arbitrarily given precision. Here by *explicit construction* we mean that all the network parameters can be computed directly. Namely, the calculated values of these parameters are as follows: $c_1 \approx 2059.373597$, $c_2 \approx -2120.974727$, $\theta_1 = -467$, and $\theta_2 = -3$. It turns out that for the above polynomial we have an exact representation. That is, on the interval $[-1, 1]$ we have the identity

$$x^3 + x^2 - 5x + 3 \equiv c_1\sigma(x - \theta_1) + c_2\sigma(x - \theta_2).$$

Let us now consider the other polynomial function

$$f(x) = 1 + x + \frac{x^2}{2} + \frac{x^3}{6} + \frac{x^4}{24} + \frac{x^5}{120} + \frac{x^6}{720}.$$

For this function we do not have an exact representation as above. Nevertheless, one can easily construct a ε-approximating network with two neurons in the hidden layer for any sufficiently small approximation error ε. Table 3 displays numerical computations of the network parameters for six different approximation errors.

At the end we consider the nonpolynomial functions $f(x) = 4x/(4 + x^2)$ and $f(x) = \sin x - x\cos(x + 1)$. Tables 4 and 5 display all the parameters of the ε-approximating neural networks for the above six approximation error bounds. As it is seen from the tables, these bounds do not alter the number of hidden neurons. Figures 3, 4 and 5 show how graphs of some constructed networks $\mathcal{N}$ approximate the corresponding target functions f.

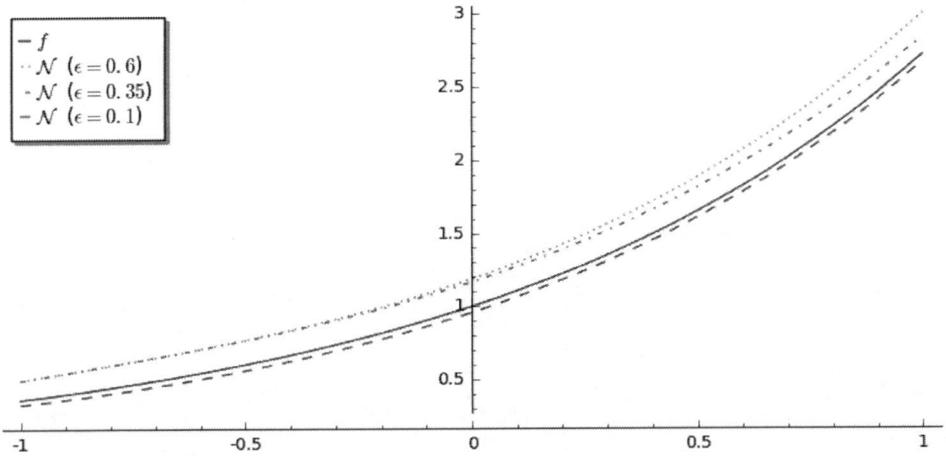

FIGURE 3. The graphs of $f(x) = 1 + x + x^2/2 + x^3/6 + x^4/24 + x^5/120 + x^6/720$ and some of its approximators ($\lambda = 1/4$)

TABLE 4. Several ε-approximators of the function $4x/(4+x^2)$

Number of neurons	Parameters of the network				Maximum error
	c_1	c_2	θ_1	θ_2	
2	1.5965×10^2	1.6454×10^2	-283	-3	0.95
2	1.5965×10^2	1.6454×10^2	-283	-3	0.60
2	-1.8579×10^3	-1.9428×10^3	-6.1840×10^{11}	-3	0.35
2	1.1293×10^4	1.1842×10^4	-4.6730×10^{34}	-3	0.10
2	2.6746×10^4	2.8074×10^4	-6.8296×10^{82}	-3	0.04
2	-3.4218×10^6	-3.5939×10^6	-2.9305×10^{4885}	-3	0.01

TABLE 5. Several ε-approximators of the function $\sin x - x\cos(x+1)$

Number of neurons	Parameters of the network				Maximum error
	c_1	c_2	θ_1	θ_2	
2	8.950×10^3	9.390×10^3	-3.591×10^{53}	-3	0.95
2	3.145×10^3	3.295×10^3	-3.397×10^{23}	-3	0.60
2	1.649×10^5	1.732×10^5	-9.532×10^{1264}	-3	0.35
2	-4.756×10^7	-4.995×10^7	$-1.308 \times 10^{180281}$	-3	0.10
2	-1.241×10^7	-1.303×10^7	-5.813×10^{61963}	-3	0.04
2	1.083×10^9	1.138×10^9	$-2.620 \times 10^{5556115}$	-3	0.01

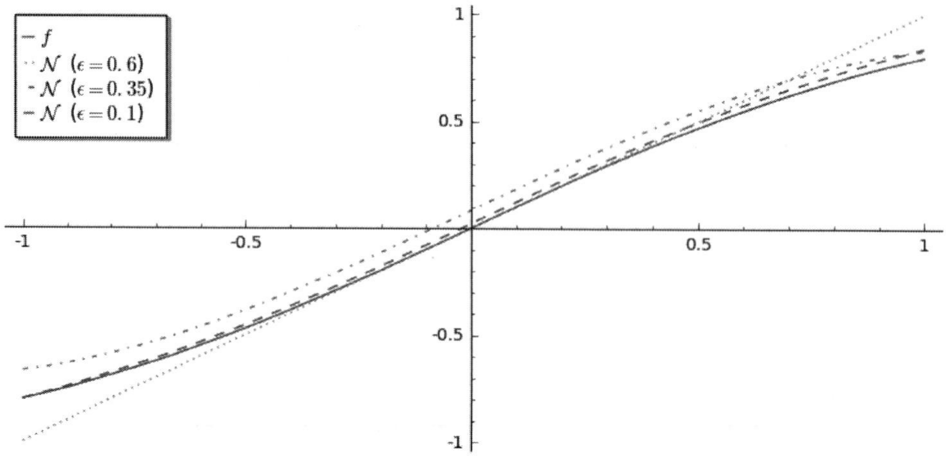

FIGURE 4. The graphs of $f(x) = 4x/(4 + x^2)$ and some of its approximators ($\lambda = 1/4$)

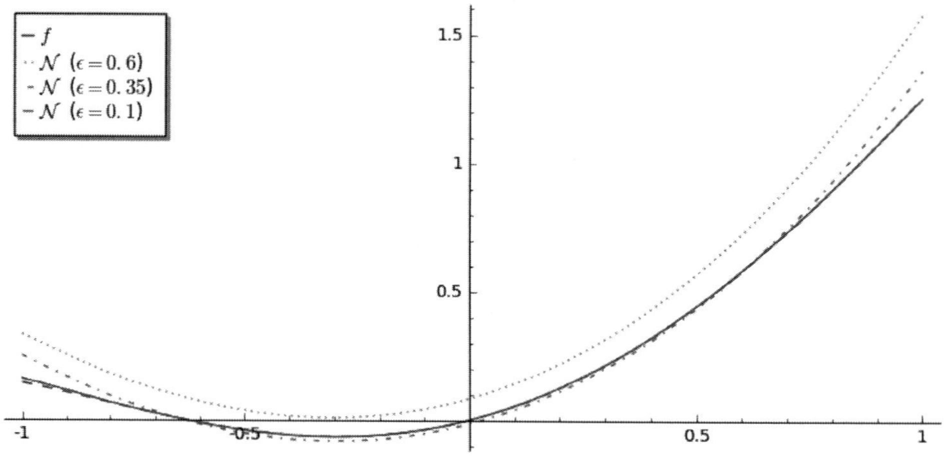

FIGURE 5. The graphs of $f(x) = \sin x - x\cos(x + 1)$ and some of its approximators ($\lambda = 1/4$)

Bibliography

[1] J. Aczél, *Lectures on functional equations and their applications*, Mathematics in Science and Engineering, Vol. 19, Academic Press, New York-London, 1966. Translated by Scripta Technica, Inc. Supplemented by the author. Edited by Hansjorg Oser. MR0208210

[2] Rashid A. Aliev, Aysel A. Asgarova, and Vugar E. Ismailov, *A note on continuous sums of ridge functions*, J. Approx. Theory **237** (2019), 210–221, DOI 10.1016/j.jat.2018.09.006. MR3868633

[3] Rashid A. Aliev and Vugar E. Ismailov, *On a smoothness problem in ridge function representation*, Adv. in Appl. Math. **73** (2016), 154–169, DOI 10.1016/j.aam.2015.11.002. MR3433504

[4] Aliev R.A., Ismailov V.E., On the representation by bivariate ridge functions, `arXiv:1606.07940`.

[5] Rashid A. Aliev and Vugar E. Ismailov, *A representation problem for smooth sums of ridge functions*, J. Approx. Theory **257** (2020), 105448, 13, DOI 10.1016/j.jat.2020.105448. MR4109084

[6] J. M. Almira, P. E. Lopez-de-Teruel, D. J. Romero-López, and F. Voigtlaender, *Negative results for approximation using single layer and multilayer feedforward neural networks*, J. Math. Anal. Appl. **494** (2021), no. 1, 124584, 11, DOI 10.1016/j.jmaa.2020.124584. MR4151572

[7] George A. Anastassiou, *Intelligent systems: approximation by artificial neural networks*, Intelligent Systems Reference Library, vol. 19, Springer-Verlag, Berlin, 2011, DOI 10.1007/978-3-642-21431-8. MR2895407

[8] V. I. Arnol′d, *On functions of three variables* (Russian), Dokl. Akad. Nauk SSSR **114** (1957), 679–681. MR0111808

[9] Aida Kh. Asgarova and Vugar E. Ismailov, *Diliberto-Straus algorithm for the uniform approximation by a sum of two algebras*, Proc. Indian Acad. Sci. Math. Sci. **127** (2017), no. 2, 361–374, DOI 10.1007/s12044-017-0337-4. MR3647158

[10] Georg Aumann, *Über approximative Nomographie. II* (German), Bayer. Akad. Wiss. Math.-Nat. Kl. S.-B. **1959** (1959), 103–109 (1960). MR0116173

[11] M.-B. A. Babaev, *The approximation of polynomials of two variables by functions of the form* $\varphi(x) + \psi(y)$ (Russian), Dokl. Akad. Nauk SSSR **193** (1970), 967–969. MR0280915

[12] M.-B. A. Babaev, *Sharp estimates for the approximation of functions of several variables by sums of functions of a lesser number of variables* (Russian), Mat. Zametki **12** (1972), 105–114. MR326243

[13] M.-B. A. Babaev, *Extremal elements and the value of best approximation of a monotone function in* $\mathbf{R}^n$ *by sums of functions of the least number of variables* (Russian), Dokl. Akad. Nauk SSSR **265** (1982), no. 1, 11–13. MR671630

[14] M.-B. A. Babaev and V. E. Ismailov, *Two-sided estimates for the best approximation in domains different from the parallelepiped*, Funct. Approx. Comment. Math. **25** (1997), 121–128. Dedicated to Roman Taberski on the occasion of his 70th birthday. MR1602374

[15] Randolph E. Bank, *An automatic scaling procedure for a D′jakanov-Gunn iteration scheme*, Linear Algebra Appl. **28** (1979), 17–33, DOI 10.1016/0024-3795(79)90114-9. MR549415

[16] Andrew R. Barron, *Universal approximation bounds for superpositions of a sigmoidal function*, IEEE Trans. Inform. Theory **39** (1993), no. 3, 930–945, DOI 10.1109/18.256500. MR1237720

[17] Helmut Bölcskei, Philipp Grohs, Gitta Kutyniok, and Philipp Petersen, *Optimal approximation with sparsely connected deep neural networks*, SIAM J. Math. Data Sci. **1** (2019), no. 1, 8–45, DOI 10.1137/18M118709X. MR3949699

[18] Dietrich Braess and Allan Pinkus, *Interpolation by ridge functions*, J. Approx. Theory **73** (1993), no. 2, 218–236, DOI 10.1006/jath.1993.1039. MR1216487

[19] R. C. Buck, *On approximation theory and functional equations*, J. Approximation Theory **5** (1972), 228–237, DOI 10.1016/0021-9045(72)90016-0. MR377363

[20] Martin D. Buhmann and Allan Pinkus, *Identifying linear combinations of ridge functions*, Adv. in Appl. Math. **22** (1999), no. 1, 103–118, DOI 10.1006/aama.1998.0623. MR1657745

[21] N. G. de Bruijn, *Functions whose differences belong to a given class*, Nieuw Arch. Wiskunde (2) **23** (1951), 194–218. MR0043870

[22] N. G. de Bruijn, *A difference property for Riemann integrable functions and for some similar classes of functions*, Nederl. Akad. Wetensch. Proc. Ser. A. **55** = Indagationes Math. **14** (1952), 145–151. MR0047120

[23] Neil Calkin and Herbert S. Wilf, *Recounting the rationals*, Amer. Math. Monthly **107** (2000), no. 4, 360–363, DOI 10.2307/2589182. MR1763062

[24] Emmanuel J. Candès, *Ridgelets: estimating with ridge functions*, Ann. Statist. **31** (2003), no. 5, 1561–1599, DOI 10.1214/aos/1065705119. MR2012826

[25] Candes E.J., *Ridgelets: theory and applications.* Ph.D. Thesis, Technical Report, Department of Statistics, Stanford University.

[26] Cao F., Lin S., Xu Z., Approximation capability of interpolation neural networks, Neurocomputing 74 (2010), 457-460.

[27] Cao F., Xie T., The construction and approximation for feedforword neural networks with fixed weights, Proceedings of the ninth international conference on machine learning and cybernetics, Qingdao, 2010, pp. 3164-3168.

[28] Carroll S.M., Dickinson B.W., Construction of neural nets using the Radon transform, in Proceedings of the IEEE 1989 International Joint Conference on Neural Networks, 1989, Vol. 1, IEEE, New York, 607-611.

[29] Chen T., Chen H., Approximation of continuous functionals by neural networks with application to dynamic systems, IEEE Trans. Neural Networks 4 (1993), 910-918.

[30] Chen T., Chen H., and Liu R., A constructive proof of Cybenko's approximation theorem and its extensions, in Computing Science and Statistics, Springer-Verlag, 1992, 163-168.

[31] Cheridito P., Jentzen A., Rossmannek F., Efficient approximation of high-dimensional functions with deep neural networks, `arXiv:1912.04310`.

[32] Charles K. Chui and Xin Li, *Approximation by ridge functions and neural networks with one hidden layer*, J. Approx. Theory **70** (1992), no. 2, 131–141, DOI 10.1016/0021-9045(92)90081-X. MR1172015

[33] C. K. Chui, Xin Li, and H. N. Mhaskar, *Limitations of the approximation capabilities of neural networks with one hidden layer*, Adv. Comput. Math. **5** (1996), no. 2-3, 233–243, DOI 10.1007/BF02124745. MR1399382

[34] Z. Ciesielski, *Some properties of convex functions of higher orders*, Ann. Polon. Math. **7** (1959), 1–7, DOI 10.4064/ap-7-1-1-7. MR109202

[35] Albert Cohen, Ingrid Daubechies, Ronald DeVore, Gerard Kerkyacharian, and Dominique Picard, *Capturing ridge functions in high dimensions from point queries*, Constr. Approx. **35** (2012), no. 2, 225–243, DOI 10.1007/s00365-011-9147-6. MR2891227

[36] Constantine P.G., del Rosario Z., Iaccarino G., Many physical laws are ridge functions, `arXiv:1605.07974`.

[37] Constantine P.G., del Rosario Z., Iaccarino G., Data-driven dimensional analysis: algorithms for unique and relevant dimensionless groups, `arXiv:1708.04303`.

[38] Danilo Costarelli and Renato Spigler, *Constructive approximation by superposition of sigmoidal functions*, Anal. Theory Appl. **29** (2013), no. 2, 169–196, DOI 10.4208/ata.2013.v29.n2.8. MR3109891

[39] Costarelli D., Spigler R, Approximation results for neural network operators activated by sigmoidal functions, Neural Networks 44 (2013), 101-106.

[40] Neil E. Cotter, *The Stone-Weierstrass theorem and its application to neural networks*, IEEE Trans. Neural Networks **1** (1990), no. 4, 290–295, DOI 10.1109/72.80265. MR1083640

[41] G. Cybenko, *Approximation by superpositions of a sigmoidal function*, Math. Control Signals Systems **2** (1989), no. 4, 303–314, DOI 10.1007/BF02551274. MR1015670

[42] Dahmen W., Micchelli C.A., *Some remarks on ridge functions*, Approx. Theory Appl. 3 (1987), 139-143.

[43] Stephen Demko, *A superposition theorem for bounded continuous functions*, Proc. Amer. Math. Soc. **66** (1977), no. 1, 75–78, DOI 10.2307/2041533. MR457651

[44] Frank Deutsch, *The alternating method of von Neumann*, Multivariate approximation theory (Proc. Conf., Math. Res. Inst., Oberwolfach, 1979), Internat. Ser. Numer. Math., vol. 51, Birkhäuser, Basel-Boston, Mass., 1979, pp. 83–96. MR560665

[45] Ronald A. DeVore, Konstantin I. Oskolkov, and Pencho P. Petrushev, *Approximation by feed-forward neural networks*, Ann. Numer. Math. **4** (1997), no. 1-4, 261–287. The heritage of P. L. Chebyshev: a Festschrift in honor of the 70th birthday of T. J. Rivlin. MR1422683

[46] Persi Diaconis and Mehrdad Shahshahani, *On nonlinear functions of linear combinations*, SIAM J. Sci. Statist. Comput. **5** (1984), no. 1, 175–191, DOI 10.1137/0905013. MR731890

[47] S. P. Diliberto and E. G. Straus, *On the approximation of a function of several variables by the sum of functions of fewer variables*, Pacific J. Math. **1** (1951), 195–210. MR43882

[48] D. Ž. Djoković, *A representation theorem for* $(X_1-1)(X_2-1)\cdots(X_n-1)$ *and its applications*, Ann. Polon. Math. **22** (1969/70), 189–198, DOI 10.4064/ap-22-2-189-198. MR265798

[49] Benjamin Doerr and Sebastian Mayer, *The recovery of ridge functions on the hypercube suffers from the curse of dimensionality*, J. Complexity **63** (2021), 101521, 29, DOI 10.1016/j.jco.2020.101521. MR4204470

[50] David L. Donoho and Iain M. Johnstone, *Projection-based approximation and a duality with kernel methods*, Ann. Statist. **17** (1989), no. 1, 58–106, DOI 10.1214/aos/1176347004. MR981438

[51] Nira Dyn, W. A. Light, and E. W. Cheney, *Interpolation by piecewise-linear radial basis functions. I*, J. Approx. Theory **59** (1989), no. 2, 202–223, DOI 10.1016/0021-9045(89)90152-4. MR1022117

[52] Leopold Flatto, *The approximation of certain functions of several variables by sums of functions of fewer variables*, Amer. Math. Monthly **73** (1966), no. 4, 131–132, DOI 10.2307/2313764. MR201888

[53] Massimo Fornasier, Karin Schnass, and Jan Vybiral, *Learning functions of few arbitrary linear parameters in high dimensions*, Found. Comput. Math. **12** (2012), no. 2, 229–262, DOI 10.1007/s10208-012-9115-y. MR2898783

[54] B. L. Fridman, *An improvement in the smoothness of the functions in A. N. Kolmogorov's theorem on superpositions* (Russian), Dokl. Akad. Nauk SSSR **177** (1967), 1019–1022. MR0225066

[55] Friedman J.H., Tukey J.W., A Projection Pursuit Algorithm for Exploratory Data Analysis, IEEE Transactions on Computers C-23 (1974), 881-890.

[56] Jerome H. Friedman and Werner Stuetzle, *Projection pursuit regression*, J. Amer. Statist. Assoc. **76** (1981), no. 376, 817–823. MR650892

[57] Funahashi K., On the approximate realization of continuous mapping by neural networks, Neural Networks 2 (1989), 183-192.

[58] Zbigniew Gajda, *Difference properties of higher orders for continuity and Riemann integrability*, Colloq. Math. **53** (1987), no. 2, 275–288, DOI 10.4064/cm-53-2-275-288. MR924073

[59] Gallant A.R., White H., There exists a neural network that does not make avoidable mistakes, in Proceedings of the IEEE 1988 International Conference on Neural Networks, 1988, vol. 1, IEEE, New York, pp. 657-664.

[60] A. L. Garkavi, V. A. Medvedev, and S. Ya. Khavinson, *On the existence of a best uniform approximation of a function in two variables by sums* $\phi(x)+\psi(y)$ (Russian, with Russian summary), Sibirsk. Mat. Zh. **36** (1995), no. 4, 819–827, ii, DOI 10.1007/BF02107327; English transl., Siberian Math. J. **36** (1995), no. 4, 707–713. MR1367249

[61] A. L. Garkavi, V. A. Medvedev, and S. Ya. Khavinson, *On the existence of a best uniform approximation of a function of several variables by the sum of functions of fewer variables* (Russian, with Russian summary), Mat. Sb. **187** (1996), no. 5, 3–14, DOI 10.1070/SM1996v187n05ABEH000125; English transl., Sb. Math. **187** (1996), no. 5, 623–634. MR1400350

[62] Andrew Glaws, Paul G. Constantine, and R. Dennis Cook, *Inverse regression for ridge recovery: a data-driven approach for parameter reduction in computer experiments*, Stat. Comput. **30** (2020), no. 2, 237–253, DOI 10.1007/s11222-019-09876-y. MR4064620

[63] M. von Golitschek and W. A. Light, *Approximation by solutions of the planar wave equation*, SIAM J. Numer. Anal. **29** (1992), no. 3, 816–830, DOI 10.1137/0729050. MR1163358

[64] Michael Golomb, *Approximation by functions of fewer variables*, On numerical approximation. Proceedings of a Symposium, Madison, April 21-23, 1958, Edited by R. E. Langer. Publication no. 1 of the Mathematics Research Center, U.S. Army, the University of Wisconsin, The University of Wisconsin Press, Madison, 1959, pp. 275–327. MR0102168

[65] Ian Goodfellow, Yoshua Bengio, and Aaron Courville, *Deep learning*, Adaptive Computation and Machine Learning, MIT Press, Cambridge, MA, 2016. MR3617773

[66] Y. Gordon, V. Maiorov, M. Meyer, and S. Reisner, *On the best approximation by ridge functions in the uniform norm*, Constr. Approx. **18** (2002), no. 1, 61–85, DOI 10.1007/s00365-001-0009-5. MR1866380

[67] Namig J. Guliyev and Vugar E. Ismailov, *A single hidden layer feedforward network with only one neuron in the hidden layer can approximate any univariate function*, Neural Comput. **28** (2016), no. 7, 1289–1304, DOI 10.1162/neco_a_00849. MR3867104

[68] Guliyev N.J., Ismailov V.E., On the approximation by single hidden layer feedforward neural networks with fixed weights, Neural Networks 98 (2018), 296-304.

[69] Guliyev N.J., Ismailov V.E., Approximation capability of two hidden layer feedforward neural networks with fixed weights, Neurocomputing 316 (2018), 262-269.

[70] Nahmwoo Hahm and Bum Il Hong, *An approximation by neural networks with a fixed weight*, Comput. Math. Appl. **47** (2004), no. 12, 1897–1903, DOI 10.1016/j.camwa.2003.06.008. MR2086107

[71] Israel Halperin, *The product of projection operators*, Acta Sci. Math. (Szeged) **23** (1962), 96–99. MR141978

[72] Hecht-Nielsen R., Kolmogorov's mapping neural network existence theorem, in Proc. I987 IEEE Int. Conf. on Neural Networks, IEEE Press, New York, 1987, vol. 3, 11-14.

[73] Hornik K., Approximation capabilities of multilayer feedforward networks, Neural Networks 4 (1991), 251-257.

[74] Hornik K., Stinchcombe M., White H., Multilayer feedforward networks are universal approximators, Neural Networks 2 (1989), 359-366.

[75] Peter J. Huber, *Projection pursuit*, Ann. Statist. **13** (1985), no. 2, 435–525, DOI 10.1214/aos/1176349519. With discussion. MR790553

[76] V. E. Ismailov, *Theorem on lightning bolts for elementary domains* (English, with English and Azerbaijani summaries), Proc. Inst. Math. Mech. Natl. Acad. Sci. Azerb. **17** (2002), 78–85, 205. Applied problems of mathematics and mechanics (Baku, 2002). MR2027426

[77] Vugar E. Ismailov, *On some classes of bivariate functions characterized by formulas for the best approximation* (English, with English and Serbo-Croatian summaries), Rad. Mat. **13** (2004), no. 1, 53–62. MR2111677

[78] Vugar E. Ismailov, *On error formulas for approximation by sums of univariate functions*, Int. J. Math. Math. Sci., posted on 2006, Art. ID 65620, 11, DOI 10.1155/IJMMS/2006/65620. MR2219166

[79] V. È. Ismailov, *On methods for computing the exact value of the best approximation by sums of functions of one variable* (Russian, with Russian summary), Sibirsk. Mat. Zh. **47** (2006), no. 5, 1076–1082, DOI 10.1007/s11202-006-0097-3; English transl., Siberian Math. J. **47** (2006), no. 5, 883–888. MR2266517

[80] Vugar E. Ismailov, *On the approximation by compositions of fixed multivariate functions with univariate functions*, Studia Math. **183** (2007), no. 2, 117–126, DOI 10.4064/sm183-2-2. MR2353881

[81] Vugar E. Ismailov, *A note on the best L_2 approximation by ridge functions*, Appl. Math. E-Notes **7** (2007), 71–76. MR2295689

[82] Vugar E. Ismailov, *Representation of multivariate functions by sums of ridge functions*, J. Math. Anal. Appl. **331** (2007), no. 1, 184–190, DOI 10.1016/j.jmaa.2006.08.076. MR2305997

[83] Vugar E. Ismailov, *Characterization of an extremal sum of ridge functions*, J. Comput. Appl. Math. **205** (2007), no. 1, 105–115, DOI 10.1016/j.cam.2006.04.043. MR2324828

[84] Vugar E. Ismailov, *On the representation by linear superpositions*, J. Approx. Theory **151** (2008), no. 2, 113–125, DOI 10.1016/j.jat.2007.09.003. MR2407861

[85] Vugar E. Ismailov, *On the approximation by weighted ridge functions*, An. Univ. Vest Timiş. Ser. Mat.-Inform. **46** (2008), no. 1, 75–83. MR2791467

[86] Vugar E. Ismailov, *On the proximinality of ridge functions*, Sarajevo J. Math. **5(17)** (2009), no. 1, 109–118. MR2527153

[87] Vugar E. Ismailov, *On the theorem of M Golomb*, Proc. Indian Acad. Sci. Math. Sci. **119** (2009), no. 1, 45–52, DOI 10.1007/s12044-009-0005-4. MR2508488

[88] V. E. Ismailov, *Approximation capabilities of neural networks with weights from two directions*, Azerb. J. Math. **1** (2011), no. 1, 122–128. MR2776103

[89] Vugar E. Ismailov, *Approximation by neural networks with weights varying on a finite set of directions*, J. Math. Anal. Appl. **389** (2012), no. 1, 72–83, DOI 10.1016/j.jmaa.2011.11.037. MR2876482

[90] Vugar E. Ismailov, *A note on the representation of continuous functions by linear superpositions*, Expo. Math. **30** (2012), no. 1, 96–101, DOI 10.1016/j.exmath.2011.07.005. MR2899658

[91] Vugar E. Ismailov, *A review of some results on ridge function approximation*, Azerb. J. Math. **3** (2013), no. 1, 3–51. MR3032796

[92] Vugar E. Ismailov, *On the approximation by neural networks with bounded number of neurons in hidden layers*, J. Math. Anal. Appl. **417** (2014), no. 2, 963–969, DOI 10.1016/j.jmaa.2014.03.092. MR3194523

[93] Vugar E. Ismailov, *Alternating algorithm for the approximation by sums of two compositions and ridge functions*, Proc. Inst. Math. Mech. Natl. Acad. Sci. Azerb. **41** (2015), no. 1, 146–152. MR3465724

[94] V. E. Ismailov, *On the uniqueness of representation by linear superpositions* (English, with English and Ukrainian summaries), Ukraïn. Mat. Zh. **68** (2016), no. 12, 1620–1628, DOI 10.1007/s11253-017-1335-5; English transl., Ukrainian Math. J. **68** (2017), no. 12, 1874–1883. MR3592387

[95] V. È. Ismailov, *Approximation by sums of ridge functions with fixed directions* (Russian, with Russian summary), Algebra i Analiz **28** (2016), no. 6, 20–69, DOI 10.1090/spmj/1471; English transl., St. Petersburg Math. J. **28** (2017), no. 6, 741–772. MR3637575

[96] Vugar E. Ismailov, *A note on the equioscillation theorem for best ridge function approximation*, Expo. Math. **35** (2017), no. 3, 343–349, DOI 10.1016/j.exmath.2017.05.003. MR3689906

[97] Vugar E. Ismailov, *Computing the approximation error for neural networks with weights varying on fixed directions*, Numer. Funct. Anal. Optim. **40** (2019), no. 12, 1395–1409, DOI 10.1080/01630563.2019.1605523. MR3953211

[98] V. E. Ismailov and A. Pinkus, *Interpolation on lines by ridge functions*, J. Approx. Theory **175** (2013), 91–113, DOI 10.1016/j.jat.2013.07.010. MR3101062

[99] Vugar E. Ismailov and Ekrem Savas, *Measure theoretic results for approximation by neural networks with limited weights*, Numer. Funct. Anal. Optim. **38** (2017), no. 7, 819–830, DOI 10.1080/01630563.2016.1254654. MR3654363

[100] Irie B., Miyake S., Capability of three-layered perceptrons, in Proceedings of the IEEE 1988 Int. Conf. on Neural Networks, 1988, Vol. 1, IEEE, New York, 641-648.

[101] Ito Y., Representation of functions by superpositions of a step or sigmoid function and their applications to neural network theory, Neural Networks 4 (1991), 385-394.

[102] Ito Y., Approximation of functions on a compact set by finite sums of a sigmoid function without scaling, Neural Networks 4 (1991), no. 6, 817-826.

[103] Ito Y., Approximation of continuous functions on $\mathbb{R}^d$ by linear combinations of shifted rotations of a sigmoid function with and without scaling, Neural Networks 5 (1992), 105-115.

[104] Jones L.K., Constructive approximations for neural networks by sigmoidal functions, Proc. IEEE 78 (1990), 1586-1589. Correction and addition, Proc. IEEE 79 (1991), 243.

[105] Fritz John, *Plane waves and spherical means applied to partial differential equations*, Interscience Publishers, New York-London, 1955. MR0075429

[106] Palle E. T. Jorgensen and James F. Tian, *Superposition, reduction of multivariable problems, and approximation*, Anal. Appl. (Singap.) **18** (2020), no. 5, 771–801, DOI 10.1142/S021953051941001X. MR4131038

[107] Paul C. Kainen and Věra Kůrková, *An integral upper bound for neural network approximation*, Neural Comput. **21** (2009), no. 10, 2970–2989, DOI 10.1162/neco.2009.04-08-745. MR2567918

[108] Kainen P.C., Kůrkova V., Vogt A., Best approximation by Heaviside perceptron networks, Neural Networks 13 (2007), no. 7, 695-697.

[109] I. G. Kazantsev, *Tomographic reconstruction from arbitrary directions using ridge functions*, Inverse Problems **14** (1998), no. 3, 635–645, DOI 10.1088/0266-5611/14/3/014. MR1630007

[110] Kazantsev I., Tomographic reconstruction using ridge functions, Proceedings of 1st World Congress on Industrial Process Tomography, Buxton, Derbishyre, UK, April 14-17, 1999, pp. 433-437.

[111] Kazantsev I., Lemahieu I., Reconstruction of elongated structures using ridge functions and natural pixels, Inverse Problems 16 (2000), 505-517.

[112] Sandra Keiper, *Approximation of generalized ridge functions in high dimensions*, J. Approx. Theory **245** (2019), 101–129, DOI 10.1016/j.jat.2019.04.006. MR3953358

[113] C. T. Kelley, *A note on the approximation of functions of several variables by sums of functions of one variable*, J. Approx. Theory **33** (1981), no. 3, 179–189, DOI 10.1016/0021-9045(81)90068-X. MR647845

[114] S. Ja. Havinson, *A Čebyšev theorem for the approximation of a function of two variables by sums* $\phi(x) + \psi(y)$ (Russian), Izv. Akad. Nauk SSSR Ser. Mat. **33** (1969), 650–666. MR0262746

[115] S. Ya. Khavinson, *Representation of functions of two variables by the sums* $\phi(x) + \psi(y)$ (Russian), Izv. Vyssh. Uchebn. Zaved. Mat. **2** (1985), 66–73, 87. MR788620

[116] S. Ya. Khavinson, *Some approximation properties of linear superpositions* (Russian), Izv. Vyssh. Uchebn. Zaved. Mat. **8** (1995), 63–73; English transl., Russian Math. (Iz. VUZ) **39** (1995), no. 8, 60–70. MR1391343

[117] S. Ya. Khavinson, *Best approximation by linear superpositions (approximate nomography)*, Translations of Mathematical Monographs, vol. 159, American Mathematical Society, Providence, RI, 1997. Translated from the Russian manuscript by D. Khavinson, DOI 10.1090/mmono/159. MR1421322

[118] A. Kłopotowski and M. G. Nadkarni, *Shift invariant measures and simple spectrum*. part 2, Colloq. Math. **84/85** (2000), no. part 2, 385–394, DOI 10.4064/cm-84/85-2-385-394. Dedicated to the memory of Anzelm Iwanik. MR1784203

[119] A. Kłopotowski, M. G. Nadkarni, and K. P. S. Bhaskara Rao, *When is* $f(x_1, x_2, \ldots, x_n) = u_1(x_1) + u_2(x_2) + \cdots + u_n(x_n)$*?*, Proc. Indian Acad. Sci. Math. Sci. **113** (2003), no. 1, 77–86, DOI 10.1007/BF02829681. Functional analysis (Kolkata, 2001). MR1971557

[120] A. Kłopotowski, M. G. Nadkarni, and K. P. S. Bhaskara Rao, *Geometry of good sets in n-fold Cartesian product*, Proc. Indian Acad. Sci. Math. Sci. **114** (2004), no. 2, 181–197, DOI 10.1007/BF02829852. MR2062398

[121] A. N. Kolmogorov, *On certain asymptotic characteristics of completely bounded metric spaces* (Russian), Dokl. Akad. Nauk SSSR (N.S.) **108** (1956), 385–388. MR0080904

[122] A. N. Kolmogorov, *On the representation of continuous functions of many variables by superposition of continuous functions of one variable and addition*, Amer. Math. Soc. Transl. (2) **28** (1963), 55–59, DOI 10.1090/trans2/028/04. MR0153799

[123] A. N. Kolmogorov and V. M. Tihomirov, *ε-entropy and ε-capacity of sets in functional space*, Amer. Math. Soc. Transl. (2) **17** (1961), 277–364. MR0124720

[124] I. M. Kolodiĭ and F. Hil′debrand, *Certain properties of the modulus of continuity* (Russian), Mat. Zametki **9** (1971), 495–500. MR281850

[125] S. V. Konyagin and A. A. Kuleshov, *On the continuity of finite sums of ridge functions* (Russian), Mat. Zametki **98** (2015), no. 2, 308–309, DOI 10.4213/mzm10787; English transl., Math. Notes **98** (2015), no. 1-2, 336–338. MR3438485

[126] S. V. Konyagin and A. A. Kuleshov, *On some properties of finite sums of ridge functions defined on convex subsets of* $\mathbb{R}^n$ (Russian, with Russian summary), Tr. Mat. Inst. Steklova **293** (2016), no. Funktsional. Prostranstva, Teor. Priblizh., Smezhnye Razdely Mat. Anal., 193–200, DOI 10.1134/S0371968516020138; English transl., Proc. Steklov Inst. Math. **293** (2016), no. 1, 186–193. MR3628479

[127] S. V. Konyagin, A. A. Kuleshov, and V. E. Maĭorov, *Some problems in the theory of ridge functions* (Russian, with Russian summary), Tr. Mat. Inst. Steklova **301** (2018), no. Kompleksnyĭ Analiz, Matematicheskaya Fizika i Prilozheniya, 155–181, DOI 10.1134/S0371968518020127; English transl., Proc. Steklov Inst. Math. **301** (2018), no. 1, 144–169. MR3841666

[128] A. Kroó, *On approximation by ridge functions*, Constr. Approx. **13** (1997), no. 4, 447–460, DOI 10.1007/s003659900053. MR1466061

[129] Marek Kuczma, *An introduction to the theory of functional equations and inequalities*, 2nd ed., Birkhäuser Verlag, Basel, 2009. Cauchy's equation and Jensen's inequality; Edited and with a preface by Attila Gilányi, DOI 10.1007/978-3-7643-8749-5. MR2467621

[130] A. A. Kuleshov, *On some properties of smooth sums of ridge functions* (Russian, with Russian summary), Tr. Mat. Inst. Steklova **294** (2016), no. Sovremennye Problemy Matematiki, Mekhaniki i Matematicheskoĭ Fiziki. II, 99–104, DOI 10.1134/S0371968516030067; English transl., Proc. Steklov Inst. Math. **294** (2016), no. 1, 89–94. MR3628495

[131] A. A. Kuleshov, *Continuous sums of ridge functions on a convex body and the class VMO* (Russian, with Russian summary), Mat. Zametki **102** (2017), no. 6, 866–873, DOI 10.4213/mzm11568; English transl., Math. Notes **102** (2017), no. 5-6, 799–805. MR3733330

[132] Svetozar Kurepa, *A property of a set of positive measure and its application*, J. Math. Soc. Japan **13** (1961), 13–19, DOI 10.2969/jmsj/01310013. MR132141

[133] Kůrková V., Kolmogorov's theorem is relevant, Neural Comput. 3 (1991), 617-622.

[134] Kůrková V., Kolmogorov's theorem and multilayer neural networks, Neural Networks 5 (1992), 501-506.

[135] Leshno M., Lin V.Ya., Pinkus A., Schocken S., Multilayer feedforward networks with a nonpolynomial activation function can approximate any function, Neural Networks 6 (1993), 861-867.

[136] Will Light, *Ridge functions, sigmoidal functions and neural networks*, Approximation theory VII (Austin, TX, 1992), Academic Press, Boston, MA, 1993, pp. 163–206, DOI 10.1144/GSL.SP.1993.071.01.08. MR1212573

[137] W. A. Light and E. W. Cheney, *On the approximation of a bivariate function by the sum of univariate functions*, J. Approx. Theory **29** (1980), no. 4, 305–322, DOI 10.1016/0021-9045(80)90119-7. MR598725

[138] W. A. Light and E. W. Cheney, *Approximation theory in tensor product spaces*, Lecture Notes in Mathematics, vol. 1169, Springer-Verlag, Berlin, 1985, DOI 10.1007/BFb0075391. MR817984

[139] Lin S., Limitations of shallow nets approximation, Neural Networks 94 (2017), 96-102.

[140] Vladimir Ya. Lin and Allan Pinkus, *Fundamentality of ridge functions*, J. Approx. Theory **75** (1993), no. 3, 295–311, DOI 10.1006/jath.1993.1104. MR1250542

[141] B. F. Logan and L. A. Shepp, *Optimal reconstruction of a function from its projections*, Duke Math. J. **42** (1975), no. 4, 645–659. MR397240

[142] G. G. Lorentz, *Metric entropy, widths, and superpositions of functions*, Amer. Math. Monthly **69** (1962), 469–485, DOI 10.2307/2311185. MR141926

[143] V. E. Maiorov, *On best approximation by ridge functions*, J. Approx. Theory **99** (1999), no. 1, 68–94, DOI 10.1006/jath.1998.3304. MR1696577

[144] Vitaly Maiorov, *Geometric properties of the ridge function manifold*, Adv. Comput. Math. **32** (2010), no. 2, 239–253, DOI 10.1007/s10444-008-9106-3. MR2581237

[145] V. E. Maiorov and R. Meir, *On the near optimality of the stochastic approximation of smooth functions by neural networks*, Adv. Comput. Math. **13** (2000), no. 1, 79–103, DOI 10.1023/A:1018993908478. MR1759189

[146] Vitaly Maiorov, Ron Meir, and Joel Ratsaby, *On the approximation of functional classes equipped with a uniform measure using ridge functions*, J. Approx. Theory **99** (1999), no. 1, 95–111, DOI 10.1006/jath.1998.3305. MR1696573

[147] Maiorov V., Pinkus A., Lower bounds for approximation by MLP neural networks, Neurocomputing 25 (1999), 81-91.

[148] Y. Makovoz, *Uniform approximation by neural networks*, J. Approx. Theory **95** (1998), no. 2, 215–228, DOI 10.1006/jath.1997.3217. MR1652888

[149] Robert B. Marr, *On the reconstruction of a function on a circular domain from a sampling of its line integrals*, J. Math. Anal. Appl. **45** (1974), 357–374, DOI 10.1016/0022-247X(74)90078-X. MR336156

[150] Donald E. Marshall and Anthony G. O'Farrell, *Uniform approximation by real functions*, Fund. Math. **104** (1979), no. 3, 203–211, DOI 10.4064/fm-104-3-203-211. MR559174

[151] Donald E. Marshall and Anthony G. O'Farrell, *Approximation by a sum of two algebras. The lightning bolt principle*, J. Funct. Anal. **52** (1983), no. 3, 353–368, DOI 10.1016/0022-1236(83)90074-5. MR712586

[152] Sebastian Mayer, Tino Ullrich, and Jan Vybíral, *Entropy and sampling numbers of classes of ridge functions*, Constr. Approx. **42** (2015), no. 2, 231–264, DOI 10.1007/s00365-014-9267-x. MR3392489

[153] Mazur S., Orlicz W., Grundlegende Eigenschaften der polynomischen Operationen I., II., Studia Math. 5 (1934), 50-68, 179-189.

[154] M. A. McKiernan, *On vanishing nth ordered differences and Hamel bases*, Ann. Polon. Math. **19** (1967), 331–336, DOI 10.4064/ap-19-3-331-336. MR221131

[155] V. A. Medvedev, *Refutation of a theorem of Diliberto and Straus* (Russian), Mat. Zametki **51** (1992), no. 4, 78–80, 142, DOI 10.1007/BF01250549; English transl., Math. Notes **51** (1992), no. 3-4, 380–381. MR1172469

[156] V. A. Medvedev, *On the sum of two closed algebras of continuous functions on a compact space* (Russian), Funktsional. Anal. i Prilozhen. **27** (1993), no. 1, 33–36, DOI 10.1007/BF01768665; English transl., Funct. Anal. Appl. **27** (1993), no. 1, 28–30. MR1225908

[157] H. N. Mhaskar, *On the tractability of multivariate integration and approximation by neural networks*, J. Complexity **20** (2004), no. 4, 561–590, DOI 10.1016/j.jco.2003.11.004. MR2068159

[158] H. N. Mhaskar and Charles A. Micchelli, *Approximation by superposition of sigmoidal and radial basis functions*, Adv. in Appl. Math. **13** (1992), no. 3, 350–373, DOI 10.1016/0196-8858(92)90016-P. MR1176581

[159] Montanelli H., Yang H., Error bounds for deep ReLU networks using the Kolmogorov–Arnold superposition theorem, Neural Networks 129 (2020), 1-6.

[160] F. Natterer, *The mathematics of computerized tomography*, B. G. Teubner, Stuttgart; John Wiley & Sons, Ltd., Chichester, 1986. MR856916

[161] K. Gowri Navada, *Some remarks on good sets*, Proc. Indian Acad. Sci. Math. Sci. **114** (2004), no. 4, 389–397, DOI 10.1007/BF02829443. MR2067701

[162] John von Neumann, *Functional Operators. II. The Geometry of Orthogonal Spaces*, Annals of Mathematics Studies, No. 22, Princeton University Press, Princeton, N. J., 1950. MR0034514

[163] Erich Novak and Henryk Woźniakowski, *Approximation of infinitely differentiable multivariate functions is intractable*, J. Complexity **25** (2009), no. 4, 398–404, DOI 10.1016/j.jco.2008.11.002. MR2542039

[164] Erich Novak and Henryk Woźniakowski, *Tractability of multivariate problems for standard and linear information in the worst case setting: Part I*, J. Approx. Theory **207** (2016), 177–192, DOI 10.1016/j.jat.2016.02.017. MR3494228

[165] Erich Novak and Henryk Woźniakowski, *Tractability of multivariate problems for standard and linear information in the worst case setting: Part II*, Contemporary computational mathematics—a celebration of the 80th birthday of Ian Sloan. Vol. 1, 2, Springer, Cham, 2018, pp. 963–977. MR3822267

[166] Ju. P. Ofman, *On the best approximation of functions of two variables by functions of the form $\varphi(x) + \psi(y)$* (Russian), Izv. Akad. Nauk SSSR Ser. Mat. **25** (1961), 239–252. MR0125381

[167] K. I. Oskolkov, *Ridge approximation, Fourier-Chebyshev analysis, and optimal quadrature formulas* (Russian), Tr. Mat. Inst. Steklova **219** (1997), no. Teor. Priblizh. Garmon. Anal., 269–285; English transl., Proc. Steklov Inst. Math. **4(219)** (1997), 265–280. MR1642280

[168] Phillip A. Ostrand, *Dimension of metric spaces and Hilbert's problem* 13, Bull. Amer. Math. Soc. **71** (1965), 619–622, DOI 10.1090/S0002-9904-1965-11363-5. MR177391

[169] Petersen P., Voigtlaender F., Optimal approximation of piecewise smooth functions using deep ReLU neural networks, Neural Networks, 108 (2018), 296-330.

[170] Pencho P. Petrushev, *Approximation by ridge functions and neural networks*, SIAM J. Math. Anal. **30** (1999), no. 1, 155–189, DOI 10.1137/S0036141097322959. MR1646689

[171] Allan Pinkus, *Approximating by ridge functions*, Surface fitting and multiresolution methods (Chamonix–Mont-Blanc, 1996), Vanderbilt Univ. Press, Nashville, TN, 1997, pp. 279–292. MR1660030

[172] Allan Pinkus, *Approximation theory of the MLP model in neural networks*, Acta numerica, 1999, Acta Numer., vol. 8, Cambridge Univ. Press, Cambridge, 1999, pp. 143–195, DOI 10.1017/S0962492900002919. MR1819645

[173] A. Pinkus, *Smoothness and uniqueness in ridge function representation*, Indag. Math. (N.S.) **24** (2013), no. 4, 725–738, DOI 10.1016/j.indag.2012.10.004. MR3124803

[174] Allan Pinkus, *The alternating algorithm in a uniformly convex and uniformly smooth Banach space*, J. Math. Anal. Appl. **421** (2015), no. 1, 747–753, DOI 10.1016/j.jmaa.2014.06.076. MR3250506

[175] Allan Pinkus, *Ridge functions*, Cambridge Tracts in Mathematics, vol. 205, Cambridge University Press, Cambridge, 2015, DOI 10.1017/CBO9781316408124. MR3559591

[176] T. J. Rivlin and R. J. Sibner, *The degree of approximation of certain functions of two variables by a sum of functions of one variable*, Amer. Math. Monthly **72** (1965), 1101–1103, DOI 10.2307/2315959. MR186995

[177] Walter Rudin, *Functional analysis*, 2nd ed., International Series in Pure and Applied Mathematics, McGraw-Hill, Inc., New York, 1991. MR1157815

[178] Marcello Sanguineti, *Universal approximation by ridge computational models and neural networks: a survey*, Open Appl. Math. J. **2** (2008), 31–58, DOI 10.2174/1874114200802010031. MR2399691

[179] Schmidt-Hieber J., The Kolmogorov–Arnold representation theorem revisited, Neural Networks 137 (2021), 119–126.

[180] Laurent Schwartz, *Théorie générale des fonctions moyenne-périodiques* (French), Ann. of Math. (2) **48** (1947), 857–929, DOI 10.2307/1969386. MR23948

[181] Shen, Z., Yang, H., Zhang, S., Neural network approximation: Three hidden layers are enough, Neural Networks 141 (2021), 160-173.

[182] Ivan Singer, *The theory of best approximation and functional analysis*, Conference Board of the Mathematical Sciences Regional Conference Series in Applied Mathematics, No. 13, Society for Industrial and Applied Mathematics, Philadelphia, Pa., 1974. MR0374771

[183] David Sprecher, *A representation theorem for continuous functions of several variables*, Proc. Amer. Math. Soc. **16** (1965), 200–203, DOI 10.2307/2033845. MR174666

[184] David A. Sprecher, *On the existence of best approximations and representations in several variables*, J. Reine Angew. Math. **234** (1969), 152–162, DOI 10.1515/crll.1969.234.152. MR257622

[185] David A. Sprecher, *An improvement in the superposition theorem of Kolmogorov*, J. Math. Anal. Appl. **38** (1972), 208–213, DOI 10.1016/0022-247X(72)90129-1. MR302838

[186] Sprecher D.A., A universal mapping for Kolmogorov's superposition theorem, Neural Networks 6 (1993), 1089-1094.

[187] J. P. Sproston and D. Strauss, *Sums of subalgebras of $C(X)$*, J. London Math. Soc. (2) **45** (1992), no. 2, 265–278, DOI 10.1112/jlms/s2-45.2.265. MR1171554

[188] Stein W.A. et al., Sage Mathematics Software (Version 7.6), The Sage Developers, 2017, `http://www.sagemath.org`.

[189] J. Sternfeld, *Dimension theory and superpositions of continuous functions*, Israel J. Math. **20** (1975), no. 3-4, 300–320, DOI 10.1007/BF02760335. MR374351

[190] Y. Sternfeld, *Uniformly separating families of functions*, Israel J. Math. **29** (1978), no. 1, 61–91, DOI 10.1007/BF02760402. MR487991

[191] Y. Sternfeld, *Superpositions of continuous functions*, J. Approx. Theory **25** (1979), no. 4, 360–368, DOI 10.1016/0021-9045(79)90022-4. MR535937

[192] Y. Sternfeld, *Dimension, superposition of functions and separation of points, in compact metric spaces*, Israel J. Math. **50** (1985), no. 1-2, 13–53, DOI 10.1007/BF02761117. MR788068

[193] Yaki Sternfeld, *Uniform separation of points and measures and representation by sums of algebras*, Israel J. Math. **55** (1986), no. 3, 350–362, DOI 10.1007/BF02765032. MR876401

[194] Stinchcombe M., White H., *Approximating and learning unknown mappings using multilayer feedforward networks with bounded weights*, in Proceedings of the IEEE 1990 International Joint Conference on Neural Networks, 1990, Vol. 3, IEEE, New York, 7-16.

[195] Bruno H. Strulovici and Thomas A. Weber, *Additive envelopes of continuous functions*, Oper. Res. Lett. **38** (2010), no. 3, 165–168, DOI 10.1016/j.orl.2010.01.004. MR2608851

[196] Xingping Sun and E. W. Cheney, *The fundamentality of sets of ridge functions*, Aequationes Math. **44** (1992), no. 2-3, 226–235, DOI 10.1007/BF01830981. MR1181270

[197] Temlyakov V.N., On approximation by ridge functions, Preprint. Department of Mathematics, University of South Carolina, 1996.

[198] V. M. Tihomirov, *The works of A. N. Kolmogorov on ε-entropy of function classes and superpositions of functions* (Russian), Uspehi Mat. Nauk **18** (1963), no. 5 (113), 55–92. MR0162910

[199] J. F. Traub, G. W. Wasilkowski, and H. Woźniakowski, *Information-based complexity*, Computer Science and Scientific Computing, Academic Press, Inc., Boston, MA, 1988. With contributions by A. G. Werschulz and T. Boult. MR958691

[200] V. N. Trofimov and L. R. Hariton, *On the error of uniform approximation of functions of two variables by a sum of functions of one variable* (Russian), Izv. Vyssh. Uchebn. Zaved. Mat. **8** (1979), 70–73. MR554506

[201] Hemant Tyagi and Volkan Cevher, *Learning non-parametric basis independent models from point queries via low-rank methods*, Appl. Comput. Harmon. Anal. **37** (2014), no. 3, 389–412, DOI 10.1016/j.acha.2014.01.002. MR3256780

[202] A. G. Vituškin and G. M. Henkin, *Linear superpositions of functions* (Russian), Uspehi Mat. Nauk **22** (1967), no. 1 (133), 77–124. MR0237729

[203] B. A. Vostrecov and M. A. Kreĭnes, *Approximation of continuous functions by superpositions of plane waves.* (Russian), Dokl. Akad. Nauk SSSR **140** (1961), 1237–1240. MR0131106

[204] Yarotsky D., Error bounds for approximations with deep ReLU networks, Neural Networks 94 (2017), 103-114.

[205] T. I. Zaitseva, Yu. V. Malykhin, and K. S. Ryutin, *On Recovery of Regular Ridge Functions* (Russian), Mat. Zametki **109** (2021), no. 2, 307–311, DOI 10.4213/mzm13000. MR4223918

Index

Selected Published Titles in This Series

263 **Vugar E. Ismailov,** Ridge Functions and Applications in Neural Networks, 2021
259 **William Heinzer, Christel Rotthaus, and Sylvia Wiegand,** Integral Domains Inside Noetherian Power Series Rings, 2021
258 **Pramod N. Achar,** Perverse Sheaves and Applications to Representation Theory, 2021
257 **Juha Kinnunen, Juha Lehrbäck, and Antti Vähäkangas,** Maximal Function Methods for Sobolev Spaces, 2021
256 **Michio Jimbo, Tetsuji Miwa, and Fedor Smirnov,** Local Operators in Integrable Models I, 2021
255 **Alexandre Boritchev and Sergei Kuksin,** One-Dimensional Turbulence and the Stochastic Burgers Equation, 2021
254 **Karim Belabas and Henri Cohen,** Numerical Algorithms for Number Theory, 2021
253 **Robert R. Bruner and John Rognes,** The Adams Spectral Sequence for Topological Modular Forms, 2021
252 **Julie Déserti,** The Cremona Group and Its Subgroups, 2021
251 **David Hoff,** Linear and Quasilinear Parabolic Systems, 2020
250 **Bachir Bekka and Pierre de la Harpe,** Unitary Representations of Groups, Duals, and Characters, 2020
249 **Nikolai M. Adrianov, Fedor Pakovich, and Alexander K. Zvonkin,** Davenport–Zannier Polynomials and Dessins d'Enfants, 2020
248 **Paul B. Larson and Jindrich Zapletal,** Geometric Set Theory, 2020
247 **István Heckenberger and Hans-Jürgen Schneider,** Hopf Algebras and Root Systems, 2020
246 **Matheus C. Bortolan, Alexandre N. Carvalho, and José A. Langa,** Attractors Under Autonomous and Non-autonomous Perturbations, 2020
245 **Aiping Wang and Anton Zettl,** Ordinary Differential Operators, 2019
244 **Nabile Boussaïd and Andrew Comech,** Nonlinear Dirac Equation, 2019
243 **José M. Isidro,** Jordan Triple Systems in Complex and Functional Analysis, 2019
242 **Bhargav Bhatt, Ana Caraiani, Kiran S. Kedlaya, Peter Scholze, and Jared Weinstein,** Perfectoid Spaces, 2019
241 **Dana P. Williams,** A Tool Kit for Groupoid C^*-Algebras, 2019
240 **Antonio Fernández López,** Jordan Structures in Lie Algebras, 2019
239 **Nicola Arcozzi, Richard Rochberg, Eric T. Sawyer, and Brett D. Wick,** The Dirichlet Space and Related Function Spaces, 2019
238 **Michael Tsfasman, Serge Vlăduţ, and Dmitry Nogin,** Algebraic Geometry Codes: Advanced Chapters, 2019
237 **Dusa McDuff, Mohammad Tehrani, Kenji Fukaya, and Dominic Joyce,** Virtual Fundamental Cycles in Symplectic Topology, 2019
236 **Bernard Host and Bryna Kra,** Nilpotent Structures in Ergodic Theory, 2018
235 **Habib Ammari, Brian Fitzpatrick, Hyeonbae Kang, Matias Ruiz, Sanghyeon Yu, and Hai Zhang,** Mathematical and Computational Methods in Photonics and Phononics, 2018
234 **Vladimir I. Bogachev,** Weak Convergence of Measures, 2018
233 **N. V. Krylov,** Sobolev and Viscosity Solutions for Fully Nonlinear Elliptic and Parabolic Equations, 2018
232 **Dmitry Khavinson and Erik Lundberg,** Linear Holomorphic Partial Differential Equations and Classical Potential Theory, 2018

For a complete list of titles in this series, visit the
AMS Bookstore at **www.ams.org/bookstore/survseries/**.